MATERIALS SCIENCE AND TECHNOLOGY

Edited by **Sabar D. Hutagalung**

Materials Science and Technology
Edited by Sabar D. Hutagalung

Published by InTech
Janeza Trdine 9, 51000 Rijeka, Croatia

Publishing Process Manager Dragana Manestar

Technical Editor Teodora Smiljanic

Cover Designer InTech Design Team

Additional hard copies can be obtained from orders@intechopen.com

Materials Science and Technology, Edited by Sabar D. Hutagalung
p. cm.
ISBN 978-953-51-0193-2

For EU product safety concerns:
IN TECH d.o.o., Prolaz Marije Krucifikse Kozulić 3, 51000 Rijeka, Croatia,
info@intechopen.com or visit our website at intechopen.com.

Meet the editor

Dr. Sabar D. Hutagalung received a Drs degree (Applied Physics) from University of North Sumatra (USU), Indonesia in 1989 and a Ph.D. degree (Materials Science) from University of Technology Malaysia (UTM), Malaysia, in 1997. He was a lecturer at the USU from 1991 to 2003. Since December 2003 he has been with the School of Materials and Mineral Resources Engineering, University of Science Malaysia (USM), Malaysia. His research interests are nanomaterials, nanoelectronics and electro-active ceramic materials areas, especially synthesis and characterization of semiconductor nanostructured materials (nanowires, nanorods, nanoparticles and quantum dots), fabrication of high-K and giant magnetoresistance ceramic, and development of ceramic-based EMI shielding materials. He has successfully guided 3 PhD, 17 MSc, and more than 50 undergraduate students. He has published more than 45 SCOPUS/ISI indexed papers. As an expert in Electronic Materials area, he has been invited as external and internal examiners to MSc and PhD theses, invited to give talk in some forum, review papers for international and national journals.

Contents

Preface

Materials are important to mankind because of the benefits that can be derived from the manipulation of their properties. Examples include electrical conductivity, dielectric constant, magnetization, optical transmittance, strength and toughness. All of these properties originate from the internal structures of the materials. Structural features of materials include their types of atoms, the local configurations of the atoms,and the arrangements of these configurations into microstructures.

Materials can be divided into various classes which may constitute separate fields such as metals, polymers, ceramics, composites, semiconductors, bio-materials and nanomaterials. Materials science is a broad field and can be considered to be an interdisciplinary area. Included within it are the studies of the structure and properties of any material, the creation of new types of materials, and the manipulation of a material's properties to suit the needs of specific application.

The basis of materials science involves relating the desired properties and relative performance of a material in a certain application to the structure of the atoms and phases in that material through characterization. Physical properties of materials usually play an important role in the selection of material for a particular application. This involves many factors such as material composition and structure, fracture and stress analysis, conductivity, optical, and thermal properties, to name a few. It also involves design, modeling, simulation, processing, and production methods. Research in the field of materials science involves many peripheral areas including crystallography, microscopy, lithography, mineralogy, photonics, and powder diffraction.

The contributors of the chapters in this book have various areas of expertise. Therefore, this book is interdisciplinary and is written for readers with a background in physical science. I believe that this book will be of interest to university students, lecturers and researchers who are interested in the fields of materials science, engineering and technology and, specifically, in advanced ceramic materials, metal, polymer composite, nanomaterials, bio-materials and optical materials. As it deals with many basic concepts that are of concern in the relevant fields, this book can also be used as a primer for studies in materials science and engineering.

The book consists of fourteen chapters that have been divided into four sections. Section one includes five chapters on advanced materials and processing. This section deals with the development of new high temperature composites based on zirconium cements for application in various consumer industries, crystal structure analysis and piezoelectricity mechanism of langasite, morphological analysis of the interface between two conducting oxides created by a single photolithography process and subsequent self-aligned pattern transfer, design and experimental analysis of composite multi leaf spring using glass fiber reinforced polymer for light passenger vehicles, modeling, simulation and experimental studies of distortions, residual stresses, and hydrogen diffusion during laser welding of as-rolled steel.

Section two includes two chapters on bio-materials which deal with the preparation of new types of collagen nanostructured bio-materials in the form of spongeous, microporous matrices, nontoxic and biocompatible with osteoblast cells, and modification of magnetite particles with different functionalizations specific for desired applications such as biomedical and industrial applications.

Section three consists of three chapters on nanomaterials, specifically on the study of carbon nanotubes, nano-machining and nanoparticles. It includes discussions on carbon nanotubes-imprinted polymers composites as useful innovative materials for the analytical determination of target analytes in complex matrices, mechanism of nano-machining and mechanical behavior of nanostructures by conducting the integrated molecular dynamics simulation of scratching and shearing with the same specimen, and development of a novel process for preparation of super-paramagnetic Fe_3O_4 nanoparticles with high crystallinity in a water system.

Section four includes four chapters on the structural investigation of crystal growth phenomena in rewritable optical recording media using an X-ray pinpoint structural measurement system, development of novel $EuMoO_4$ microstructures synthesized via a facile simple and mild hydrothermal route, a review on plasma switching by laser ablation (PLASLA) including introduction to PLASLA phenomenon, PLASLA formation method, PLASLA properties, the interaction of PLASLA with a magnetic field and the significance of PLASLA in materials science, and the development of a method to determine the molecular weight (MW) and molecular weight distribution (MWD) prediction of cellulose by means of a rheology-based method.

As Editor, I would like to thank all the contributors of the chapters in this book for their tremendous efforts in producing excellent work. Last but not least, I would like to express my sincere gratitude to Ms Dragana Manestar, publishing process manager, for the effective communication and assistance given during the preparation of this book.

Sabar D. Hutagalung
University Sains Malaysia
Malaysia

Part 1

Advanced Materials and Processing

1

New Cements and Composite Materials Based on Them for Atomic Industry

Nickolay Ilyoukha, Valentina Timofeeva and Alexander Chabanov
Academic Ceramic Center, Kharkov, Ukraine

1. Introduction

New cements contains double oxides and aluminates of calcium, barium, stroncium. The date obtained allow to classify new cements as high-property fire-proof, quick-hardening and high-strength binders. The synthesis of cement clinker takes place through solid phase reactions. The newer developed cements of the higest fire resistance show only insignificant loss of strength when being heated (15% to 20% by weight), which can be accounted for by formation of high temperature - proof epitaxial contacts between the new hydrate formations and double oxides grains.

The aim of this work is to enlarge our knowledge about this new cements and hightemperature composites on them and find answers to questions about the optimum application of this product in severe industrial conditions and to analyse how this material performs in high temperatures.

The composite before trials investigated with the help of roentgen - structural and roentgen - spectral of the analyses, optical and raster electronic microscopy on standard techniques.

2. The using of high temperature composites based on zirconia's cement in recycling and catcher of fuel

For coating of high temperature gas - dynamic of channels (> 2000 °C) used high temperature composites based on zirconia's cements. The properties and structure of a material are adduced in table 1, 2. The erosive and temperature stability of a material on a bench, special energies heat treatment channel a bench was previously evaluated. The test specifications are adduced in table 3.

Complex three-component source structure, durables maintenance at temperatures exceeding not only temperature of a sintering, but also the melting aluminates of barium (-2100K) and the sintering of a dioxide of a zirconium, intensive of component's evaporation making of cement in a combination to large gradients of temperature result in steep changes on width of lining of a structure and structure of a material. It, naturally, is reflected in strength and reliability of maintenance of lining, that stipulates necessity of a detail research of a formed structure, in particular for detection of zones destruction.

No	Water to solids, % by weight	Setting time, min		Strength, N/mm2		
		start	end	2nd	4 th	8 th
				day	day	day
1	15	25	55	550	880	1050
2	15	55	90	400	540	900
3	15	65	98	190	400	520
4	13	70	105	170	280	330
5	13	86	120	150	200	240
6	12	29	63	180	280	480
7	12	62	98	150	250	290
8	12	75	110	140	240	350
9	12	79	130	120	240	320
10	15	30	98	600	910	1100
11	15	60	90	450	600	950
12	15	70	105	210	450	600
13	13	78	115	200	310	350
14	12	91	125	180	250	270

Table 1. Properties of zirconia's cements.

Properties	Content of binder / mass - %		
	10	20	30
Baddeleyite (ZrO2)	90	80	70
Refractoriness, K	2673	2473	2073
Compression strength cold, MPa	38	45	65
Compression strength, MPa			
burnt at 120°C	29	38	57
burnt at 300°C	17	34	39
burnt at 500°C	19	32	46
burnt at 800°C	23	37	44
burnt at 1000°C	36	39	63
burnt at 1200°C	38	44	52
burnt at 1400°C	39	45	50
burnt at 1600°C	40	47	56
burnt at 1750°C	42	45	58
Linear change, 1750°C, mass-%	1.1	1.7	2.0
Temperature of initial deformation under load, °C, no less	1750	1660	1560
Thermal shocr resistance 1300 - water heat changes, no less	16	15	11

Table 2. Physical and technical properties of composites materials based on zirconiur cements.

The composite material from a dioxide of a zirconiumipn cement contains a filler from stabilized cubic ZrO_2 and cement including 30 mass-% of monoaluminates and 70 % of zirconat. A dioxide of a zirconium making 80 % of concrete, is represented by three fractions: large (0.2 – 2.5 mm), average (0.1 – 0.5 mm), small-sized (are more sma!l-sized (more finely) than 0.1 mm), including with the linear sizes of partials 1-5 microns. Used in experiments a dioxide of a zirconium was stabilized 6-7 mass-% oxide yttrium.

Parameter	Tank of the combustion chamber (CC)	Site a channel
Expenditure of products of combustion of natural gas, kg/s	2-5	
Volumetric share of oxygen in an oxidizer, mass - %	40	
Temperature of a heating of an oxidizer, K	1100	
Coefficient of excess oxidizer	0,85-1,05	
Temperature of products of combustion, K	2600 - 2700	2500 - 2650
Flow velocity, m/s	80	700
Pressure (absolute), MPa	0,25	0,25-0,15

Table 3. Specification of a test of lining.

The zirconium cement in an outcome of interaction with a water is subjected hydratation with derivation of a number of connections ($BaAl_2O_4 \cdot 6H_2O$, $BaAl_2O_4 \cdot 7H_2O$, $BaAl_2O_4 \cdot 2H_2O$, $Ba_2Al_2O_5 \cdot 5H_2O$), carrying on which among are two and six-water aluminates of barium. The accretion of chips of hydrat phases results in creation of a frame ensuring strength of a material at room temperature. The developed link between partial's of cement and filler at this stage is absent. The porosity of a material after hardening makes 15-2 mass-%.

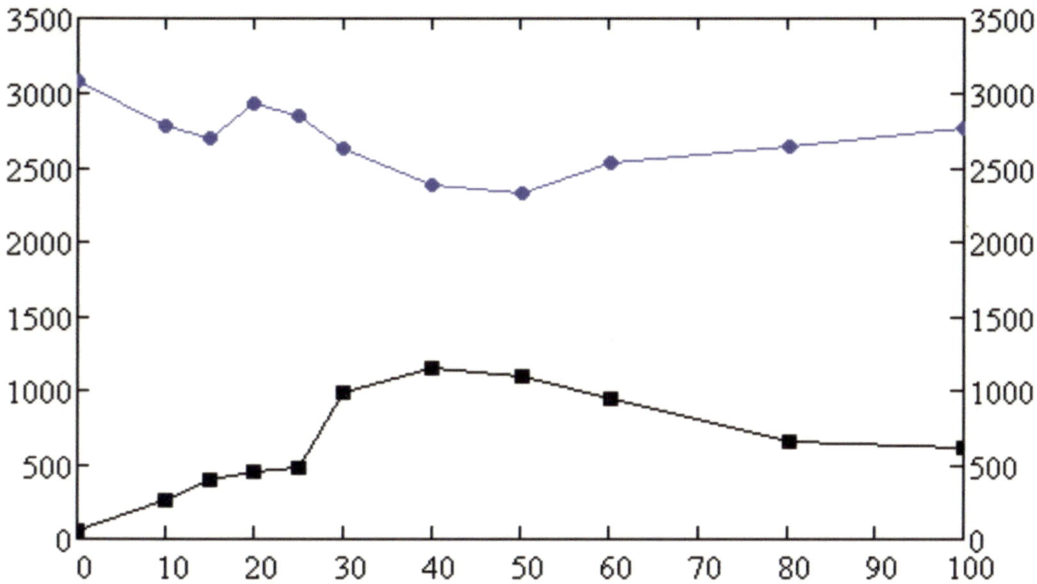

Fig. 1. Dependence of refractoriness and strengh from contents.

At temperatures is lower 400 K a material does not undergo changes. In remaining sites happens dehydratation knitting. Above 1400 T is observed a sintering of a material. Thus in a zone, which hot boundary corresponds to temperature 1700 K, the sintering happens mainly at the expense of development of contacts between partials of cement, in a zone up to 2150 K - at the expense of link of partials of cement among themselves and to grains of a filler, in a more high-temperature layer the direct contacts of partials dioxide of a zirconium develop. In intermediate area (between the poorly changed zone and by a sintering layer) the structure of a material is determined by a degree dehydratation knitting, accompanying by destruction source hydro aluminates of barium. In a zone, from the cold side to sintering to a layer, where has a place practically full loss of a water, the destruction of a frame derivates, and, as a corollary it, recession of strength of a material happens. It is necessary to mark, that while in service destructions of lining on this zone, as a rule, does not happen, probably, by virtue of design features of a wall.

The width sintering of a layer depends on a level of temperature on a hot surface. At the temperature of about 2200 K the width sintering of a layer makes 20-25 mass-% from width of lining (for linings by width hS =12-60 mm), at 2300 - 2350 K 0,45 hS - 0.6 hS (on the profile of temperature partial transparency of a material) essentially influences. In a hot part sintering of a layer a sintering of a material increasing at the expense of derivation of a melt aluminates of barium. The rnelt partially quits on a firing surface of lining. The selection of a melt on a firing surface of a material during first 0,5-1 h of unisothermal heating immediately was observed during trials.

In bounds of a sintering layer it is possible to select some zones. Immediately zone adjoins to a hot surface, which distinctive feature is the presence only in the extremely minor amount (less than 1 %) components of cement. The cold boundary of this zone approximately coincides an isotherm 2200 K. by a Dominating phase a sintering layer is trie cubic hard solution because of dioxide of a zirconium. Aluminates and zirconate of barium register as separate of inclusions. Aluminates making is represented in main inclusions Al_2O_3. The average content BaO in a sintering layer makes 0.2 – 0.4 mass-%, Al_2O_3 0.3 – 0.4 mass-%.

The structure of a material of a hot layer, going to a firing surface, shows, that during trials it in main loses components of cement. Zirconat of barium is partial in other zones, is partially decomposed with selection as an independent phase Al_2O_3. The loss of components of cement carries on to growth of a porosity of a material in a hot layer on 5 mass-% - 8 mass-% in comparison with source.

The structure of a hot layer of lining by width 0.14 hS- 0.16 hS (here and further at the temperature of of working surface is higher 2300 K) represents sintering porous ceramics because of dioxide of a zirconium. The character of a structure, formed in this layer, is determined by a level of temperature and amount of a liquid phase at the first phase of a sintering. In to a going firing surface a layer (the hot part of a zone of porous ceramics) happens heavily division of source grains of a filler into more small-sized fragments. Present here at initial stage of trials the melt aluminates of barium, which amount can be increased at the expense of arrival from more cold sites, promotes both process dispergation, and reallocation of products dispergation and partials of a small-sized fraction on size. The partials of a small-sized fraction dioxides of a zirconium are intensive decristallisation. In an outcome in to a hot surface a layer the material loses obviously expressed division into large and small-sized fractions and acquires a rather homogeneous structure.

In more cold sites of a zone of porous ceramics an amount of a liquid phase present to not initial stage of trials and capable to ensure a regrouping, of a capable, much below, than in of surface area. Dispergation of grains of a filler with derivation of free interfaces between blocks is not accompanied by essential reallocation of derivate partials on size.

Because of grains of a filler will be derivates of partials dioxide of a zirconium, inclu a grain and going to it a partial of a small-sized fraction.

In accordance with transition to even more cold sites the decrease of the size of grains, of the filler which has undergone dispergation is observed. Dispergation of large grains of a filler happens in a hot part, which width makes approximately 60 - 65 mass-% from width of a zone of porous ceramics. In underlying zones in large grains of a filler the increase S of a porosity registers in comparison with source. In more small-sized grains the growth of a porosity is combined with splitting on blocks. The latter will be realized(sold) mainly near to boundaries of a grain of a filler. As well as in hotter sites, in a material the developed contacts of grains of a filler with decristallisation by partials of a small-sized fraction are observed.

The link between large by dispergation grains of a filler implements derivations of a complex configuration generated from partials of average and small-sized fractions dioxide of a zirconium. A structure of these derivations, number and quality of contacts, explicating between them and large grains of a filler, largely determine strength of a material in a zone of porous ceramics because of dioxide of a zirconium. Their most characteristic variants of allocation of structural units in intervals between large grains it is necessary to mark a structure, in which the primary orientation is absent.

The hot part of a zone of porous ceramics differs from hot sites less by rectangular distribution of a material and large size of vacuum in intervals between grains of a filler. Thus the fair quantity porosity, oriented along a hot surface is observed. In some cases the concentration such nop on approximately one horizon happens, the depth makes 0.12 hS - 0.14 hS. The rise of concentration porosity carries on to lowering strength of a material on appropriate horizon. At availability change, of temperature in this area the cracks, carrying on to destruction linings can develop.

As already it was marked, a dominating phase in a zone of porous ceramics is the cubic hard solution because of dioxides of a zirconium. The concentration of the stabilizing " component Y_2O_3 in a layer, which adjoins to a hot surface, makes 3 - 4 mass-% (molar shares) ambassador 50 hours of trials. The same level of concentration of the component registers through 200 - 250 hours after a beginning of trials. In accordance with transition to more steep layers of this zone concentration Y_2O_3 the gradually is increased and near to cold boundary of a zone of porous ceramics receives a reference value. The not stabilized dioxide of a zirconium either is absent, or registers in an amount less than 1 %. The absence of the not stabilized dioxide of a zirconium shows, that in a hot part of this zone the reallocation of the component Y_2O_3 between a source cubic hard solution and not stabilized ZrO_2, selected happens at decomposition zirconat barium.

The link of particles of a dioxide of a zirconium in a zone of porous ceramics implements mainly direct contacts ZrO_2 - ZrO_2. The part of contacts will be realized by means of films by width a micron and less, containing oxide of a aluminum.

Near to cold boundary of a zone of porous ceramics the increase of number of * inclusions zirconat of barium is observed. On depth 0.14hS – 0.16hS the sharp increase of a content zirconat of barium up to 20 - 25 mass-% (figure 1) happens, that considerably exceeds the source value - about 14 mass-%. Zirconat of barium in this area will derivate congestions on a surface of particles dioxide of a zirconium, fills in the porosity and crack in grains of a filler. The particles zirconat of barium forming on boundaries of grains dioxide of a zirconium, have the linear sizes 2-50 microns and differ by dense stacking. Sometimes such are be surrounded by a film Al_2O_3. The content Al_2O_3 in a zone of synthesis zirconat of barium makes 1.5-1.7 mass-%. Thus an amount aluminates of barium makes 1 %.

The useful increase of an amount zirconat of barium allows to assume, that the part a pair(vapour.couple) BaO selected at decomposition zirconat of barium in a zone of porous ceramics, acts to space in more cold zones, where is condensed and enters response with a dioxide of a zirconium with derivation secondary zirconat of barium. It is necessary to marRrTRaT the derivation $BaZrO_3$ in a considered zone is also in case of absence zirconat of barium in source concrete. Thus the amount synthesized during trials zirconat of barium does not exceed 3-4 %.

The material of a zone enriched zirconat of barium, after trials is saturated with cracks. The development of cracks should be promoted by the volumetric extension accompanying synthesis zirconat of barium. The link of particles of a dioxide of a zirconium implements both by means of direct contacts, and through inclusions zirconat of barium.

In a cold part of a zone of synthesis zirconat of barium the number of inclusions aluminates of barium accrues. The amount of the latter spasmodically increases on depth 0,24hS - 0,26hS. The underlying layer by width about 0,07hS differs by an increased content aluminates of barium (12 - 15 mass-%), essentially exceeding a source level (6 mass-%). In a hot part of a zone enriched $BaZrO_3$, aluminates of barium in a combination to particles $BaZrO_3$ will derivate interlayer up to 150 microns on boundaries of grains of a filler. The content zirconat of barium is close a reference value. The particles of cement differ by rather dense stacking. Their linear sizes make in main 3-15 microns. The direct contacts of particles dioxide of a zirconium here practically are absent. At the same time in this area the development of contacts of particles dioxide of a zirconium both with zirconat and with aluminates of barium is provided. The material of lining in a considered layer differs in increased density in comparison with the poorly changed zone. It is possible to assume, that the derivation of this packed zone is stipulated by arrival of a melt aluminates of barium from hot layers of lining and it a chip.

From the cold side the layer adjoins to a zone with increased density, in which the content of components of cement is identical source. The sintering of cement in this zone flows past rather actively and is accompanied of the linear change. In an outcome in intervals between grains of a filler the including also small-sized particles of a dioxide of a zirconium will be derivated densely stacked of particles of cement. The structure of a material here is identical to a structure of concrete, in isothermal conditions at 1900 -2000 K.

In accordance with deleting from a zone saturated aluminates of barium, the size of particles of components of cement of a gradually decreases, the quality of link between a filler and particles of cement is worsened. At the same time the linear change within the limits of a small-sized fraction hinders with derivation of a continuous frame of cement, that in a combination to not enough developed contacts a filler - the cement carries on to lowering strength of a material in this zone. At cooling of lining the explicating in a material of power result in derivation of trunk cracks in a cold part of a circumscribed above zone on depth, making 0.3hs - 0.4hs.

In more cold sites down to cold boundary of a layer the sintering of cement happens without obviously expressed the linear change. Recristallisation of particles of cement is expressed poorly. In a material the continuous frame of cement is formed. Thus the link between a filler and cement is advanced poorly. As already it was marked, the cold boundary sintering of a layer passes on depth 0.45hs - 0.6hs.

The circumscribed above zones in a sintering layer are present at the rather developed form in 2 hours after a beginning of trials. The more durables trials do not result in increase of width of a sintering layer and zones, composing it. With rise of duration of trials the zone of porous ceramics because of dioxides of a zirconium in a greater degree is cleared of inclusions containing barium and aluminum. Besides the more expressed character is acquired with concentration of the poores in a cold part of a zone of ceramic porous.

Is conducted about 20 launches of installation for trial of materials by duration from 50 till approximately 400 o'clock. During trials temperature on a surface of lining (both in the combustion chamber is fixed, and in gas-dynamic a channel), as a rule, 2250 -2400 K, in separate experiments it was possible to reach temperature 2600 K.

The rise of temperature of a surface of lining Ts is precisely fixed at decrease of a share of a firing surface occupied by copper edges and at increase of width of lining h (from Ts≈2000 K at x≈0,15 and h=10 mm up to Ts≈2400 K at x≈0 and h=25-30 mm). Is marked, that at Ts >2200 K defining a role in creation of a thermal mode of a firing surface begins to play of walls among themselves (density of a heat flow ≥1 MW/m2 at density of a convective heat flow ≤0.4 MW/m2).

At temperatures a surface of concrete is higher 2150 T concrete lining on zirconia's cement in a upper layer well conglomerated in first 2-6 hours of installation, that provided to lining high erosive stability purely erosion of a material during trials was less than 1 micron / hour.

Is clarified, that the lining on zirconia's cement allows conduct fast (for 20 - under 60 min.) heating of an outline of installation. Moreover, such heating even is preferable to concrete (acceleration of a sintering of a high layer results in decrease of initial erosion of concrete).

3. The using of zirconia's cement composite in coating of quartz glass tanks

The new cements and composite based on then can be used in the coating of quartz glass tanks. In contact with glass there are used refractories made of natural baddeleyite without stabilizing oxides. The coating of this material showed good characteristics under thermal shock and quartz glass penetration resistance properties.

3.1 Red mud

Red mud is a waste material which results from alumina production during the Bayer process. Approx. 35-40% of the bauxite ore processed goes into waste as red mud. This is a cheap source of raw material for the manufacture of low-cost ceramic products such as building blocks, floor and wall tiles, sanitary ware and as an additive for cement. The chemical composition of red mud is as follows (Table 4).

Oxidic Compounds	Content [Mass - %]
Fe_2O_3	38.1
$A1_2O_3$	27.01
SiO_2	14.15
TiO_2	5.01
Na_2O	6.03
K_2O	0.36
CaO	2.26
MgO	1.01
CO_2	4.02
SO_3	2.05

Table 4. Cemical composition of the calcined red mud.

Red mud is a very complex material. It is a mixture of several oxides and minerals such as hematite, sodium aluminum silicates, and rutile. These make the mud a potential raw material for the production of additives for cements.

The particle size analysis of the calcined red mud showed that 85% of the particles are <10um (Figure 2). The calcined red mud ranged in size from 35 to 5 urn average particle diameter. The particle size analysis of the calcined red mud is given in Figure 3.

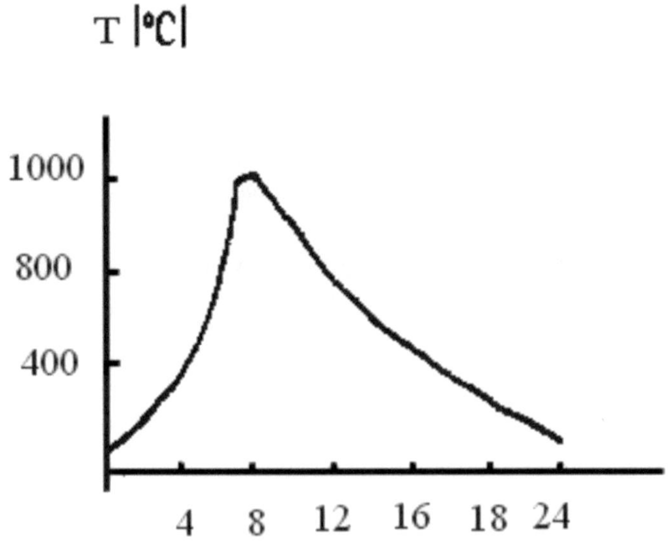

Fig. 2. The firing treatment of red mud. The treatment of red mud n dependent on temperature and time.

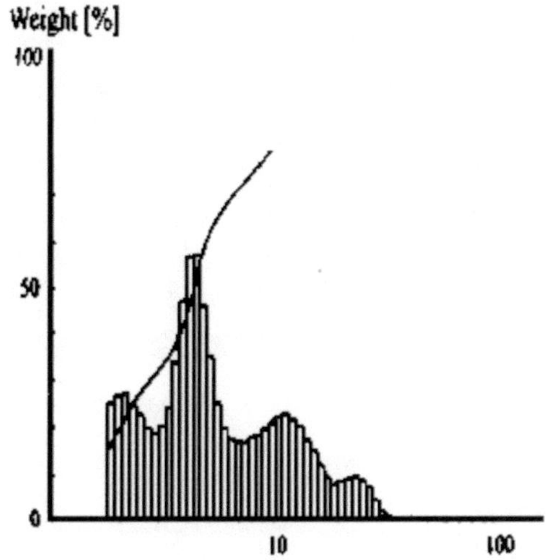

Fig. 3. The particle size analysis of calcined red mud. X-Ray diffraction results of the calcined red mud.

3.2 Clays nonconditional caolins

Ukraine has a great number of clay deposits. These clays can be used for traditional ceramics production and for production as an additive to cement. The physicoceramic date revealed that, in general, the clays sensitive to drying from this area, have significant plasticity (between 30 and 58%) and relatively high total shrinkage (between 6 and 21%) (Figure 4), have good to very good capability, average water adsorption (between 9 and 16%) good to very good mechanical resistance (compressive strength between 325 and 355 daN/cm^2) (Figure 5), and bending strength between 90 and 460 daN/cm^2.

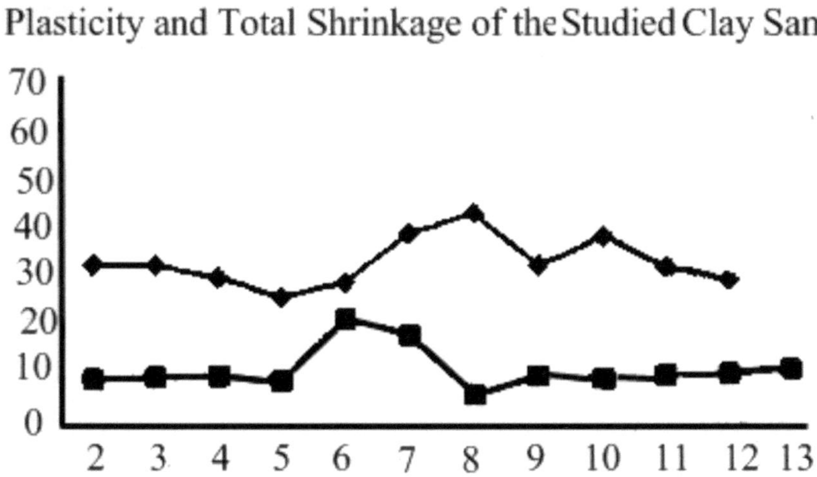

Fig. 4. The Augustinic diagram of use of the clay deposits.

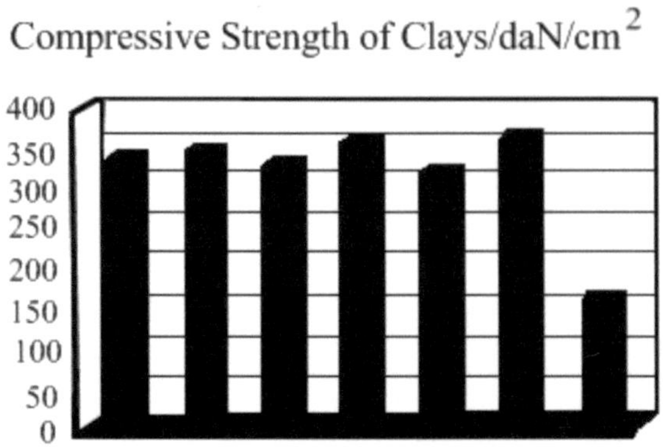

Fig. 5. Comparative diagrams of physico-mechanical characteristics of clay deposits.

The red mud sample was supplied by Nicolayew Plant(Ukraine), Kaolin wastes(region of Donetsk), Nepheline wastes(Atschinsk, Russia).

The chemical analysis of the sample was accomplished by using an atomic absorption spectrophotometer or digital photometer and wet chemical methods (Table 5). X-Ray diffraction studies on the sample were undertaken using a Philips X-Ray diffraction unit. The particle size distribution of the sample was measured by using a Sedigraph (5000D Micrometrics). The photographs of the samples were taken with an optical microscope (Olympus BH2-IMA).

Working zones		Distance from hot face, mm			Contents		
	SiO_2	$A1_2O_3$	Fe_2O_3	MgO	Na_2O	CaO	K_2O
1	68.11	14.85	5.6	0.91	0.14	0.96	0.57
2	69 52	15	4.6	0.6	3.75	1.68	0.75
3	52	31	8	0	2.1	0	2.3
4	54.6	23 19	2.1	1.6	0.3	0.6 2.03	2.6
5	58.38	19	6.6	1.85	0.95	2.03	
6	64	14.4	6.7	1.1	1.13	2.14	1.27
7	57.03	18.57	7.37	2.22	1.04	2.16	2.88
8	64.88	12.51	5.13	2.29	1.63	5.99	2.1
9	65.07	16.48	6.08	1.18	0.74	1.18	1.66
10	51.62	27.97	3.92	1.09	1.59	0.76	1.59
11	14.76	26.94	37.6	1.01	5.83	2.06	0.34
12	57.21	20.02	1.44	0.94	0.55	1.96	3.2
13	53.43	32.07	5.82	2.06	0.05	2.14	0.19
14	59.1	25.64	7.1	1.28	0.06	2.27	0.85
15	50.12	29.78	5.04	1.00	0.04	0.45	0.7
16	43.02	27.68	3.09	0.6	11.62	2.04	9.34
17	6.86	35.9	0.9	4.39	0.15	48.9	0.06

Table 5. Trend of contents of coating after working (153 days).

The obtaining of additives from alumina wastes is based on a theoretical study of the system CaO - $AI2O3$ – $SiO2$ – $Fe2O3$, MgO - CaO - $AI2O3$ – $SiO2$ [Figures 6, 7].

The obtaining of cements is based on a theoretical study of the system CaO, SrO, BaO - Al_2O_3 - SiO_2 - Fe_2O_3, MgO-CaO-Al_2O_3-SiO_2.

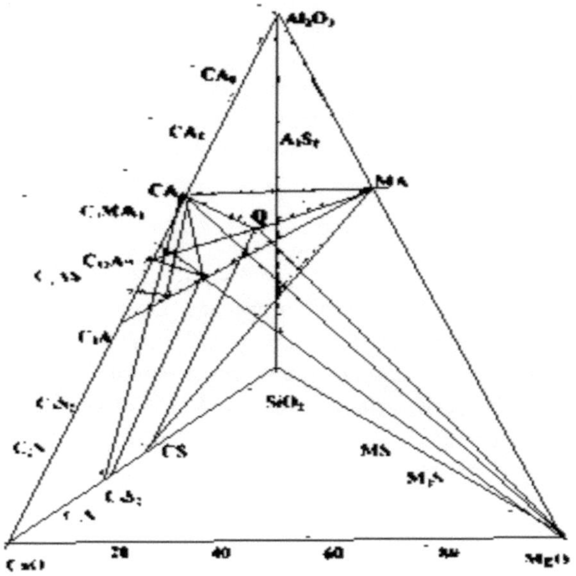

Fig. 6. Optimum region for obtaining of special cements. Subsolidus structure of the system CaO - Al_20_3 - Fe_20_3 - $Si0_2$.

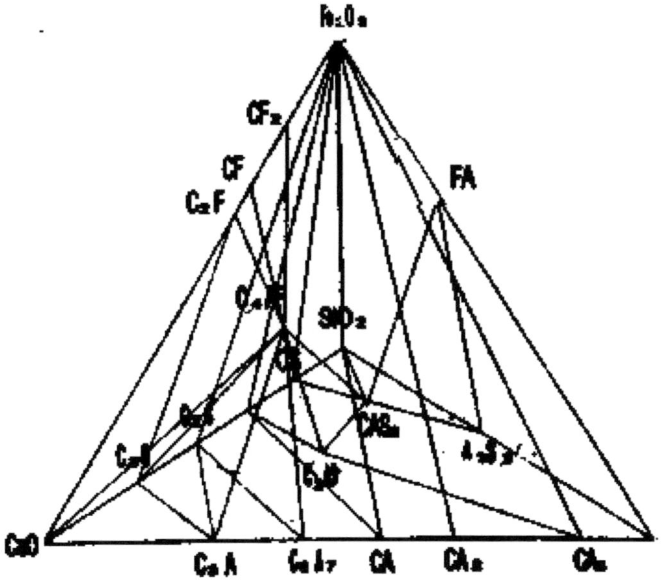

Fig. 7. Optimum region of obtaining of special cements.

It was found that pseudocuts containing double oxides and binder compounds (C_2S, C_3S, CA) have greater binder activity. The initial stage of hydration begins with leaching the surface atoms and active chemical adsorption of water molecules due to presence at active centres followed by their relation through forming OH hydroxide groups and a surfactant layer having specific surface area (to 2*10A6 sq.m/kg) and consisting of aluminum hydroxide. At the next stage the surfactant layer of double oxides adsorbs ions Ca^{+2}. Chemical adsorption of the cations decelerates nucleation and promotes formation of epitaxy contacts on the surface of the double oxides as well as oriented growth of hydrate formations of hydroaluminates hydrosilicates of calcium. As a result of chemical adsorption process and oriented crystallization 01 hydrates around the double oxides grains there is formed a contact zone that is significantly higher than that of pure cement stone. This determines the high strength of the double oxides containing cement.

4. Conclusion

The development of new high temperature composites based on zirconium cements for the application in various consuming industries has been illustrated and is well documented in terms of performance improvements. They are meant to protect units from influence of temperature more than 2073 K. They are used for coating of high temperature headtreatment, in coating of fuel - construction, in coating of furnace for making fuel, carbon-reactor, H_2 - Furnace, petrochemistry reactors.

5. References

[1] Ilyoukha N., 1984, *Refractory cements,* Ukraine Publishing Company, Kiev.

[2] Ilyoukha N.,Timofeeva W., 1995, UNITEGR'95, Japan, 10 November,1995, Ilyoukha N.,Timofeeva W., *Refractory cements.*

[3] Ilyoukha N.,Timofeeva W., 1997, UNITEGR'97, USA, New-Orlean, 15 November, 1997, Ilyoukha N.,Timofeeva W., *Development of zirconia hydraulic cements.*

[4] Ilyoukha N. , Lebedenko E., Refractories clinker, *Chemistry and Chemical Technology of Ukraine.* V.№ 5., 2007, P. 34-39.

[5] N. Iloukha, Z. Barsova, I. Cwhanovskaya, V. Timofeeva, Kinetic investigations of phaseformation processes in the system BaO - Al_2O_3 - Fe_2O_3, *Chemistry and chemical technology of Ukraine.* V.4, № 2, 2010, P.91-93.

2

Origin of Piezoelectricity on Langasite

Hitoshi Ohsato[1,2,3]
[1]*Hoseo University,*
[2]*Nagoya Institute of Technology,*
[3]*Nagoya Industrial Science Research Institute,*
[1]*Korea*
[2,3]*Japan*

1. Introduction

Piezoelectric materials produce polars in the crystal structure and charges on the surface of the crystal, when the crystals are stressed mechanically as shown in Figure 1(a). The surface charges leads to a voltage difference between the two surfaces of the crystal. On the contrary, when the crystals are applied with an electric field, they exhibit mechanical strain or distortion as shown in Figure 1(b).

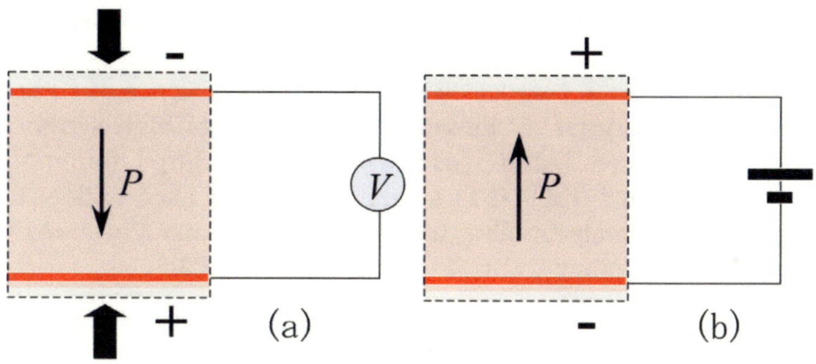

Fig. 1. The piezoelectric effects. (a) generated *V* by an applied force. (b) compressed crystal by an applied voltage.

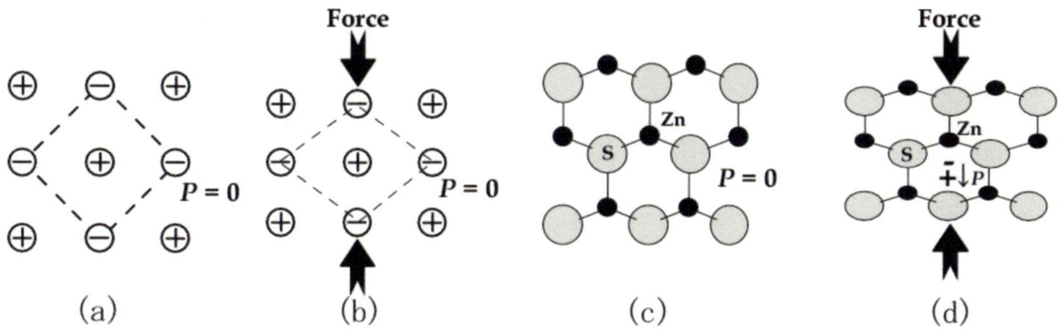

Fig. 2. (a) and (b) NaCl type crystal with *i*. (c) and (d) Hexagonal unit cell without *i*.

The crystal structure of piezoelectric materials should be no center of symmetry *i* that is inversion symmetry. Figure 2(a) and 2(c) shows the crystal structure with *i* and without *i*, respectively. The first one is NaCl structure with *i*, in which the centers of mass of positive charges and negative charges are in the same position. This case, a net dipole moment *P* does not appear in the crystal structure. Under mechanical stress, also the *P* moment does not appear, though the polarity appears under the electric field. The second one is ZnS zincblende without *i*, in which the centers of mass of positive and negative charges are in the same position as shown in Figure 2(c). However, under stress as shown in Figure 2(d), the centers of mass of positive charges and negative charges are in different positions. In this case, a net dipole moment *P* appears in the crystal structure, which is piezoelectricity.

These piezoelectric materials without *i* are included in point groups (except *O* = 432) of 2nd to 7th columns as shown in Table 1. Here, the 1st column is Laue group with *i*, and 3rd to 6th for optical activity, 5th to 7th for pyroelectricity, and 4th and 5th for enantiomorphism. Ferroelectric materials are ones with spontaneous polarization in pyroelectricity 5th to 7th. All ferroelectric materials show piezoelectricity, but the reverse is not true, that is, not all piezoelectric materials show ferroelectricity.

Piezoelectric materials such as quartz, topaz, Rochelle salt, and tourmaline and so on were discovered in 1880 by the Curie brothers. Quartz single crystal is one of the most useful piezoelectric materials. The point group is 32 of the 4th group in Table 1 which is none polar piezoelectric material. As the electromechanical coupling factor is small but temperature coefficient of resonant frequency (*TCf*) is near zero ppm/°C, quartz has been used for bulk transducer and SAW devices. Artificial quartz single crystals fabricated by hydrothermal synthesis have been used for these devices because natural single crystals are deteriorated by Brazil-type twin weaken the piezoelectricity due to co-existence of right and left hand crystals (Figure 3(a)) in a crystal as shown. The twin brings high symmetry producing mirror symmetry by screw axes 3_1 and 3_2 as shown in Figure 3(b). Lithium Niobate $LiNbO_3$ (LN) and Lithium Tantalate $LiTaO_3$ (LT) single crystals are used widely for SAW filter, resonator et al., which are treated poling for adjusting the polar directions because of the ferroelectrocity based on the point group 3*m* and space group *R*3*c*.

Lead-zirconate-titanate $Pb(Zr,Ti)O_3$ (PZT) ceramics located on the morphtropic phase boundary (MPB) co-existing trigonal and tetragonal phases has been used for bulk piezoelectricity transducer, resonator, and SAW filter, as it shows excellent piezoelectricity that has a huge electromechanical coupling factor. Though the PZT is standing at a critical moment because of toxins for health, still it is being used for industrial applications. As PZT ceramics composed by tetragonal and trigonal crystals, point group 4*mm* (space group: *P*4*mm*) and 3*m* (*R*3*m*), respectively, belonging No.7th group in Table 1, they are ferroelectrics having polar. So, poling treatments are useful for improving the piezoelectricity. The properties of the binary PZT compound between $PbZrO_3$ and $PbTiO_3$ are improved more adding $Pb(Mg_{1/3}Nb_{2/3})O_3$. The three components perovskite compounds are used IF SAW filter on 10 to more MHz region. Furthermore, the properties of these perovskite compounds are improved by applying single crystals by Hosono & Yamashita (2004) as shown in Figure 4.

Recently, Pb-free piezoelectric materials have been researched because of the toxin of Pb for health. It was applied as restriction of hazardous substances (RoHs) from 1 July 2006 based on directive 2001/95/EC of the European Parliament and of the Council of 27 January 2003,

	1	2	3	4	5	6	7	Crystal system
1	$C_i = \bar{1}$				$C_1 = 1$			Triclinic
2	$C_{2h} = 2/m$				$C_2 = 2$	$C_s = m$		Monoclini
3	$D_{2h} = mmm$			$D_2 = 222$		$C_{2v} = mm2$		Orthorhombic
4	$C_{4h} = 4/m$		$S_4 = \bar{4}$		$C_4 = 4$			Tetragonal
5	$D_{4h} = 4/mmm$		$D_{2d} = \bar{4}2m$	$D_4 = 422$			$C_{4v} = 4mm$	
6	$C_{3i} = \bar{3}$				$C_3 = 3$			Hexagonal
7	$D_{3d} = \bar{3}m$			$D_3 = 32$			$C_{3v} = 3m$	
8	$C_{6h} = 6/m$	$C_{3h} = \bar{6}$			$C_6 = 6$			
9	$D_{6h} = 6/mmm$	$D_{3h} = \bar{6}m2$		$D_6 = 622$				
10	$T_h = m3$			$T = 23$				Cubic
11	$O_h = m3m$	$T_d = \bar{4}3m$		$O = 432$				
	Laue group			Enantiomorphism				
					Pyroelectricity			
			Optical activity					
		Piezoelectricity (except 432)						

Table 1. Point groups and properties.

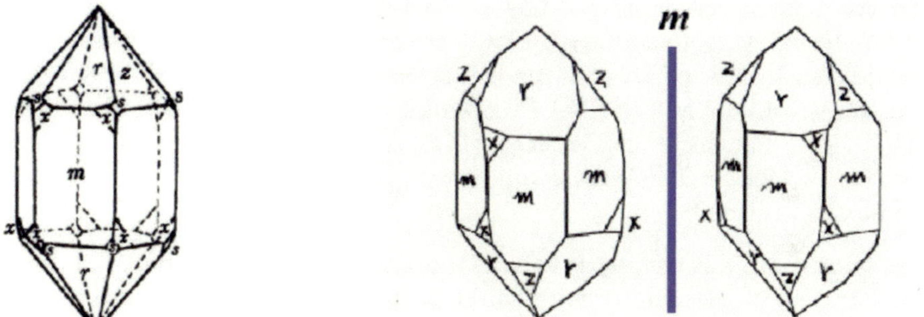

Natural crstal with Brazil-type twin.
***m*-symmetry is added.**
(a)

Left-hand and Right hand crystals
Enantiomorphism
(b)

Fig. 3. Natural quartz crystal (a) including Brazil-type twin of right and left hand crystals (b). *m*: mirror plane.

that new electrical and electronic equipment put on the market does not contain lead, mercury, cadmium, hexavalent chromium, polybrominated biphenyls (PBB) or polybrominated diphenyl ethers (PBDE) as appeared in Article 4 of the directive [web site 1]. Lead in electronic ceramic parts (e.g. piezoelectric devices) has been excluded from the RoHs directive as appeared in the Annex of the directive.

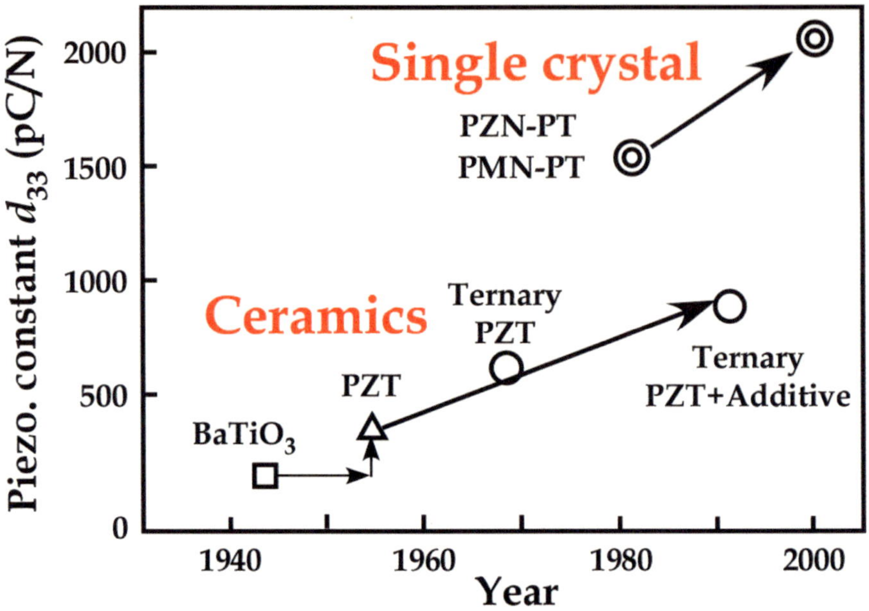

Fig. 4. Trend of piezoelectric materilas.

There are candidate materials for the Pb-free piezoelectric materials such as niobate based ceramics group, tungsten-bronze group, and perovskite relative materials. In the author's laboratory, KNO_3 being basically hard sintering materials was successfully formed as high density ceramics co-doping of La_2O_3 and Fe_2O_3 by Kakimoto et al. (2004), and Guo et al. (2004) found $[Li_{0.06}(Na_{0.5}K_{0.5})_{0.94}]NbO_3$ ceramics with excellent piezoelectric constant d_{33} values reach 235 pC/N on the MPB. Saito (2004) also fabricated highly orientated $(Na_{0.5}K_{0.5})NbO_3$-$LiTaO_3$ (NKN-LT) ceramics by template method. With the d_{33} values reach 400 pC/N. Langasite $La_3Ga_5SiO_{14}$ (LGS) group crystals are also Pb-free piezoelectric material, which are being reviewed just in this chapter, presented by Mill et al. (1982) and Kaminskii et al., (1983).

The target of in this chapter is to state mechanism of piezoelectricity on Langasite. So, some mechanisms of piezoelectricity should be presented here. Figure 5 shows mechanism of piezoelectricity in the case of BeO. Figure 5(a) is crystal structure of BeO in the absence of an applied force, here, +: Be, -:O. Under an applied force, the center of mass for positive and negative ions are different positions producing a net dipole moment P which generate negative and negative charges on the surfaces as shown in Figure 5(b). In the case of quartz SiO_2 which is the most important piezoelectric material, the structure is also constructed by six membered SiO_4 rings. The mechanism of piezoelectricity is similar with BeO, which is produced by the deformation of the six membered ring. Perovskite structures which are very important crystal structure for electro materials show very great piezoelectricity. The structure has some polymorphs such as cubic, tetragonal, orthorhombic and hexagonal. Cubic structure has center of symmetry i, and transform to tetragonal without i at curie temperature about 120 ^{o}C for $BaTiO_3$.

In this chapter, LGS, $Pr_3Ga_5SiO_{14}$ (PGS) and $Nd_3Ga_5SiO_{14}$ (NGS) single crystals grown by Czochralski method were analysed the crystal structure by X-ray single crystal diffraction

(XRSD) and clarified the differences of crystal structure. The mechanism of the piezoelectricity was clarified based on the crystal structure, which was confirmed by the deformation of the crystal structure under high pressure. Also relationships between the crystal structure and the properties are discussed, and it was also clarified that LGS has higher piezoelectric properties than NGS.

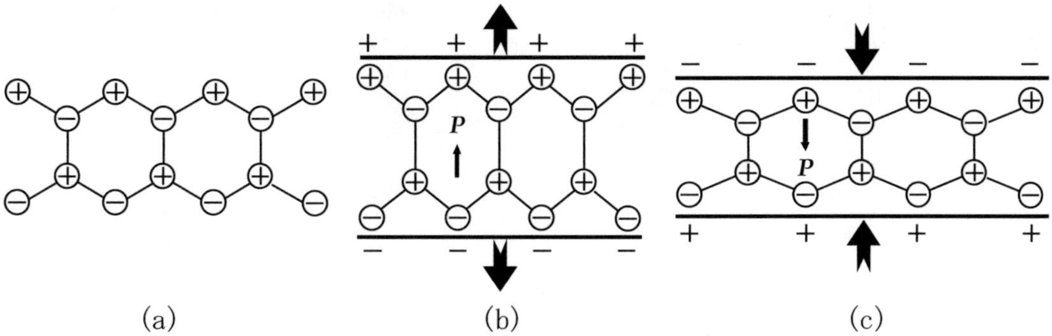

Fig. 5. Mechanism of piezoelectricity of BeO. +: Be ion, -: O ion.

2. Crystal growth of langasite

Langasite group single crystals have been grown by many growing methods such as Czochralski (Cz) technique, Bridgeman method, floating zone (FZ) method, micro-pulling down (μ-PD) technique, as these crystals grow easily because of a low melting point around 1470 ^{o}C being able to use stably Pt-crucible, no phase transition and congruent melting. The most useful method is Cz-method which is pulling up a single crystal using the seed crystal from melts in a crucible heated (Figure 6(a)-(c)). Bridgeman method is easy, which is grown in a crucible with gradient temperature spontaneously nucleating in the bottom of the crucible. FZ-method has a melting and crystallizing zone between seed crystal and sintered ceramic rod with same composition as grown crystal, which is grown without contamination because of no crucible. The viscosity of langasite crystals is considered suitable for FZ-method, because the ratio between oxygen ions and cations in tetrahedra *C* and *D*-site in the crystal structure $A_3BC_3D_2O_{14}$ is 2.8, locating near one dimensional or ring framework with 3. μ-PD method is special which grows long size thin single crystal with around 1mm ϕ by few cm meters to over one meter (Figure 6(d)). A single crystal grows during pulling down from the bottom small hole of a container made by a heater such as Pt. LGS, PGS and NGS single crystals as shown used in this chapter were grown by a conventional radiation frequency (RF)-heating Cz-method with output power 60kW by Sato et. al. (1998). Starting melts formed from single-phase powder of langasite sintered 1300 ^{o}C using high purity 99.99 % low materials were applied a height of 40 mm in platinum and iridium crucible with 50 mm in diameter and 50 mm in height. The growth atmosphere was a mixture of Ar and 1 vol % of O_2 gases in order to avoid the evaporation of gallium oxide from the melt during growth. The heating of melts was performed by Pt-crucible itself by induction heating. The crucible was isolated by ZrO_2 granules. Before seeding, the melts was clarified during 1 h at least. The pulling velocity and the crystal rotation rates were 1.0-1.5 mm/h and 10 rpm, respectively. The seeds were used a small <001> oriented LGS single crystal rods. Growing crystal was kept the temperature by a passive double after-heating system made of alumina ceramics.

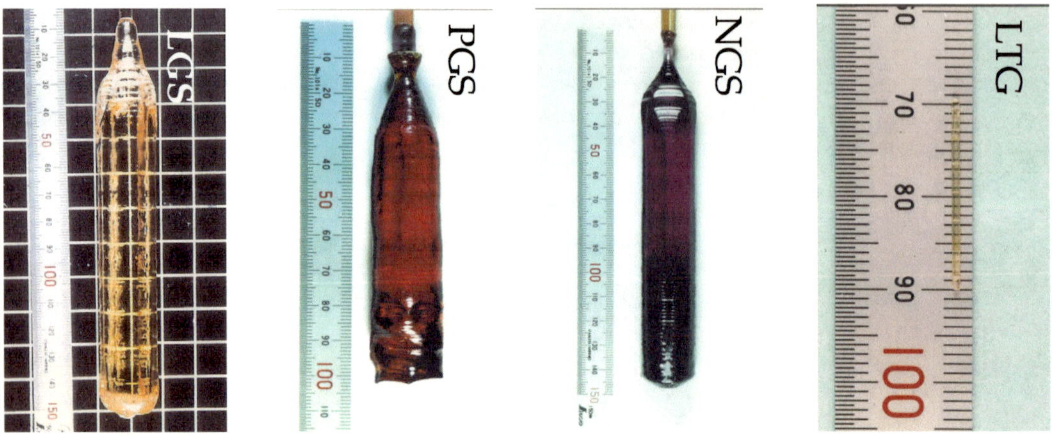

(a) (b) (c) (d)

Fig. 6. Single crystals of Langasite (a)-(c) and (d) grown by Cz-method, and by μ-PD technique, respectivery.

Defect-free LGS, PGS and NGS single crystals with constant diameter of 22 mm and lengths up to 145 mm were grown as shown in Figure 6(a), (b) and (c), respectively (Sato et al., 1998). These ingot diameters were high constancy over the whole length. The optimum pulling rates are not exceeding 1.5 mm/h for inclusion-free perfect single crystals, and a higher temperature gradient at the growing interface controls the growth preventing a distinct facet enlargement and asymmetrical growth leading to spiral morphology.

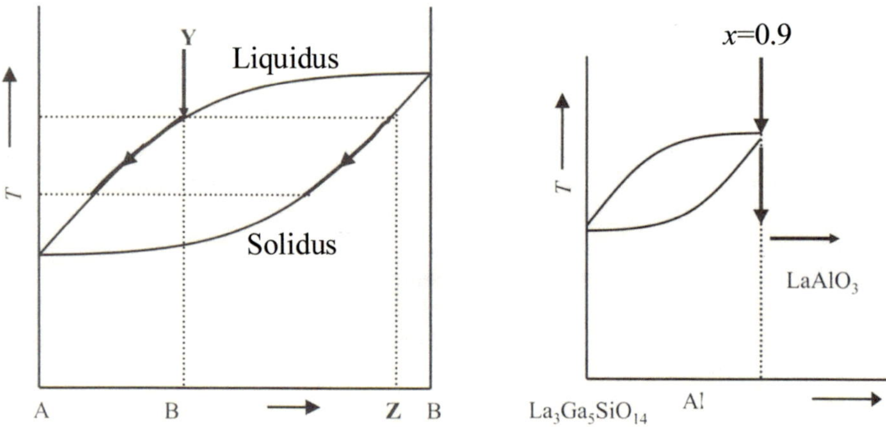

(a) (b)

Fig. 7. Crystal growth from quasi-congruent melt. (a) Crystal growth of solid solutions: composition of precipitaed should be changed gradually. (b) Formatin of quasi-congruent melt at $x = 0.9$ due to precipitated secondary phase $LaAlO_3$.

LGS compounds with congruent melting are grown easily because of same composition of growth crystals and liquid. In the case of solid solutions, composition of precipitated crystal gradually changes during crystal growth as shown in Figure 7(a). Takeda & Tsurumi (2011) presented crystal growth with homogeneous composition from quasi-congruent melt (Takeda, 1998, Kumatoriya et al., 2001). In the case of Al-substituted $La_3Ga_{5-x}Al_xSiO_{14}$ (LGASx), as the limitation of the solid solutions is located at $x = 0.9$ as shown in Figure 7(b), endmember of the solid solutions makes congruent melt. At $x = 0.9$ composition, good quality single crystals of LGAS0.9 are grown as named quasi congruent melt growth.

	NGS	PGS	LGS
Composition	$Nd_3Ga_5SiO_{14}$	$Pr_3Ga_5SiO_{14}$	$LaGa_5SiO_{14}$
Formula weight	1033.401	1023.415	1017.409
Crystal system	Trignal	Trignal	Trignal
Space group	*P*321	*P*321	*P*321
Point group	32	32	32
Lattice parameter *a* (Å)	8.0674(5)	8.0944(5)	8.1674(4)
c (Å)	5.0636(9)	5.0724(9)	5.0964(8)
Unit cell volume (Å^3)	285.40(5)	287.81(6)	294.41(5)
Formula number *Z*	1	1	1
Calculated density D_x (g/cm^3)	6.0127	5.9047	5.7384
Linear absorption coefficient μ (cm^{-1})	258.603	247.274	228.496
R	0.0320	0.0292	0.0346
Rw	0.0293	0.0264	0.0317
GOF	1.2132	1.3497	1.2169

Table 2. Crystallographic data and reliability factors for crystal structure analysis of NGS, PGS and LGS.

3. Crystal structure of langasite

Langasite crystal structure was analyzed originally by Mill et al. (1982). The crystal structure is isostructural to $Ca_3Ga_2Ge_4O_{14}$ presented by Belokoneva et al. (1980). The crystal system is trigonal, point group 32, space group *P*321 (No.150), lattice constants approximately $a = 8.1$, $c = 5.1$ Å, $Z = 1$, which is similar as quartz SiO_2. We determined the three langasite-group LGS, PGS and NGS crystals using the initial atomic parameters presented by Mill et al. (1982). Table 2 shows the crystallographic data and experimental conditions for X-ray single crystal diffraction (XRSD) analysis. Sphere single crystals around 0.04 mm diameter grounded a cut single crystal were used for the single-crystal structure analysis performed by a four-circle diffractometer with graphite monochrometer. After X-ray intensity data collected with MoKα radiation were corrected based on Lorentz and polarization (*Lp*) correction and absorption, the refinements of the crystal structure were performed by full-matrix least-squares program RADY (Sasaki, 1982). The site occupancy of *D*-site Ga:Si was obtained from multiplicity *g* determined by the linear constrain as follows:

$$g(\mathrm{Ga}) = \mathrm{cal.} \quad (1)$$

$$g(\mathrm{Si}) = 1/3 - g(\mathrm{Ga}). \quad (2)$$

On the procedure of crystal structure analyzing, a scale factor, coordinates of each atoms, and anisotropic temperature factors were refined, and at final step of refinement, anisotropic extinction corrections were performed.

Table 3(a), 3(b), and 3(c) show atomic coordinates of LGS, PGS and NGS, respectively (Iwataki, 2002, Master thesis). The equivalent isotropic temperature factors (*B*eq.) were calculated using anisotropic temperature factors by following equation:

$$B_{eq.} = 1/3\ \Sigma_i \Sigma_j B_{ij}\, a^*_i\, a^*_j\, a_i .\ a_j. \tag{3}$$

The $B_{eq.}$ values are reasonable as around 0.7 for cations and 1.5 for oxygen ions. And site occupancy ratios of Ga and Si ions in *D*-site are almost 1:1. The final reliability factors: *R* and R_w values are fine around 0.03.

The crystal structure figures projected from [001] and [120] are shown in Figure 8(a) and 8(b), respectively. The structure represented by the structural formula, $A_3BC_3D_2O_{14}$, is constructed four sites: *A*-, *B*-, *C*, and *D*-site projected from two ways as shown in Figure 8. *A*-site is decahedron with eight coordination number (*c.n.*) named as twisted Thomson cube, *B*-site octahedron with six *c.n.*, and *C*- and *D*-sites tetrahedra with four *c.n.* as shown in Figure 8(c). The size of *D*-site is slightly smaller than that of *C*-site. Rare earth La^{3+}, Pr^{3+} and Nd^{3+} occupy the *A*-site, Ga^{3+} occupies the *B*, *C* and half of the *D*-sites, and Si^{4+} half of the *D*-sites, respectively. This structure is constructed by framework layer structure: *B*-*C*-*D*-*C*-*D*-*C* six-membered rings around *A*-site as shown in Figure 8(a) projected from [100]. Tetrahedra *C*- and *D*-site, and decahedra, octahedra and open-space form layer structure as shown in Figure 8(b). Large cation sites *A*- and *B*-sites, and open-spaces makes one layer. The open-space plays important role for piezoelectric properties as described in section 5.

(a) Atomic parameter of LGS

atom	site	occupancy	*x*	*y*	*z*	$B_{eq.}$ (Å^2)
La	3*e*	1	0.41865(3)	0	0	0.625(3)
Ga1	1*a*	1	0	0	0	0.844(8)
Ga2	3*f*	1	0.76517(6)	0	1/2	0.660(7)
Ga3 Si	2*d*	0.507(3) 0.493	1/3	2/3	0.5324(2)	0.557(8)
O1	2*d*	1	1/3	2/3	0.198(1)	1.16(5)
O2	6*g*	1	0.4660(4)	0.3123(4)	0.3186(6)	1.41(7)
O3	6*g*	1	0.2208(4)	0.0811(4)	0.7629(5)	1.56(8)

(b) Atomic parameter of PGS

atom	site	occupancy	*x*	*y*	*z*	$B_{eq.}$ (Å^2)
Pr	3*e*	1	0.41787(3)	0	0	0.639(2)
Ga1	1*a*	1	0	0	0	0.762(7)
Ga2	3*f*	1	0.76473(5)	0	1/2	0.656(6)
Ga3 Si	2*d*	0.498(3) 0.502	1/3	2/3	0.5346(2)	0.548(7)
O1	2*d*	1	1/3	2/3	0.1967(9)	1.16(5)
O2	6*g*	1	0.4668(4)	0.3167(4)	0.3151(5)	1.53(7)
O3	6*g*	1	0.2215(4)	0.0777(4)	0.7614(4)	1.50(6)

(c) Atomic parameter of NGS

atom	site	occupancy	*x*	*y*	*z*	$B_{eq.}$ (Å^2)
Nd	3*e*	1	0.41796(3)	0	0	0.660(3)
Ga1	1*a*	1	0	0	0	0.782(8)
Ga2	3*f*	1	0.76466(6)	0	1/2	0.669(7)
Ga3 Si	2*d*	0.498(3) 0.502	1/3	2/3	0.5351(2)	0.557(8)
O1	2*d*	1	1/3	2/3	0.198(1)	1.33(6)
O2	6*g*	1	0.4674(4)	0.3181(4)	0.3131(7)	1.63(8)
O3	6*g*	1	0.2218(4)	0.0762(4)	0.7597(5)	1.50(7)

Table 3. Atomic paremeters of LGS, PGS and NGS.

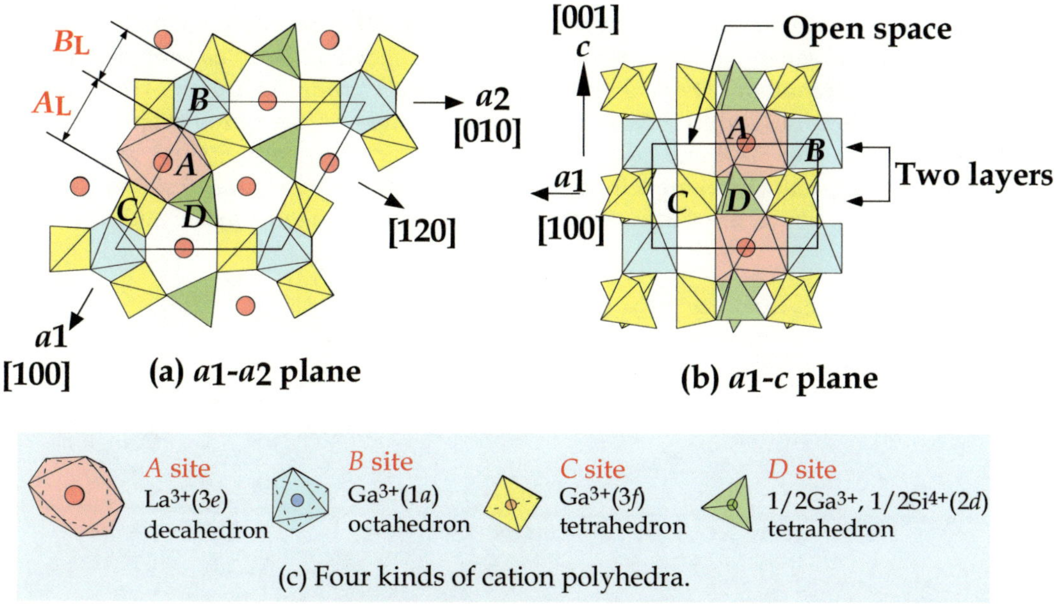

Fig. 8. Crystal structure of Langasite. (a) and (b) are viewed from [001] and [120], respectively. (c) is four kinds of cation polyhedra.

The crystal structures among LGS, PGS and NGS differ mostly in shape of each site. In particular, the change of the *A*-site is remarkable. The decahedral *A*-site expands with the increase of ionic radius of rare earth (*R*) that occupies the *A*-site. The *A*-site expands greatly in [100] directions compared to the expansion in [120], which is perpendicular to [100], with the increase of the ionic radius of *R*.

4. Piezoelectric properties of langasite

4.1 langasite and properites

Langasite shows piezoelectricity but none ferroelectricity, based on crystallographic point group 32 belonging the 4th group in Table 1. This point group is the same with that of quartz showing excellent piezoelectricity. As langasite is none polar piezoelectric crystal, poling treatment is not necessary. However ceramics that is polycrystals show isotropic properties as a whole because each orientation of grains turns to every direction. So, non-polar piezoelectric materials should be used as a single crystal. For a single crystal, the knowledge of the directions of piezoelectricity is very important. As the directions are the same one of polar, they could be derived based on the point group. The Piezoelectric constants of the point group 32 for langasite is as following tensor:

$$\begin{pmatrix} d_{11} & -d_{11} & 0 & d_{14} & 0 & 0 \\ 0 & 0 & 0 & 0 & -d_{14} & -2d_{11} \\ 0 & 0 & 0 & 0 & 0 & 0 \end{pmatrix} \tag{4}$$

Figure 9 shows the stereographic projection of general positions on point group. Figure 9(a) shows [001] direction without polarity because of the same number positions on the opposite directions of [001] that is same number ○ and x. [210] direction (Figure 9(c)) also is non-polarity by the same manner. Only [100] direction shows polarity as shown in Figure 9(b). The configurations of a typical crystal with a point group 32 are shown along the stereographic projections, which crystal surfaces are plotted on the stereo projections. The crystal structures along [120] and [100] as shown in Figure 9(d) and (e) show asymmetry, and symmetry, respectively, along left and right directions. Now, Figure 10 shows *Y*-cut of crystal. Here, *X*, *Y*, and *Z* are Cartesian coordination, and hexagonal axis *a* and *c* also are shown.

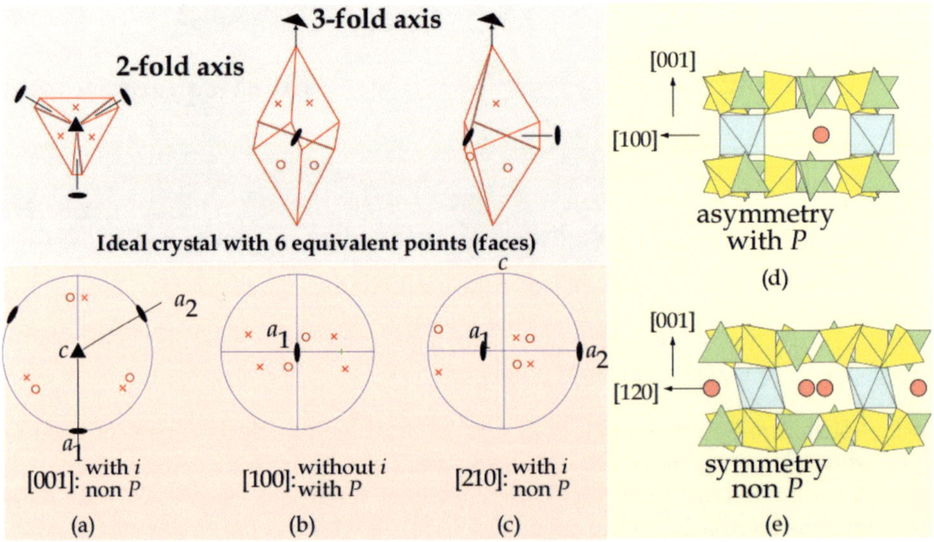

Fig. 9. Determination of piezoelectricity direction based on point group 32. Stereo graphs (a), (b) and (c) with equivalent points are projected from [001], [100], and [210], respectively. X: upper points, o: opposite points. Configurations of a crystal with point group 32 also are drown for supporting the stereo projections. (d) and (e) show the crystal structure along [100] and [120] showing asymmetry and symmetry, respectively. Dipole moment will be appeared in (d).

Figure 11(a) shows equivalent series resistance as a function of vibration modes of resonators on the LGS and quartz single crystals (Shimamura, 1996). The resistance of LGS is one order smaller than that of quartz. So, as if the surface roughness of LGS is large, high frequency oscillation is easy. Moreover, as the equivalent series resistances at high vibration mode as 7th and 9th are small, LGS filter is useful for high frequency wave area filter. Figure 11(b) shows electromechanical coupling factor k_{12} as a function of piezoelectric constant -d_{11} on the langasite group such as LGS. PGS, NGS and so on, comparing quartz single crystal. Figure 12(a) shows electromechanical coupling factor k^2 as a function of *TCf* on the piezoelectric materials. The value of langasite is near zero. Figure 12(b) shows temperature dependence of frequency and equivalent series resistance of filter made of Y-cut LGS single crystal (Shimamura, 1996). The temperature dependence of frequency shows a secondary curve with good values of 1-2 ppm/°C. In the range of -20 to 70 °C, the dependence of temperature is good value of 100 to 150 ppm/°C. Table 4 shows the properties comparing some piezoelectric crystals such as

$LiTaO_3$, LGS, quartz, and $La_3Ga_{5.5}Nb_{0.5}O_{14}$ (LGN) (Fukuda et al. 1998). The properties of LGS locate between $LiTaO_3$ and quartz. Electro-mechanical coupling factor k of LGS is 15 to 25 % locating between that of $LiTaO_3$ 43 % and quartz 7 %. The temperature frequency variation of LGS is 100 to 150 ppm/°C locating between that of $LiTaO_3$ 200 to 400 ppm/°C and quartz 50 to 80 ppm/°C. Here, LGN single crystal substituted Nb^{5+} and Ga^{3+} for Si^{4+} has superior properties for piezoelectric properties. Figure 13 shows Pass band characteristic of filter made of Y-cut LGS single crystal (Fukuda, 1995). Y-cut LGS single crystal has a very wide pass band characteristic width of 45 KHz at 3 dB attenuation which is 3-times that of quartz with 15 KHz band width. This means the electromechanical coupling constant K_{12} of LGS is about 3-times larger than that of quartz.

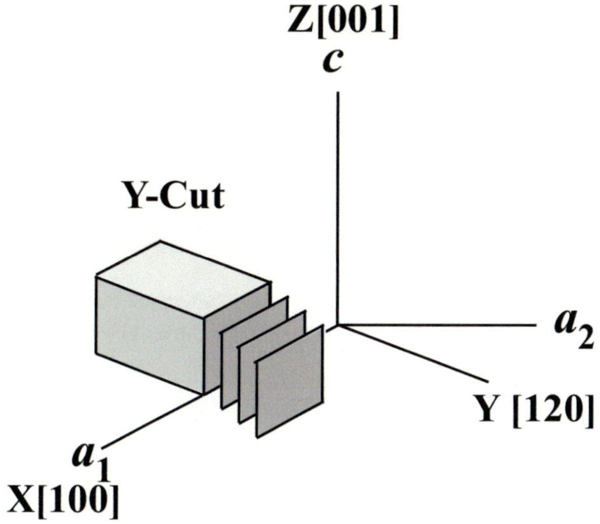

Fig. 10. Y-cut for langasite single crystal for piezoelectric measurements.

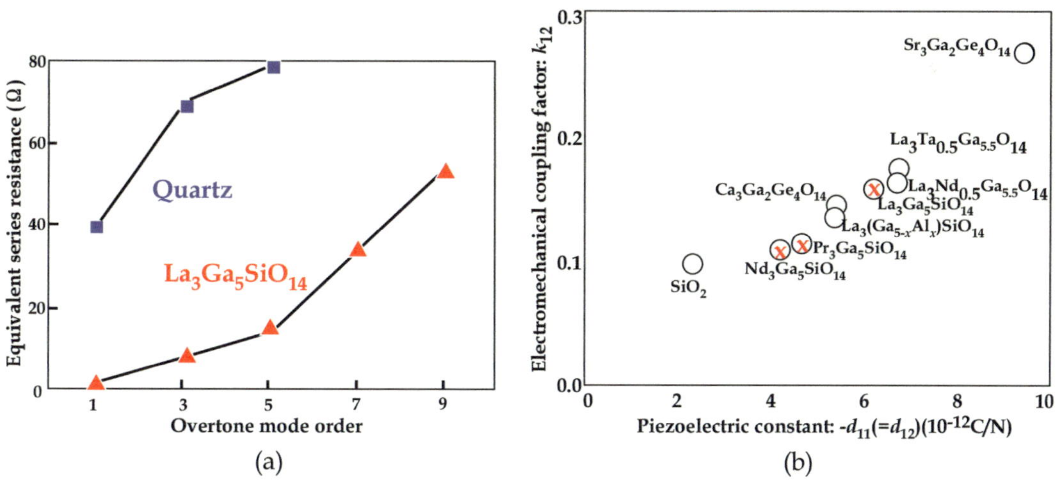

Fig. 11. (a) Equivalent series resistance of quartz and langasite single crystals as a function of resonator vibration modes. (b) electromechanical coupling factor of langasite series as a function of piezoelectric constant.

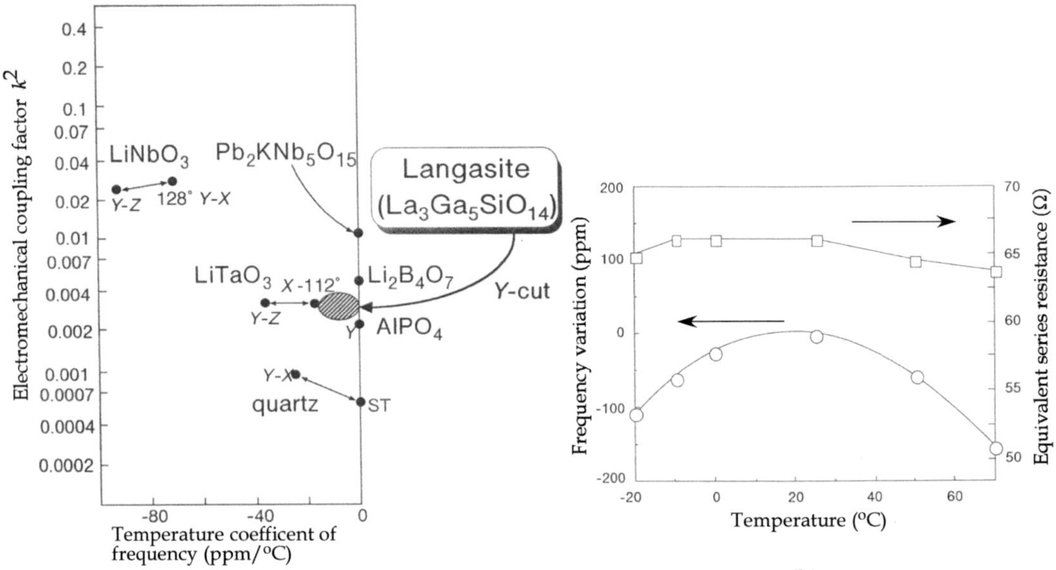

(a) (b)

Fig. 12. (a) Electromechanical coupling factor vs. temperature coefficent of frequency of piezeelectric materials. (b) Frequency variation/equivalent series resistance as a function of temperature on the $La_3Ga_5SiO_{14}$ filter.

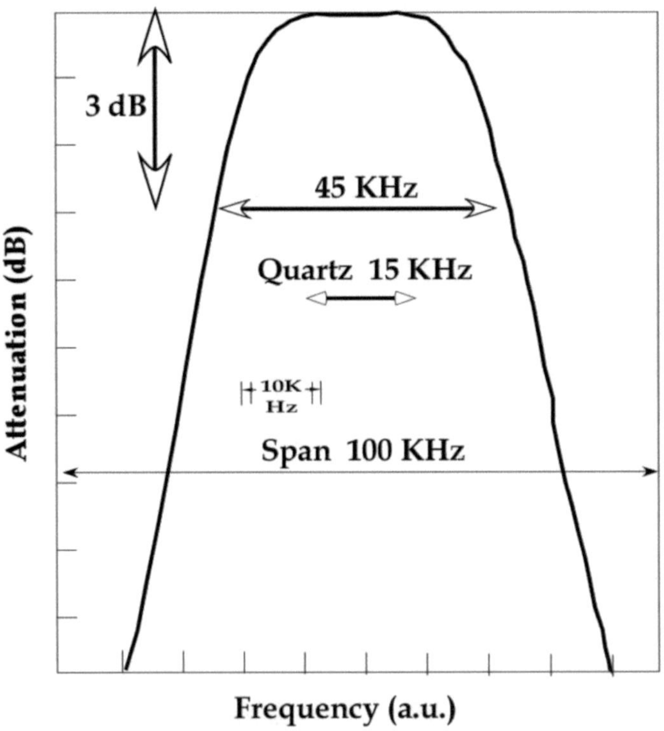

Fig. 13. Filter properties of $La_3Ga_5SiO_{14}$ single crystal.

	$LiTaO_3$	$La_3Ga_5SiO_{14}$	$La_3Ga_{5.5}Nb_{0.5}O_{14}$	Quartz
phase transition	665°C	none	none	573°C
melting temp. (°C)	1650	1470	1470	——
mohs hardness	5~6	6~7	6~7	7
electro-mechanical coupling factor k [%]	43	15~25	~30	7
Q value	5000	30000~40000	60000~120000	100000~200000
equivalent series resistance [Ω]	——	2~5	0.7~1.7	20~40
frequency variation [ppm] (−20~70°C)	200~400	100~150	——	50~80

Table 4. Comparison of properties of each crystal.

4.2 Ordered crystal structure and properties

LGS, PGS and NGS having been described here are compositionally disordered crystals. The structural formulae are $[R_3]_A[Ga]_B[Ga_3]_C[GaSi]_DO_{14}$ (R = La, Pr and Nd). Here, as D-site is occupied disorderly by Ga and Si, these crystals are disordered. An ordered langasite structural formula such as $Sr_3TaGa_3Si_2O_{14}$ (STGS), $Sr_3TaGa_3Ge_2O_{14}$ (STGG), $Sr_3NbGa_3Si_2O_{14}$ (SNGS), $Ca_3NbGa_3Si_2O_{14}$ (CNGS), and $Ca_3TaGa_3Si_2O_{14}$ (CTGS) are presented and characterized by Mill et al., (1998) and Takeda et al., (2000). The structural formula is $[Sr/Ca_3]_A[Nd/Ta]_B[Ga_3]_C[Ge/Si]_DO_{14}$: large A-decahedron is occupied by Sr or Ca cations, middle size B-octahedron by Nd or Ta cations, and C- and D-tetrahedra by the larger Ga and the smaller Ge or Si cations, respectively. The ordering should be called as "compositional ordering" compared with ordering based on the order-disorder transition.

Table 5 shows characterization of disordered and ordered langasite-type piezoelectric single crystals at room temperature and 500 °C (Zhang, 2009). Though the ordered crystals posess lower piezoelectric coefficients than disordered ones at room temperature, they posess much higher mechanical quality factor and electrical resistivity at elevated temperature of 500 °C. The high mechanical quality factor and the high electrical resistivity has been expected for a high-temperature bulk acoustic wave (BAW) and SAW resonator and ignition pressure sensor, respectively. Density and dielectric constant of ordered cyrstals are lower than those of disordered ones, which contribute to the high acoustic velocity on the high frequency devices. LTGA and LNGA disordered crystals in the Table 5 are substituted Al for Ga on the $La_3R_{0.5}Ga_{5.5-x}Al_xO_{14}$ (LRGAx, R=Ta or Nb), which are contributed to the low raw material cost. Takeda et al. (2005) presented LTG, LTGA0.3 and 0.5 in which d_{14} values are increased 3.68, 4.03 and 4.19 pC/N in the order, and those resistivity increased 2.2 x 10^7, 4.6 x 10^7 and 7.1 x 10^8 Ω•cm in the order at 400 °C as shown in Figure 14 (Takeda et al., 2005). The resistivity of LTGA0.5 increased about 30 times of that of LTG. Also Al-substituted LGS ($La_3Ga_{5-x}Al_xSiO_{14}$: LGASx) are studied for high resistivity at elevated temperature and low cost, which are presented by Kumatoriya et al. (2001), Takeda et al. (2002). The piezoeoectric properties d_{11} and resistivity of LGAS0.9 was improved from 6.075 to 6.188 pC/N and 5.9 x 10^7 to 7.6 x 10^8 Ω•cm, respectively. The Al-substitution is effective for high resistivity and also reduce the raw material cost. CTAS in the Table 5 is substituted Al for Ga compleatly which has high resistivity of 2.7x10^9 Ω•cm. On the other hand, Fe-substituted langasite-type crystals are expected for multiferroic materials (C. Lee, et al., 2010).

								500 °C		
Material	Structure	Density (g/cm^3)	$□_{11}$	Loss	K_{12}	s^E_{11} (pm^2/N)	d_{11} (pC/N)	Q	□ (□ cm)	RC (ms)
LGS	Disordered	5.85	18.0	>0.001	0.16	8.86	6.20	···	$9.0x10^6$	0.02
LTG	Disordered	6.12	19.6	>0.001	0.17	9.07	7.10	···	$1.5x10^7$	0.03
LNG	Disordered	5.95	20.7	>0.001	0.18	9.27	7.40	···	$5.0 x10^7$	0.10
LTGA	Disordered	6.07	21.0	>0.001	0.16	9.15	6.60	1000	$2.2 x10^7$	0.05
LNGA	Disordered	5.90	19.5	>0.001	0.16	9.90	6.90	1000	$1.1 x10^8$	0.22
SNGS	Ordered	4.65	12.4	<0.001	0.17	8.80	5.40	7500	$6.3 x10^7$	0.08
STGS	Ordered	5.12	11.8	<0.001	0.16	8.69	4.90	11000	$3.7 x10^8$	0.42
CNGS	Ordered	4.15	17.5	<0.001	0.11	8.75	4.00	8000	$6.9 x10^7$	0.11
CTGS	Ordered	4.63	16.5	<0.001	0.11	8.95	4.00	19000	$1.7 x10^9$	2.36
CTAS	Ordered	4.04	13.0	<0.001	0.14	8.51	4.30	20000	$2.7 x10^9$	3.25

Table 5. Characterization of disordered and ordered langasite-type piezoelectric single crystals at room temperature and 500 °C. LTG:$La_3Ta_{0.5}Ga_{5.5}O_{14}$; LNG:$La_3Nb_5GaO_{14}$; LTGA: $La_3Ta_{0.5}Ga_{5.3}Al_{0.2}O_{14}$; LNGA: $La_3Nb_{0.5}Ga_{5.3}Al_{0.2}O_{14}$;SNGS: $Sr_3NbGa_3Si_2O_{14}$; STGS: $Sr_3TaGa_3Si_2O_{14}$; CNGS: $Ca_3NbGa_3Si_2O_{14}$; CTGS: $Ca_3TaGa_3Si_2O_{14}$;CTAS: $Ca_3TaAl_3Si_2O_{14}$.

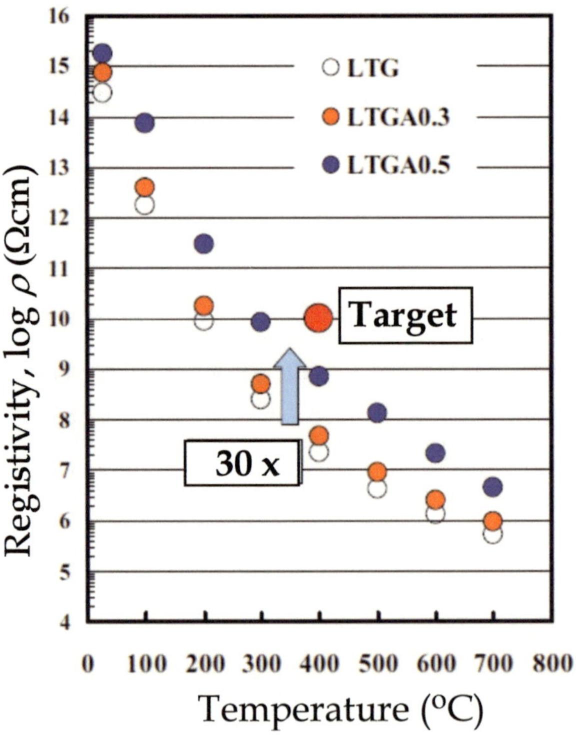

Fig. 14. Resistivity of LTG, LTGA0.3 and 0.5 at elevated temperature. Al-substituted LTG 0.5 was improved about 30 times at 400 °C. More high resistivity is expected.

5. Mechanism of piezoelectricity of langasite based on the crystal structure

In this section, mechanism of piezoelectricity of langasite will be presented based on the crystal structure. Fukuda et al. (1995) and Sato et al. (1998) presented an excellent relationship between the piezoelectric properties of LGS, PGS and NGS, and the lattice parameters and the ionic radius of *R*-ion in *A*-decahedron as shown in Figure 15(a) and (b), respectively. Now, we will consider the reason why LGS has the best piezoelectric properties among LGS, PGS and NGS. The volumes of *A*-site increase depending on the ionic size, and the lattice parameter of *a*-axis elongates by 0.100 Å larger than that 0.023 Å of *c*-axis based on the crystal structure as described section 3. Deformation of *A*-site along [100] direction might bring the piezoelectricity. The direction is just direction generating the piezoelectricity as shown in Figure 9 described in the Section 4. So, we will consider the deformation of *A*-site based on the crystal structure obtained. Figure 16 shows the deformation of *A*-decahedron. This decahedron is a twisted Thomson cube with 8 coordination numbers composed of three kinds of oxygen ions: O1, O2 and O3 as shown in Table 3 in Section 3. Here, important oxygen ions for the deformation are named by I, II and III as shown in Figure 16. The I anion is O3, and the II and III anions are equivalent of O2 by 3-fold axis. Here, O1 on the top oxygen of *D*-tetrahedron just located on the 3-fold axis. These I, II and III anions are shifted toward arrow direction in Nd, Pr and La order based on the crystal structure obtained. For the III anion, coordinate x increased, and y also increased. On the contrary, for the II anion, coordinate x decreased and y also decreased. As a result, the *A*-site deforms anisotropically as expanding to [100] and shrinking to [120] direction. Though other sites also deform a little, the explanations will be elsewhere because it is not so important for piezoelectric properties in this crystal structure. This *A*-site deformation compared with *B*-site: A_L/B_L as shown in Figure 17(a) has relationships with piezoelectric constant d_{11} and electromechanical coupling factor k_{12} as shown in Figure 17(b). Here, A_L and B_L are lengths of *A*-site and *B*-site along [100], respectively. Three components compounds: LGS, PGS and NGS show just linear relationship with d_{11} and k_{12}.

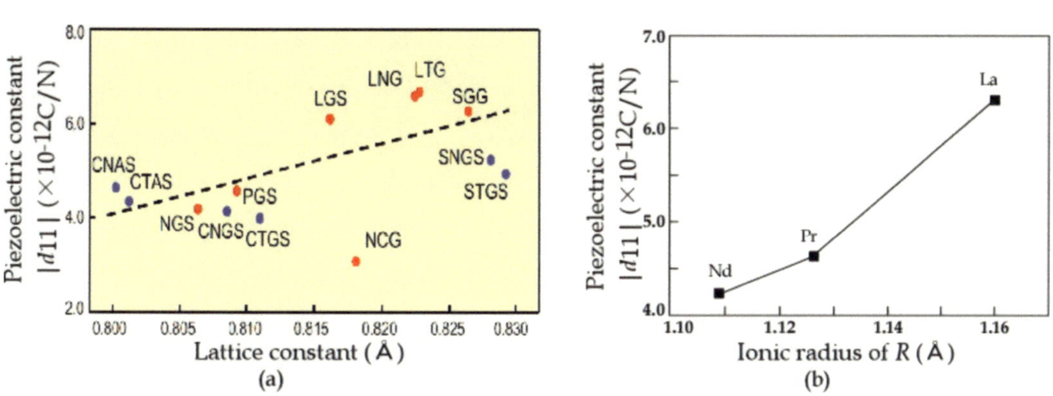

Fig. 15. (a) Piezoelectric constant d_{11} as a function of lattice constant for Langasite series crystals. (b) d_{11} of LGS, RGS and NGS as a function of ionic radius.

Iwataki et al. (2001) also shows relationship between A_{L1}/A_{L2} and piezoelectric modulus $|d_{11}|$ as shown in Figure 18. Here, A_{L1} is same length with A_L as shown in Fig.17, and A_{L2} is length along [120] of *A*-polyhedron. As seen from Figure 18(b), the ratio A_{L1}/A_{L2} increases with the ionic radius of R and the piezoelectric modulus also increases. The ratio shows distortion of A-polyhedron, and large distortion brings high d_{11}. LGS with large La ion brings larger d_{11} more than PGS and NGS with Pr and Nd, respectively.

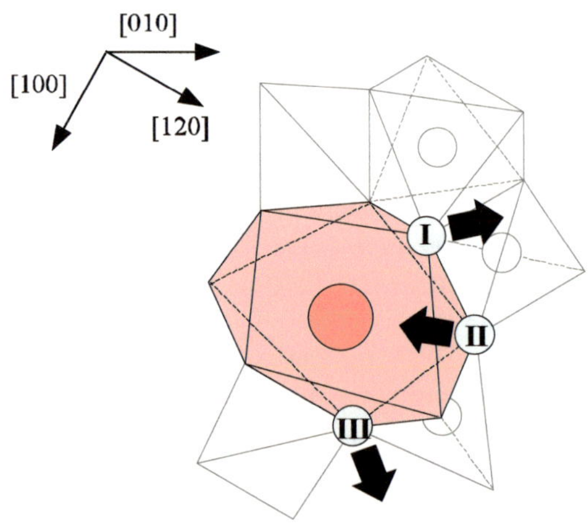

Fig. 16. Deformation of A-decahedron. I:O3, II and III:O2. When A-site was occupied by large ion such as La, the A-site will be deformed to arrow direction of oxygen ions anisotropically.

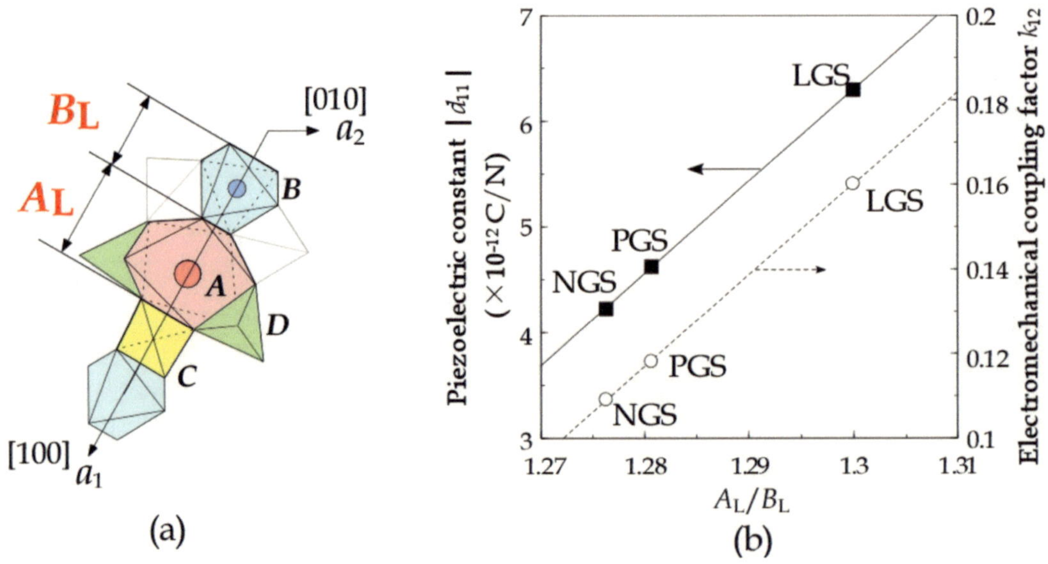

Fig. 17. (b) Piezoelectric constant d_{11}/ electromechanical coupling factor k_{12} of LGS, PGS and NGS as a function of A_L/B_L which is shown in (a).

We will consider a mechanism for piezoelectricity based on the information obtained before. The direction for piezoelectricity is [100] direction based on point group 32 as shown in Figure 9 described in Section 4. Along the [100] direction, there are two cation polyhedra: A-decahedron and B-octahedron, and open-space located among the polyhedra, which play an important role for piezoelectricity as sown in Figure 19. B-ions in octahedron locate on the

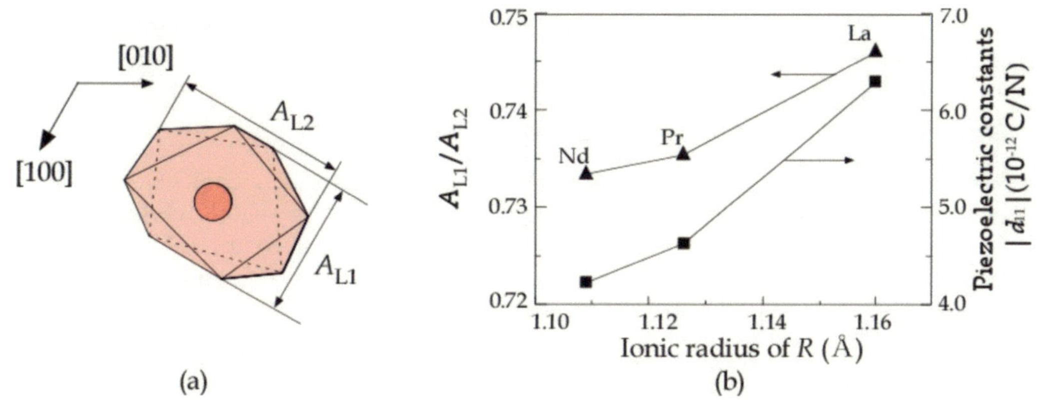

Fig. 18. (a) Size A_{L1} and A_{L2} of A-polyhedron along [100] and [120], respectively. (b) The ratio A_{L1}/A_{L2} and piezoelectric modulus $|d_{11}|$ as a function of ionic radius of R.

origin of unit cell, and A-ion locates in the decahedron contacted with B1-octahedron and open-space.

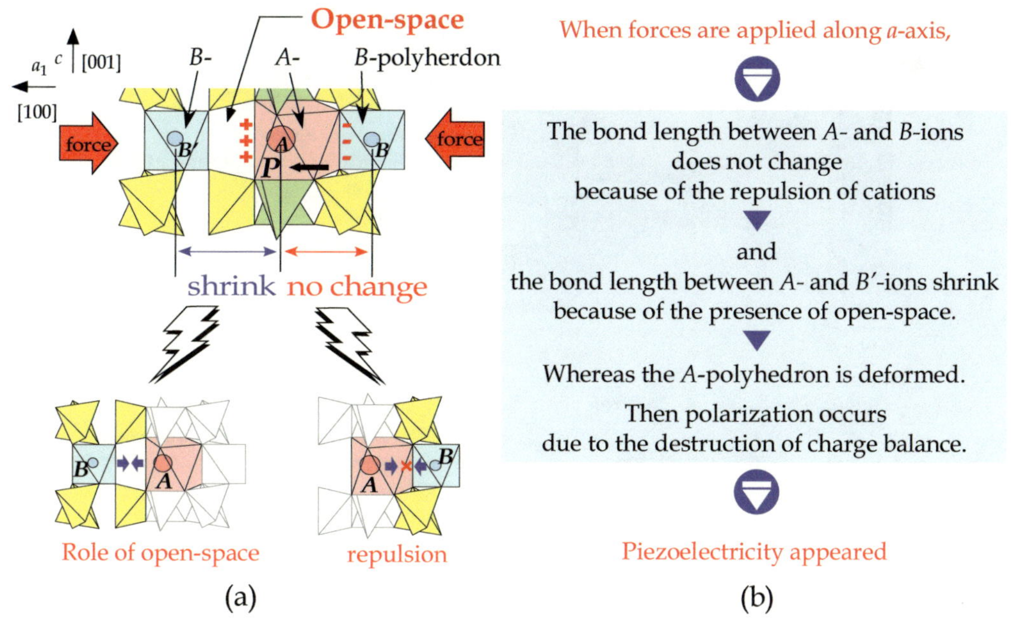

Fig. 19. Mechanism of piezoelectricity presented imaginary based on role of open-space in the crystal structure.

With an induced pressure, though all polyhedra are compressed, distances between cations show different movements. The distance between A and B shows no change, and that between A and B' shows shrinkage. This phenomenon could be explained by the existence of open-space, that is, no change in the distance between A and B should be generated by the repulsion between the cation charges and by the role of shock absorption of the open-space. The shrinkage between A and B' is generated by the shock absorption of the open-space. As a result, the position of A-cation is shifted from the center of the decahedron, so, it brings the piezoelectricity that the centers of mass of positive charges and negative charges are in different positions.

6. Crystal structure and piezoelectricity under pressure

In this section, the mechanism of the piezoelectricity presented imaginary in section 5 will be confirmed under applied pressure. Pressures applied to the single crystals using a diamond anvil cell (DAC) as shown in Figure 20 were calibrated by the ruby fluorescence technique. The DAC was putted on the four-circle diffractmeter. The unit cell parameters were refined by using the 2θ-ω step scan technique.

Table 6 shows crystallographic data of LGS and NGS including the lattice parameters measured at various pressures. The pressure dependence of lattice parameters (a_1 and c) and unit cell volume are plotted in Figure 21. Both of lattice parameters along a_1- and c-directions shrunk linearly with an increase of applied pressure. Noteworthy is that the a_1-axis is preferentially shrunk direction compared to the c-axis. This indicates that the compression of langasite crystals occurs preferentially in the a_1-a_2 plane.

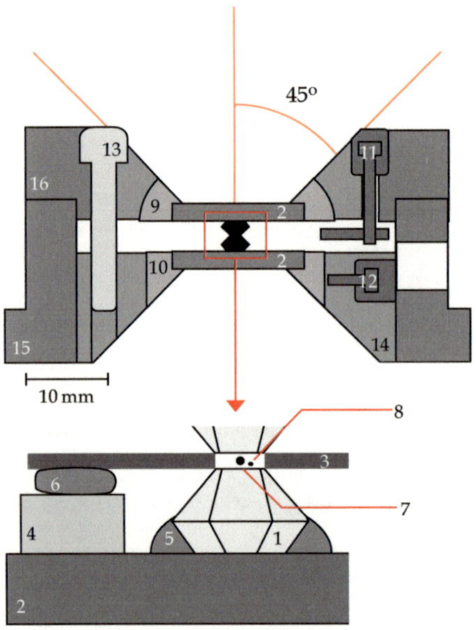

(1) diamond anvils
(2) beryllium diamond support disc
(3) gasket
(4) spacer ring
(5) epoxy resin
(6) adhesive
(7) single crystal
(8) ruby
(9) hemispherical plate
(10) disc-shaped plate
(11) inclined adjusting screw
(12) lateral-position adjusting screw
(13) driver screw
(14) lower sliding piston
(15) cylindrical sleeve
(16) upper stationary disc

Fig. 20. Schematic cross section of diamond anvil cell.

	LGS			NGS		
Pressure (GPa)	Atm.	3.3	6.1	Atm.	3.5	6.8
Lattice parameter a (Å)	8.1674(4)	8.103(1)	8.046(3)	8.0674(5)	7.999(1)	7.926(1)
c (Å)	5.0964(8)	5.070(1)	5.052(1)	5.0636(9)	5.041(1)	5.011(1)
Unit cell volume (Å^3)	294.41(5)	288.3(1)	283.2(1)	285.40(5)	279.4(1)	272.6(1)
Calculated density D_x (g/cm^3)	5.7384	5.860	5.965	6.013	6.143	6.295
Linear absorption coefficient m (cm^{-1})	228.496	233.339	237.505	258.608	264.210	270.751
R	3.45	4.70	6.01	3.17	5.88	4.48
wR	3.60	5.59	7.03	3.30	6.78	5.17

Table 6. Crystallographic data of LGS and NGS under pressure.

(a) Atomic parameter of LGS at atmospheric pressure

atom	site	occupancy	*x*	*y*	*z*	Beq. (Å^2)
La	3*e*	1	0.41862(3)	0	0	0.632(3)
Ga1	1*a*	1	0	0	0	0.849(8)
Ga2	3*f*	1	0.76514(6)	0	1/2	0.665(7)
Ga3 Si	2*d*	0.5 0.5	1/3	2/3	0.5324(2)	0.553(7)
O1	2*d*	1	1/3	2/3	0.196(1)	1.14(5)
O2	6*g*	1	0.4655(5)	0.3114(5)	0.3165(7)	1.43(7)
O3	6*g*	1	0.2218(5)	0.0816(5)	0.7639(6)	1.57(8)

(b) Atomic parameter of LGS at 3.3 GPa

atom	site	occupancy	*x*	*y*	*z*	Beq. (Å^2)
La	3*e*	1	0.4212(3)	0	0	1.00(4)
Ga1	1*a*	1	0	0	0	1.4(1)
Ga2	3*f*	1	0.7671(5)	0	1/2	1.08(7)
Ga3 Si	2*d*	0.5 0.5	1/3	2/3	0.531(2)	1.1(1)
O1	2*d*	1	1/3	2/3	0.199(7)	1.4(5)
O2	6*g*	1	0.456(4)	0.306(3)	0.306(5)	2.2(4)
O3	6*g*	1	0.224(3)	0.084(3)	0.765(4)	1.6(3)

(c) Atomic parameter of LGS at 6.1 GPa

atom	site	occupancy	*x*	*y*	*z*	Beq. (Å^2)
La	3*e*	1	0.4216(4)	0	0	0.94(6)
Ga1	1*a*	1	0	0	0	1.4(2)
Ga2	3*f*	1	0.7675(8)	0	1/2	0.96(10)
Ga3 Si	2*d*	0.5 0.5	1/3	2/3	0.531(2)	0.8(2)
O1	2*d*	1	1/3	2/3	0.205(8)	1.0(7)
O2	6*g*	1	0.459(5)	0.308(4)	0.312(6)	2.2(5)
O3	6*g*	1	0.217(5)	0.080(5)	0.757(5)	2.2(6)

(d) Atomic parameter of NGS at atmospheric pressure

atom	site	occupancy	*x*	*y*	*z*	Beq. (Å^2)
Nd	3*e*	1	0.41796(3)	0	0	0.658(3)
Ga1	1*a*	1	0	0	0	0.776(7)
Ga2	3*f*	1	0.76460(6)	0	1/2	0.667(6)
Ga3 Si	2*d*	0.5 0.5	1/3	2/3	0.5352(2)	0.559(7)
O1	2*d*	1	1/3	2/3	0.196(1)	1.31(6)
O2	6*g*	1	0.4673(5)	0.3174(5)	0.3104(7)	1.60(7)
O3	6*g*	1	0.2234(5)	0.0771(5)	0.7607(5)	1.46(7)

(e) Atomic parameter of NGS at 3.5 GPa

atom	site	occupancy	*x*	*y*	*z*	Beq. (Å^2)
Nd	3*e*	1	0.4193(4)	0	0	1.02(6)
Ga1	1*a*	1	0	0	0	1.1(2)
Ga2	3*f*	1	0.7667(9)	0	1/2	1.12(10)
Ga3 Si	2*d*	0.5 0.5	1/3	2/3	0.535(2)	0.8(2)
O1	2*d*	1	1/3	2/3	0.189(9)	2.2(10)
O2	6*g*	1	0.469(5)	0.322(5)	0.316(5)	2.1(6)
O3	6*g*	1	0.227(5)	0.076(5)	0.770(5)	2.5(6)

(f) Atomic parameter of NGS at 6.8 GPa

atom	site	occupancy	*x*	*y*	*z*	Beq. (Å^2)
Nd	3*e*	1	0.4227(3)	0	0	0.79(5)
Ga1	1*a*	1	0	0	0	1.2(1)
Ga2	3*f*	1	0.7688(7)	0	1/2	0.90(8)
Ga3 Si	2*d*	0.5 0.5	1/3	2/3	0.534(1)	1.0(1)
O1	2*d*	1	1/3	2/3	0.181(8)	2.3(9)
O2	6*g*	1	0.467(6)	0.319(4)	0.312(5)	2.6(5)
O3	6*g*	1	0.224(5)	0.075(4)	0.768(5)	3.1(6)

Table 7. Atomic parameter of LGS (a) to (c) and NGS (d) to (f) under the pressure.

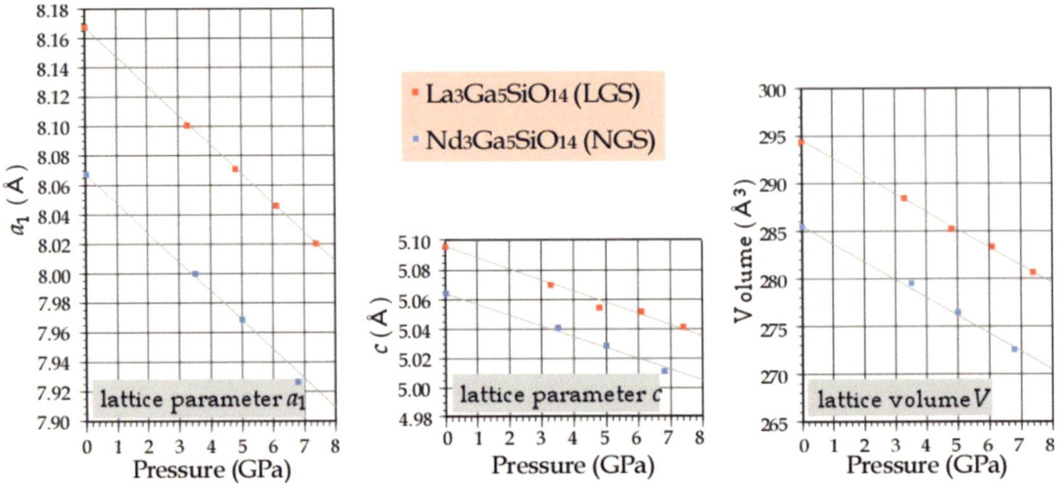

Fig. 21. Lattice parameters and volume of LGS and NGS as a function of pressure.

Table 7 shows atomic parameters of LGS and NGS under pressure. The site occupancies of Ga and Si ions on the *D*-site are fixed to 0.5 according to the data as shown in Table 3. The temperature coefficients are calculated from anisotropic temperature coefficients obtained. Bond lengths and volumes of this structure are calculated from these data. Table 8 compares the variation of A_L, B_L and S_L, presented in Figure 22, along [100] direction between atmospheric pressure, around 3 and 6 GPa. Lattice parameter a_1 equals the sum of $A_L+B_L+S_L$. Therefore, we can consider that preferential shrinkage observed along a_1-axis is also divided into three kinds of length. As seen in Table 8, the change in S_L is much larger than the other length of A_L and B_L, indicating a larger contribution of shrinkage of an open-space. The open-space forms corner shares with *A*- and *B*-polyhedra. In contrast, *A*- and *B*-polyhedra make shared edges with each other. When the pressure induced in the [100] direction, it can be speculated that the corner-shared open-space is easily distorted compared with the edge-shared *A*- and *B*-polyhedra.

	LGS			NGS		
	Pressure (GPa)			Pressure (GPa)		
(Å)	Atm.	3.3	6.1	Atm.	3.5	6.8
A_L	3.516	3.541	3.536	3.411	3.324	3.413
B_L	2.956	2.949	2.848	2.982	3.024	2.869
S_L	1.694	1.612	1.661	1.674	1.652	1.644
A_L/B_L	1.19	1.20	1.24	1.14	1.10	1.19

Table 8. *A*- and *B*-polyhedra size A_L and B_L, and open-space size S_L along *a*-axis and A_L/B_L for LGS and NGS. Atm.: atmosphere pressure.

M_L, N_L, O_L, Dipole moment (M_L-O_L) and a_1 (Å)	LGS			NGS		
	Pressure (GPa)			Pressure (GPa)		
	Atm.	3.3	6.1	Atm.	3.5	6.8
M_L	3.419	3.413	3.392	3.372	3.354	3.350
N_L	4.748	4.690	4.654	4.696	4.645	4.576
O_L	3.271	3.249	3.215	3.229	3.203	3.172
M_L- O_L	0.148	0.164	0.177	0.143	0.151	0.178
a_1	8.167	8.103	8.046	8.067	7.999	7.926

Table 9. Atomic distances (M_L, N_L), center position X-*B* ion distance (O_L), dipole moment (M_L - O_L) and lattice constant of a_1 of LGS and NGS along [100] direction under the pressure. Atm.: atmosphere pressure.

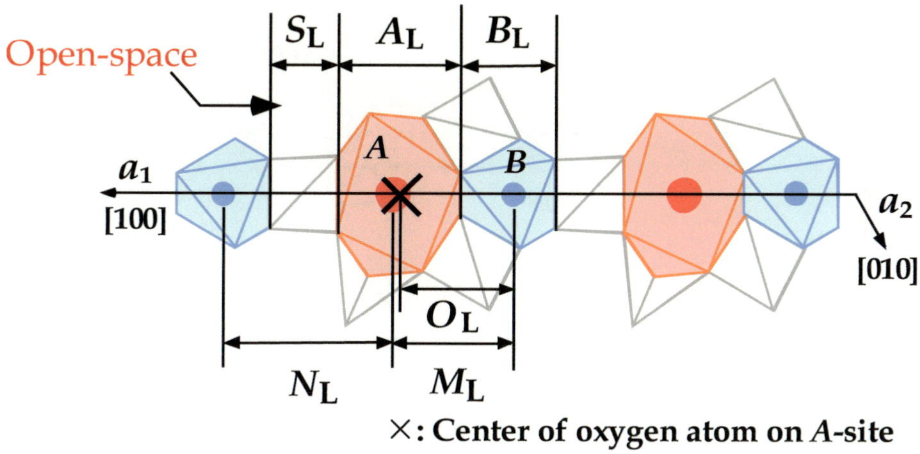

Fig. 22. Sizes of *A*-, *B*-site and open space *S*, and atomic distances of M_L, N_L and O_L between atoms. Position X is the center of oxygen atoms on *A*-polyhedron.

Piezoelectricity is originated from polarization caused by the destruction of charge balance depending on the displacement of ions when pressure is induced. The piezoelectricity of langasite is known to be generated in the [100] direction. From the rule of symmetrical operation, langasite should have polarization in *A*- and *C*-polyhedra in contrast with no polarization in *B*- and *D*-polyhedra. We have already stated the mechanism of piezoelectricity imaginary on Langasite as shown in Figure 19 in section 5. In this section, evidence of the mechanism is clarified based on crystal structure analysis under the pressure. And also, the reason of La-langasite having larger piezoelectricity than Nd-langasite is clarified. With an induced pressure, distance M_L between *A* and *B* cations as shown in Figure 22 and Table 9 is not changed around 3.4 Å, and distances of N_L (*A* ion-[open-space]-*B* ion) in LGS and NGS are shrunk from 4.75 to 4.65, and 4.70 to 4.58 Å, respectively. On the other hand, *A*-polyhedron is distorted to $[\dot{1}00]$ direction. The distortion is clarified by shortage of distance O_L between center position X on *A*-polyhedron and *B*-ion as shown in Figure 22 and 23. The O_L lengths of LGS and NGS are shortened from 3.27 to 3.21 Å, and 3.23 to 3.17 Å, respectively. As a result, a dipole moment *P* appeared according to moving the centers of mass of positive and negative charges to produce piezoelectricity.

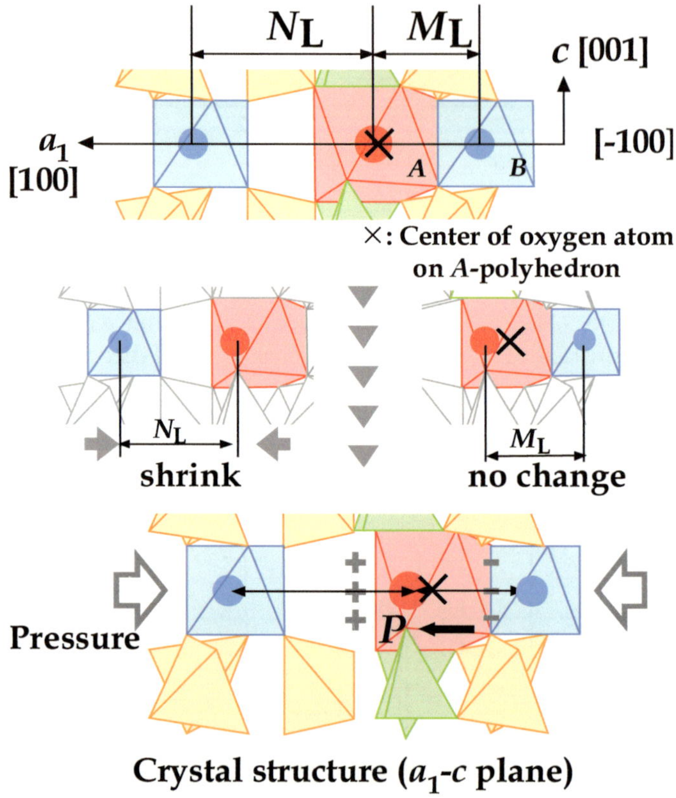

Fig. 23. Origin of piezoelectricity on Langasite projected from [120]. Position X is the center of oxygen atoms on *A*-polyhedron. Under pressure, *ML* does not chang and *NL* shrinks *A*-ion.

Here, differences (M_L-O_L) between center positions X and *A*-ion positions on the LGS and NGS calculated based on crystal structure are equivalent to polarization as shown in Table 9, and Figure 24. The values of LGS and NGS as a function of pressure are increased from 0.147 to 0.177 Å (at 6.1 GPa) and 0.143 to 0.178 Å (at 6.8 GPa), respectively. The value increases with pressure. So, the mechanism of piezoelectricity on the langasite structure series is clarified as follows: though the *A*-polyhedron is deformed to $[\dot{1}00]$ with applied force, *A*-ion stay on same position by repulsion force from *B*-ion under the existence of open-space which has no atoms in the center and is working as damper. The difference between *A*-cation and center position X should make a net dipole moment *P* along *a*-axis. Dipole moment should be enhanced if the distance between the charge centers of cations and anions becomes large. Therefore, the enhancement of piezoelectric properties is related with the shrinkage of open-space in langasite crystal structure, and it is clear that the increase of polarization is caused by the induced pressure in [1 0 0] direction of langasite structure.

The reason of La-langasite having larger piezoelectricity than Nd-langasite is clarified by the difference of the M_L-O_L (dipole moments *P*). As shown in Figure 24, the dipole moments *P* for LGS are larger than that for NGS.

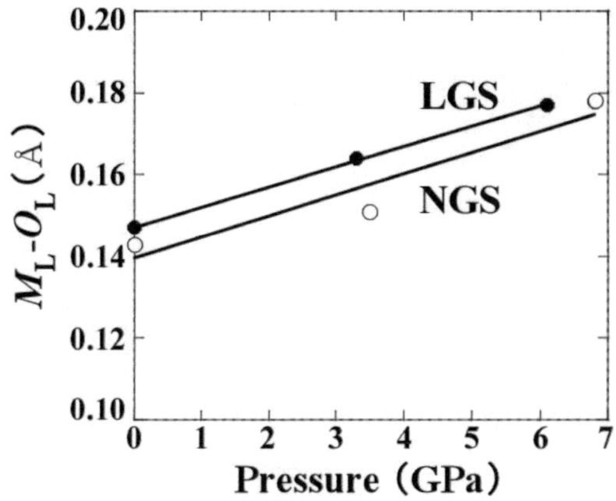

Fig. 24. The difference M_L-O_L (dipole moment P) for LGS and NGS as a function of pressure.

7. New piezoelectric materials with framework structure

Langasite is a kind of materials with framework structure without inversion symmetry *i*. quartz and BeO are also framework structure which is formed mainly by a covalent bond such as SiO_4, AlO_4, ZnO_4. Especially, silicates including langasite make many framework structures by connection of SiO_4 tetrahedra as cyclo-, ino-, phyllo-, and tecto-silicates. These framework structures are noticed recently for applications on many kinds of properties such as zeolite for optical properties by absorption of special compounds.

Here, we will present a candidate for piezoelectric materials. Nepheline ($KNa_3Al_4Si_4O_{16}$) is one of the alumino-silicates with framework as shown in Figure 25(a). The crystal structure has hexagonal ring framework without *i* based on the space group $P6_3$ (No. 173), the point group 6. When stress will be added to the ring, the ring will deform and cations located in tetrahedra and near the center of ring will shift. If the center of cations and anions will become different, piezoelectricity will occur.

Recently, Hosono (2010) presented that $Ca_{12}Al_{14}O_{32}$ ($C_{12}A_7$) clinker compound with big cages including O^{2-} in a $[Ca_{24}Al_{28}O_{64}]^{4+}$ framework shows specific properties such as electride, transparent electride, transparent p-type conducting oxides, transparent semiconductor, super conductor, etc. The crystal structure has 12 cages of 4.4 Å in diameter in a unit cell of 12 Å cube, and 2 cages of 12 cages include oxygen ion (O^{2-}) as shown in Figure 25(b). As this O^{2-} is bonding weakly with the framework, this ion could be removed or exchanged with other anions easily. Transparent metal oxide ($C_{12}A_7$:H^-) transformed to electro conductor by photon induced phase transition, $C_{12}A_7$ compound including much active oxygen O^- atoms (Hayashi et al., 2002), and $C_{12}A_7$: e^- electride stable in room temperature and air-condition (Matsuishi et al., 2003) are presented by Hosono group.

This $C_{12}A_7$ crystal grown by Cockayne & Lent (1979) is expected for SAW, because of the high SAW velocity and reasonable bulk electromechanical coupling by Whatmore (to be published). Two space groups are reported such as $I\bar{4}3d$ by Cockayne & Lent (1979) and $I\bar{4}2m$ by Kurashige et al. (2006). The point groups are $\bar{4}3m$ and $\bar{4}2m$, respectively. Both point groups without *i* show piezoelectricity and the latter shows additional rotatory power. Moreover, additional atoms group in the cages of the frame structure are expected to be designed for SAW suitable properties.

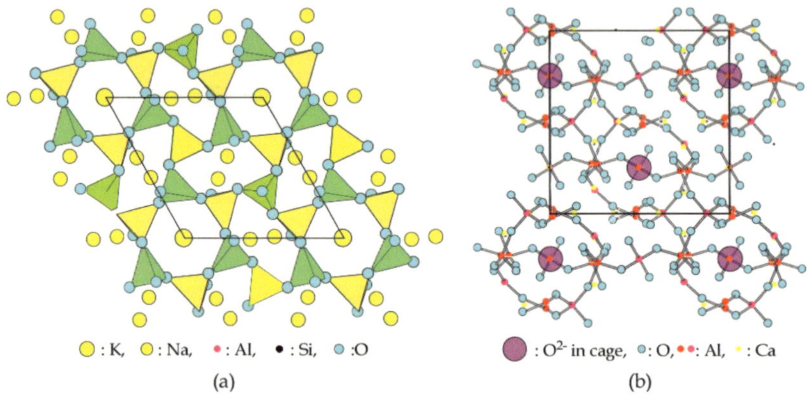

Fig. 25. (a) Nepheline structure as an example of alumino-silicates with framework. (b) Crystal structure of aluminate calcium $C_{12}A_7$. New-type superior properties such as super-conductor were designed from this structure.

8. Conclusion

Origin of piezoelectricity on Langasite has been explained based on the crystal structure, and clarified by crystal structure analysis under high pressure. In the introduction, the principle of piezoelectricity and present condition required such as Pb-free piezoelectricity are stated. Next, crystal growth of langasite, crystal structure analysis and piezoelectric properties of langasite are stated, and mechanism of piezoelectricity are presented based on the crystal structure and properties. And the mechanism is confirmed by crystal structure analysis under high pressure. Finally, a search of new piezoelectric materials are proposed on the direction of framework compound such as nepheline, and introduced the $C_{12}A_7$ compounds with framework also for piezoelectricity.

9. Acknowledgment

The author would like to thank Dr. Katsumi Kawasaki, Jun Sato and Hiroki Morikoshi of TDK Co. for the presenting of single crystal LGS, PGS and NGS. Professor Hiroaki Takeda of Tokyo Institute of Technology and Dr. Hiroki Morikoshi for supporting writing this chapter. Also, M. Eng. Tsuyoshi Iwataki and Nobukazu Araki of NIT for studying crystal structure analysis and experiments under high pressure. And Professors Yasuhiro Kudo and Takahiro Kuribayashi of Tohoku University for supporting crystal structure analysis under high pressure, Professors Cheon Chae-Il and Kim Jeong-Seog of Hoseo University Korea for discussion of the contents, Professors Ken-ichi Kakimoto and Isao Kagomiya of NIT for supporting experimental conditions. Moreover, my wife Keiko Ohsato for supporting my health conditions.

10. References

Araki, N., (2004). "Crystallographic Study for the piezoelectric mechanism of langasite under high pressure, electric field & high & low temperature." *Master-thesis of Nagoya Institute of Technology*.

Araki, N., Ohsato, H., Kakimoto, K., Kuribayashi, T., Kudoh, Y., & Morikoshi, H., (2007). "Origin of Piezoelectricity for Langasite $A_3Ga_5SiO_{12}$ (A = La & Nd) under high Pressure" *J. Eur. Ceram. Soc.*, 27, pp. 4099-4102.

Belokoneva, E. L., Simonov, M. A., Butashin, A. V., Mill, B. V., & Belov, N. V., (1980). "Crystal structure of calcium gallogermanate $Ca_3Ga_2Ge_4O_{14}$ = $Ca_3Ge[(Ga_2Ge)Ge_2O_{14}]$ and its analog $Ba_3Fe_2Ge_4O_{14}$ = $Ba_3Fe[(FeGe_2)Ge_2O_{14}]$" *Sov. Phys. Dokl.*, 25, pp. 954-957.

Bussen, W., & Eitel, A., (1936). "Die Struktur des Pentacalciumtrialuminats" *Z. Kryst.*, 95, pp. 175-188. (Germany).

Cockayne, B., & Leat, B., (1979). "SINGLE CRYSTAL GROWTH OF 12 CaO 7 $A1_2O_3$" *J. Crystal Growth*, 46(2) pp. 467-473.

European Commission - Environment - Waste – WEEE. Date of access: 2nd, November, 2011, [web1]http://ec.europa.eu/environment/waste/weee/index_en.htm [web2]http://ec.europa.eu/environment/waste/weee/pdf/hazardous_substances_ report.pdf

Fukuda, T., Shimamura, K., Kohno, T., Takeda, H., & Sato, M., (1995). "New Piezoelectric Crystal "Langasite" *J. Jap. Asso. Crystal Growth*, 22(5), 358-363. (Japanese)

Fukuda, T., Takeda, H., Shimamura, K., Kawanaka, H., Kumatoriya, M., Murakami, S., Sato, J., Sato, M., (1998). "Growth of New Langasite Single Crystals for Piezoelectric Applications" *Proceedings of the Eleventh IEEE International Symposium*, ISBN:0-7803-4959-8, Montreux, Switzerland, August 1998.

Guo, Y., Kakimoto, K., & Ohsato, H., (2004). "Phase transitional behaviour and piezoelectric properties of $(Na_{0.5}K_{0.5})NbO_3$-$LiNbO_3$ ceramics" *App. Phys. Lett.*, 18, pp. 4121-4123.

Hayashi, K., Hirano, M., Matsuishi, S., & Hosono, H., (2002). "Microporous Crystal $12CaO\bullet7Al_2O_3$ Encaging Abundant O-Radicals" *J. Am. Chem. Soc.*, 124, pp. 738-739.

Hosono, H., (2010). Chap.10, in Handbook of Transparent Conductors, Edited by Ginley, D., Hosono, H., & Paine, D., Springer.

Hosono, H., Hayashi, K., Kamiya, T., Atou, T., & Susaki:, T., (2011). "New functionalities in abundant element oxides: ubiquitous element strategy" *Sci. Technol. Adv. Mater.*, 12, PP. 034303 (22pp).

Hosono, Y., & Yamashita, Y., (2004). "High-Efficiency Piezoelectric Single Crystal", *Toshiba Review*, 59(10), pp. 39-42.

Iwataki, T., (2002). "Study for the relationship between crystal structure and piezoelectric properties of langasite-type piezoelectric crystals" *Master-thesis of Nagoya Institute of Technology*. (Japanese)

Iwataki, T., Ohsato, H., Tanaka, K., Morikoshi, H., Sato, J., & Kawasaki, K., (2001). "Mechanism of the piezoelectricity of langasite based on the crystal structures" *J. Eur. Ceram. Soc.*, 21, pp. 1409-1412.

Kakimoto, K., Masuda I., & Ohsato, H., (2004). "Ferroelectricity and Solid-Solution Structure of $KNbO_3$ Ceramics Doped with La and Fe" *Key Eng. Mater.*, 269, pp. 7-10.

Kaminskii, A. A., Mill, B. V., Khodzhabagyan, G. G., Konstantinova, A. F., Okorochkov, A. I., & Silvestrova, M., (1983). "Investigation of trigonal $(La_{1-x}Nd_x)_3Ga_5SiO_{14}$ crystals. I. Growth and optical Properties" *Physica Status Solidi (a)*, 80(1), pp. 387–398.

Katsuro, H., Matsuishi, S., Kamiya, T., Hirano, M., & Hosono, H., (2002). "Light-induced conversion of an insulating refractory oxide into a persistent electronic conductor" *Nature*, 419, pp. 462-465.

Kazuhisa, K., Toda, Y., Matstuishi, S., Hayashi, K., Hirano, M., & Hosono, H., (2006). "Czochralski Growth of $12CaO\bullet7Al_2O_3$ Crystals" *Cryst. Growth Design*, 6, pp. 1602-1605.

Kumatoriya, M., Sato, H., Nakanishi, J., Fujii, T., Kadota M., & Sakabe, Y., (2001). "Crystal growth and electromechanical properties of Al substituted langasite ($La_3Ga_{5-x}Al_xSiO_{14}$)" *J. Cryst. Growth*, 229, pp. 289-293.

Lee, C., Kan, E., Xiang, H., & Whangbo, M. H., (2010). "Theoretical Investigation of the Magnetic Structure and Ferroelectric Polarization of the Multiferroic Langasite $Ba_3NbFe_3Si_2O_{14}$" *Chem. Mater.*, 22(18), pp. 5290-5295.

Matsuishi, S., Toda, Y., Miyakawa, M., Hayashi, K., Kamiya, T., Hirano, M., Tanaka, I., & Hosono, H. (2003). "High Density Electron Anions in a Nano-porous Single Crystal: $[Ca_{24}Al_{28}O_{64}]^{4+}(4e^-)$" *Science*, 301, pp. 626-629.

Mill, B. V., Belokoneva, E. L., & Fukuda, T., (1998). "New compounds with a $Ca_3Ga_2Ge_4O_{14}$-Type Structure: $A_3XY_3Z_2O_{14}$ (A = Ca, Sr, Ba, Pb; X = Sb, Nb, Ta; Y = Ga, Al, Fe, In; Z = Si, Ge)" *Russian J. Inorg. Chem.* 43, pp. 1168-1175. Translated from *Zh. Neorganicheskoi Khimii*, 43, pp. 1270-1277.

Mill, B. V., Buntashin, A. V., Khodzhabagyan, G. G., Belokoneba, E. L., & Belov, N. V. (1982). "Modified rare-earth gallates with a $Ca_3Ga_2Ge_4O_{14}$ structure" *Dokl. Akad. Nauk SSSR*, 264, pp. 1385-1389.

Mill, B. V.; Butashin, A. V.; Khodzhabagyan, G. G.; Belokoneva, E. L.; Belov, N. V., (1982). "Modified rare-earth gallates with a $Ca_3Ga_2Ge_4O_{14}$ structure" *Sov. Phys. Dokl.*, 27, pp. 434-437.

Saito, Y., Takao, H., Tani, T., Nonoyama, T., Takatori, K., Homma, T., Nagaya, T., & Nakamura, M., (2004). "Lead-free piezoceramics" *Nature*, 432, pp. 84-87.

Sasaki, S., (1982). "A Fortran program for the least-squares refinement of crystal structure" *XL Report, ESS, State Univ. of New York*, pp.1-17.

Sato, J., Takeda, H., Morikoshi, H., Shimamura, K., Rudolph, P. & Fukuda, T. (1998). "Czochralski growth of $RE_3Ga_5SiO_{14}$ (RE=La, Pr, Nd) single crystals for the analysis of the influence of rare earth substitution on piezoelectricity" *J. Crystal Growth*, 191, pp. 746-753.

Shimamura, K., Kato, T., Sato, J., & Fukuda, T., (1999). "Crystal Growth and Characterization of New Langasite-type Compounds for Piezoelecteic Applications" *Proceedings of The 9th US-Japan on Dielectric and Piezoelectric Ceramics*, held November 2-5, Rizzan Sea Park Hotel, Tancha Bay, Okinawa, Japan.

Shimamura, K., Takeda, H., Kohno, T., & Fukuda, T., (1996). "Growth and characterization of lanthanum gallium silicate $La_3Ga_5SiO_{14}$ single crystals for piezoelectric applications" *J. Crystal Growth*, 163, pp. 388-392.

Takeda, H., Izukawa, S., Shimizu, H., Nishida, T., Okamura, S., & Shiosaki, T.,(2005)."Growth, Structure and Piezoelectric Properties of $Ln_3Ga_5SiO_{14}$ (Ln= La, Nd) Single Crystals"Trans. *Mater. Res. Soc. Jpn.* 30, pp.63-66.

Takeda H., & Tsurumi, T., (2011). "Crystal Growth of High Temperature Piezoelectric Crystals from Quasi-Congruent Melt", *Bulletin of the Ceramic Society of Japan*, 46(8), pp. 657-661.

Takeda, H., (1998). "A Study of the Growth, Structure and Properties of $A_3BC_3D_2O_{14}$-type Complex Oxide Crystals" *Ph. D. Thesis, Tohoku University*.

Takeda, H., Kumatoriya, M., & Shiosaki, T., (2002). "Structure and Piezoelectric Properties of Al-substituted Langasite ($La_3Ga_{5-x}Al_xSiO_{14}$) Crystals" *Key Eng.*, 216, pp. 43-46.

Takeda, H., Sato, J., Kato, T., Kawasaki, K., Morikoshi, H., Shimamura, K., & Fukuda, T., (2000). "Symthesis and characterization of $Sr_3TaGa_3Si_2O_{14}$ single crystals" *Mater. Res. Bull.*, 35, pp.245-252.

Whatmore, R. W., (1980). "New Polar Materials: Their Application to SAW and Other Devices" *J. Crystal Growth*, 48, pp. 530-547.

Zhang, S., Zheng, Y., Kong, H., Xin, J., Frantz E., & Shrout, T. R., (2009). "Characterization of high temperature piezoelectric crystals with an ordered langasite structure" *J. Appl. Phys.*, 105, pp. 114107-1-6.

3

Photolithography and Self-Aligned Subtractive and Additive Patterning of Conductive Materials

Gert Homm, Steve Petznick, Torsten Henning and Peter J. Klar
Justus Liebig University,
Institute of Experimental Physics I, Gießen
Germany

1. Introduction

Quantum theory predicts a number of phenomena for materials scaled down to a size where confinement effects occur in one or more dimensions. Numerous devices that are based on these effects have been developed, as for example tunnel diodes, quantum well lasers, etc. An essential component in many of these device concepts are interfaces between conductive materials. To make the devices as efficient as possible in a reproducible way, the interfaces need to be controllable and tunable in their shape, morphology, and transport properties. For multilayer growth, investigations have already shown that a proper control of the quality of the interfaces between the stacked layers is of major importance for the device performance (Fasol et al., 1988; Hillmer et al., 1990). For in-plane interfaces, however, a proper characterization is still missing. To date and to our knowledge, only investigations of grain boundaries have been reported, in which the interfaces were arranged randomly (Schwartz, 1998; Watanabe, 1985; 1993; Watanabe & Tsurekawa, 1999).

A typical example are thermoelectric materials. Theory predicts that the thermoelectric figure of merit of a material

$$ZT = \frac{S^2\sigma}{\kappa}T, \tag{1}$$

where S is the Seebeck coefficient, σ the electrical conductivity and κ the thermal conductivity, could be significantly improved by reducing the dimensions of this material, namely by artificial structuring (Dresselhaus et al., 2007; Hicks & Dresselhaus, 1993a;b). As can be seen from equation (1), an improvement can be achieved by either increasing $S^2\sigma$ (the so called power factor) or by decreasing κ, without affecting the other parameters in an unwanted way. Assuming a constant Seebeck coefficient, the equation (1) implies that materials with a high electrical conductivity and a low thermal conductivity are desired, a design goal which, for metals, is somewhat contradicted by the Wiedemann-Franz law (Franz & Wiedemann, 1853; Lorenz, 1872). This directly leads to semiconductors as the materials class of choice for thermoelectrics. Here, numerous advantages over the metals can be used: The free carrier concentration can be adjusted by doping such that the electric conductivity is still fairly high, but the thermal conductivity is dominated by phonon transport. Simultaneously, the thermal conductivity can be reduced further by phonon blocking almost without affecting the free carrier transport, an approach for which various methods have been developed.

One fairly simple and effective way of improving the thermoelectric figure of merit is to take a thermoelectric material, e.g. PbTe, and then try to enhance it by ball milling and subsequent compacting. This method turned out to further improve the thermoelectric figure of merit of the bulk material due to a lowering of the thermal conductivity, while the electrical conductivity and the Seebeck coefficient remained high (Sootsman et al., 2009). This procedure has been applied to ZnO as well, with similar results, namely the reduction of the thermal conductivity by growing a highly disordered polycristalline material, i. e., by introducing interfaces (Alvarez-Quintana et al., 2010; Huang et al., 2011; Igamberdiev et al., 2010). A more defined way of introducing interfaces is to grow multilayers. The thermoelectric properties are then measured in cross-plane direction, i. e. perpendicular to the sample surface. High figures of merit could be achieved this way (Venkatasubramanian, 2000; Venkatasubramanian et al., 2001), but the sample preparation is comparatively complex, and for some applications, an in-plane geometry is more desirable. An elegant way to introduce well defined interfaces in-plane with a relatively simple process is the lateral structuring of thin films with alternating materials. Calculations show that for materials with incompatible phonon dispersion relations, the propagation of phonons of certain energies across the interfaces is suppressed (Bachmann et al., 2011; Bies et al., 2000; Capinski et al., 1999; Chen, 1998; Daly et al., 2003; Müller et al., 1994; Yang & Chen, 2003; Yao, 1987).

Motivated by this prediction we developed the sample preparation process that uses photolithography and self-aligned pattern transfer for defining lateral interfaces in thin-film structures and that will be detailed in the following. In fig. 1, a lateral structure with interfaces between undoped ZnO and Al-doped ZnO is shown as the typical result of this process. A

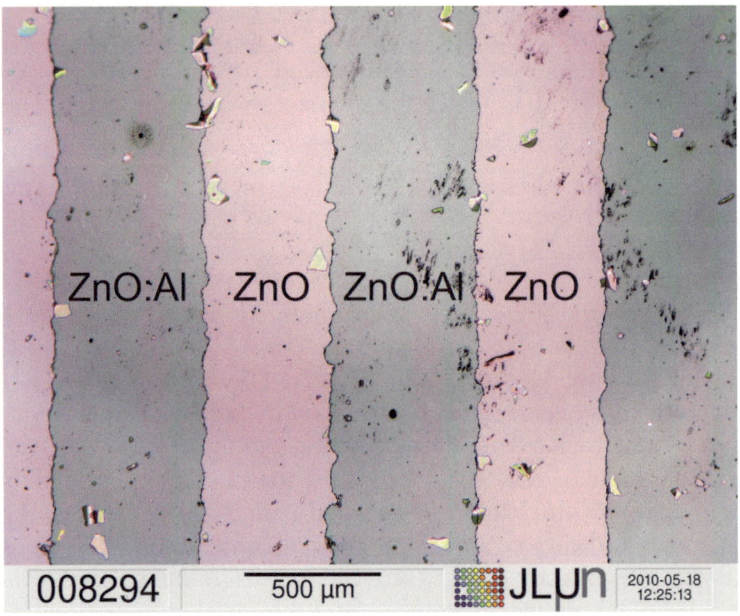

Fig. 1. Optical micrograph of structured alternating ZnO and ZnO:Al bars prepared by photolithography and self-aligned pattern transfer. The roughness of the interfaces is defined by the edge roughness of the structures on the photomask, which had been ink-jet printed in this case.

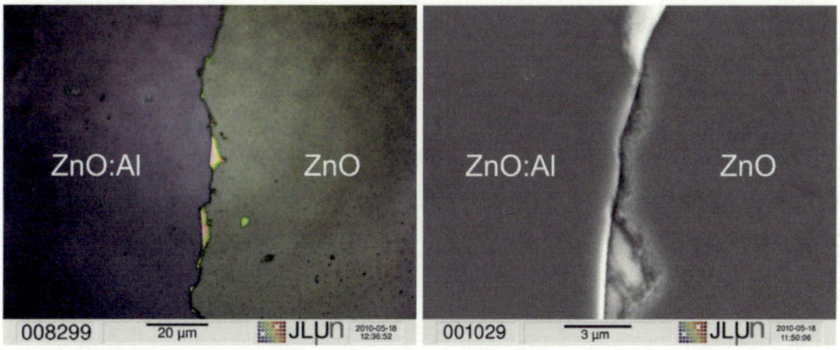

(a) Optical micrograph. (b) Scanning electron micrograph.

Fig. 2. Images of an interface between two conductors with a point-wise contact.

series of four such samples with different numbers of interfaces (2, 4, 8, 16) was prepared. Measurements of the Seebeck coefficient as well as of the electrical conductivity showed that the resistivity decreased with an increasing number of interfaces. In accordance with these results, the absolute values of the Seebeck coefficients decreased with increasing electrical conductivity. Details of the underlying physical effects have been published elsewhere (Homm et al., 2011).

Obviously, there are two length scales at which the properties of the interfaces between the two conductors are determined. On the one hand, there is the roughness of the interface that is determined by the roughness of the edges of the printed structures on the photomask. In a high quality photomask, this roughness is below the wavelength of light and hence negligible, unless an artificial roughness is created intentionally. This roughness would typically be on the micrometer scale, and it determines the length of the resulting interface and hence the area on which transport between the two conductors can occur. On the other hand, there is the length scale of the grain size of the respective conductors and of the average width of the gap between both conductors as a result of the pattern transfer process. At this length scale, the quality of the interface is determined. It can range from an ideal full-contact interface via an interface with an average density of point-wise contacts per unit interface length to a non-contact interface, in which the gap is wider than the typical grain size. An example of a point-wise contacting interface is shown in fig. 2.

In the following sections, we will first describe the sample fabrication procedure, namely the self-aligned pattern transfer. We will then describe how the properties of the interface depend on variations in the process parameters, and which degrees of freedom one has when using this fabrication method. Finally, possible variations and extensions of the technique will be pointed out. It should be noted that the process has been tailored to conductive oxide thin films, but is in principle transferable to other thin film materials.

2. Self-aligned pattern transfer process

Figure 3 shows a somewhat idealized process flow. It starts with the blanket deposition of material A by virtually any thin film deposition method onto a substrate (a). The sample is then coated with photoresist, which is exposed (b) and developed (c). The photoresist masking

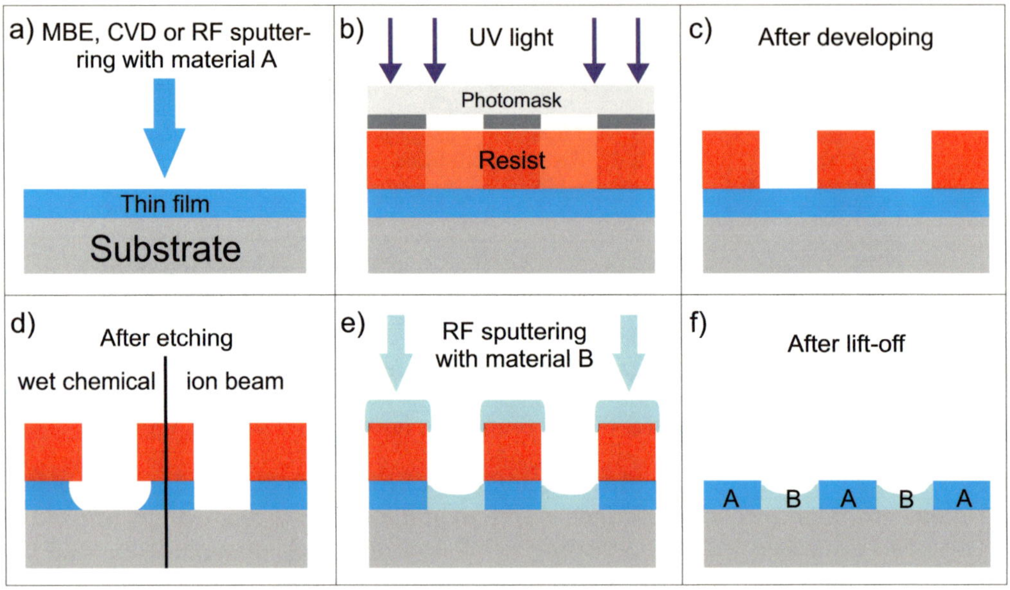

Fig. 3. Lateral patterning of alternating films of two materials by self-aligned pattern transfer.

layer is then used for subtractive patterning of material A, e. g. by wet chemical etching or by ion beam etching (d). The same photoresist masking serves as a lift-off mask in the subsequent additive patterning of material B, consisting of the two steps of thin film deposition (e) and the final lift-off of superfluous material B by dissolving the photoresist masking (f).

In panels (b) and (c) of fig. 3, the assumption has been made that a positive photoresist is used, i. e. a photoresist whose exposed portions are removed during development. As we will see, negative resists can be used as well. In this case, the transparent and opaque structures on the photomask have to be reversed.

The parts of the thin film of material A that are not covered by the patterned photoresist masking are now removed by etching. The etching method may be either wet-chemical etching or any of the wide variety of dry etching methods, such as ion beam etching or reactive ion etching. In the selection of the etching method, two properties have to be considered especially. Firstly, the different etching methods differ in their selectivity, that is the ratio of the etching rate of the thin film (that is to be etched) and of the resist masking (which, ideally, should not be etched at all). Secondly, the different etching methods will deliver differing degrees of anisotropy. In wet-chemical etching, and given that the thin film will be polycrystalline or at best oligocrystalline, etching will be isotropic, i. e., the rate at which the thin film is being removed in horizontal direction will be equal to the rate at which it is etched in vertical direction (Madou, 2002). Ion beam etching, on the other hand, can deliver a certain degree of anisotropy, resulting in steep sidewalls of the etched film and a low degree of underetching (removal of material under the resist masking). The different shapes of the etched thin film are, again somewhat idealized, shown in panel (d) of fig. 3.

The selectivity of the resist masking can be improved by a hardbake procedure, in which the developed sample is heated up to a temperature slightly below the glass point of the resist.

Temperature control must be precise, because at a temperature even slightly too high, the resist masking may lose its shape, resulting in a failure of the remaining processing steps.

While the deposition method for material A can be practically any out of the multitude of methods available for thin films, including the vast class of chemical vapor deposition (CVD) methods or molecular beam epitaxy (MBE), the deposition method for material B has to be compatible with the resist masking still present at this step in the process flow (panel (e) of fig. 3). Namely, the temperature at the sample surface has to remain below the glass point temperature of the photoresist, which usually means a maximum temperature in the range between 120 °C and about 200 °C, dependent on the precise type of resist used. This excludes all CVD and MBE methods and leaves only physical vapor deposition methods (PVD) available. These may be evaporation coating, pulsed laser deposition, or a variety of sputter deposition methods. In the experimental work presented here, material B was always deposited by sputter deposition. The advantages of sputter deposition over evaporation coating are that the resulting films will have a higher degree of crystallinity, and that sputter deposition tends to give a better coverage of sidewalls (conformal coating) than does evaporation coating. This is important since it is usually desirable that the vertical sidewalls of material A make contact to material B. The coverage of the sidewalls may, however, not be too high, since that would mean covering the sidewalls of the resist masking as well, and that would prevent the solvent in the lift-off step (panel (f) of fig. 3) from reaching and hence from dissolving the photoresist. The solvent has to be chosen for good selectivity to the thin film materials and can be either a general purpose solvent such as acetone, or, especially in the case of hardbaked photoresist, a special organic formulation known as "remover".

3. Freedom of pattern choice and limitations

As mentioned above, the interface shape on a length scale larger than the wavelength of light may be chosen rather arbitrarily. This is demonstrated by the five different interface shapes which were realized and which are shown schematically in fig. 4. The bottom images show optical micrographs of the interface structures transferred into photoresist after the development step. It can be seen that defining these interface shapes on the length scale of a few micrometers is easy with photolithography. Many of the results discussed in the following are based on samples where pattern (d), a wavy interface shape, was used. The alternating bars of the two conductors were designed to cover an area of $5 \times 5\,\mathrm{mm}^2$ between two contact bars each 1 mm wide and extending over the 5 mm perpendicular to the direction of electrical transport. Two orientations of the interfaces to the direction of current flow were implemented, as shown in fig. 5.

The pitch of the conducting bars, or, in other words, their packing density is limited (among others) by the thickness of the thin film of material A. Since, as mentioned above, wet-chemical etching is in many cases nearly perfectly isotropic, the resulting distance between two conductor bar edges at the end of the process flow has to be more than twice the thickness of the thin film. If the structures were too close together, they would either not be separated in the depth of the film (here the surface of the substrate), or they would be damaged near the top film surface once etching has been performed long enough to separate the conducting parts. It should be kept in mind that in order to ensure process safety, given a slight uncertainty in

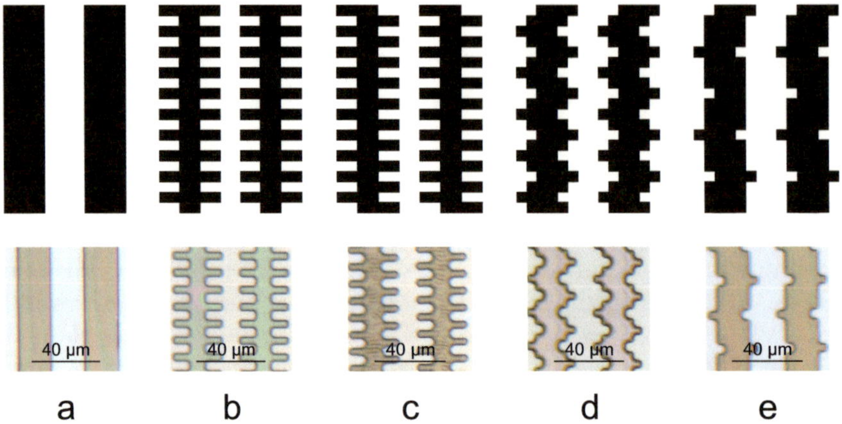

Fig. 4. Schematic drawing of the five different interface shapes designed for structuring of lateral interfaces: (a) straight interfaces, (b) teeth in-phase, (c) teeth in counter-phase, (d) wavy and (e) toothed interfaces with constant distance. The length of a tooth is typically on the order of 5 µm, the pitch (center-to-center-distance of the conducting bars of the same material) on the order of 40 µm. The bottom images are optical micrographs of the pattern transferred into the photoresist after the development step (step c) of fig. 3).

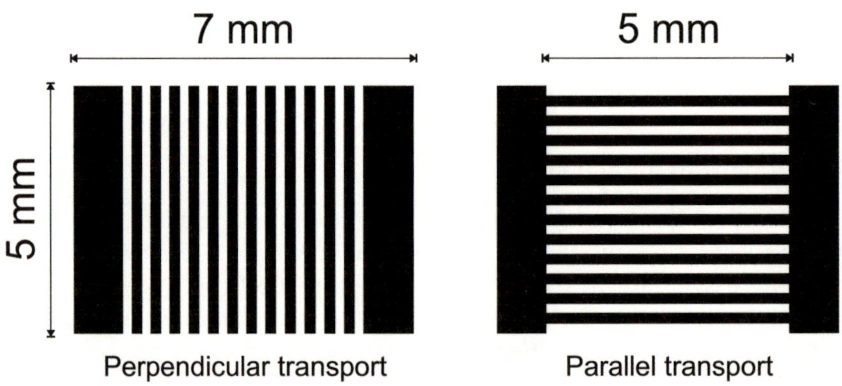

Fig. 5. Schematic drawing of the positioning and orientation of alternating conductor bar structures between the electrical contact bars (to the left and right, respectively).

actual etching rates, etching always has to be carried out for slightly more than the minimum time required theoretically (so-called "overetching").

The etching rate will, among others, depend on the crystallinity of the material being etched. This means that the etching rate of material A will depend on whether it has been deposited with a method like MBE, which gives a high degree of crystallinity and usually lower etch rates, or sputter deposition, which leads to polycrystalline material with a higher etch rate, since etching along the grain boundaries is faster than the etching of the individual grains. As a second-oder effect, the morphology of the etched edge will depend on the deposition method as well. In the case of polycrystalline samples with samples growing predominantly in vertical direction, the conductor edge will be lined with vertical isolated columns, whereas the etched edge of an oligocrystalline film will tend to be rather smooth. Figure 6 depicts

AFM images of etched edges of three ZnO thin-films. The images a) and c) show sputtered ZnO films etched by wet-chemical and ion-beam etching, respectively. Image b) shows an AFM image of a MBE grown ZnO film after wet-chemical etching: The different edge morphologies can be clearly distinguished. The combination of wet-chemical etching and granular sputtered material yields a nanostructured interface region with columns of original material protruding from the closed interface. In contrast, ion-beam etching of sputtered material yields comparatively sharp interfaces with rather steep edges. Wet-chemical etching of epitaxial MBE-grown ZnO shows a closed surface of the edges, but with a smoother slope than for ion-beam etching.

For material B, an upper limit of the film thickness is given by a certain fraction of the thickness of material A. If this fraction exceeds a value around $\frac{3}{4}$, the conformal coating of the resist sidewalls will prevent a succesful lift-off of material B. While the upper limit of the film thickness is governed by the pattern transfer process as described above, the lower limit of the film thickness is given by the requirement that the film be continuous. The limit at which a film becomes continuous varies with the deposition method, and in general, higher deposition temperatures require a larger film thickness to pass the threshold to continuity. Evaporation coating can result in continuous films for thicknesses as low as 10 nm (Völklein & Zetterer, 2006), while for sputter deposition of conducting oxides, the limit is on the order of 50 nm. In addition, a certain minimal film thickness is desirable since the conductivity, which, in first order, is inversely proportional to the film thickness, should not be too low for transport properties to be reliably measurable.

4. Control over the interface properties

In this section we will discuss different combinations of process parameters, namely of the resist and the etching method used, will present the resulting interface morphologies and will correlate the properties to the fabrication process in order to demonstrate the range over which the interface properties can be controlled and tuned.

4.1 Examples of different interface morphologies obtained by a single photolithography step followed by self-aligned pattern transfer

Very often, ***interfaces with point-wise contacts*** are obtained. Two examples are shown in fig. 7 and 8 which were obtained after processing using different photoresists and wet-chemical etching. In the first case the material A consisted of a layer of ZnO with a thickness of 200 nm, grown by molecular beam epitaxy on top of a 300 μm thick sapphire substrate. Photolithography was first performed with a 1.5 μm thick layer of positive photoresist (ma-P 1215 from micro resist technology, Berlin). Exposure was done with a UV broadband source in a Suss MA 56 mask aligner at doses of 67.2 mJ cm^{-2} at 365 nm wavelength and 122.4 mJ cm^{-2} at 405 nm wavelength. The resist was developed in ma-D 331 developer (also from micro resist technology, Berlin) for 15 s and hardbaked for 90 minutes in a convection oven. The zinc oxide was wet-chemically etched in a mixture of 84 % orthophosphoric acid (H_3PO_4), 100 % acetic acid ($C_2H_4O_2$), and water (H_2O) in a ratio of 1:1:100 parts by volume (Elm et al., 2008). Etch end detection could be performed by monitoring the electrical resistance in situ, since ZnO is intrinsically electrically conducting. As material B, a 150 nm thick layer of Ga-doped ZnO was deposited by radio frequency (RF) sputter coating. This thickness is very close to the

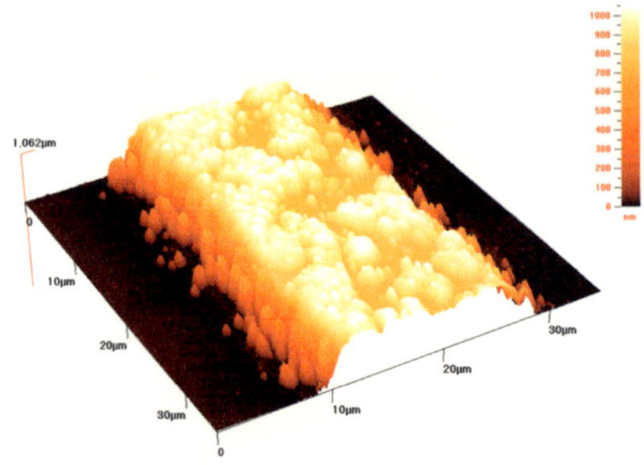

(a) Atomic force micrograph.

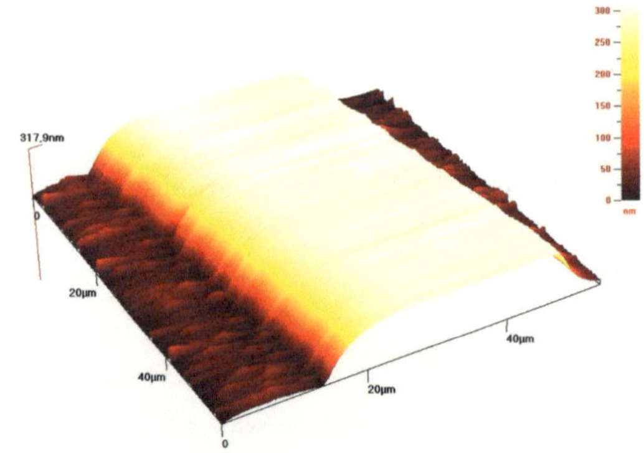

(b) Atomic force micrograph.

(c) Atomic force micrograph.

Fig. 6. Atomic force microscopy (AFM) images of etched edges of three ZnO thin-films. The images (a) and (b) show a sputtered (a) and an MBE grown (b) film after pattern transfer with wet-chemical etching. Image (c) shows an AFM image of a sputtered ZnO film after pattern transfer by ion-beam etching: The different edge morphologies can be clearly distinguished.

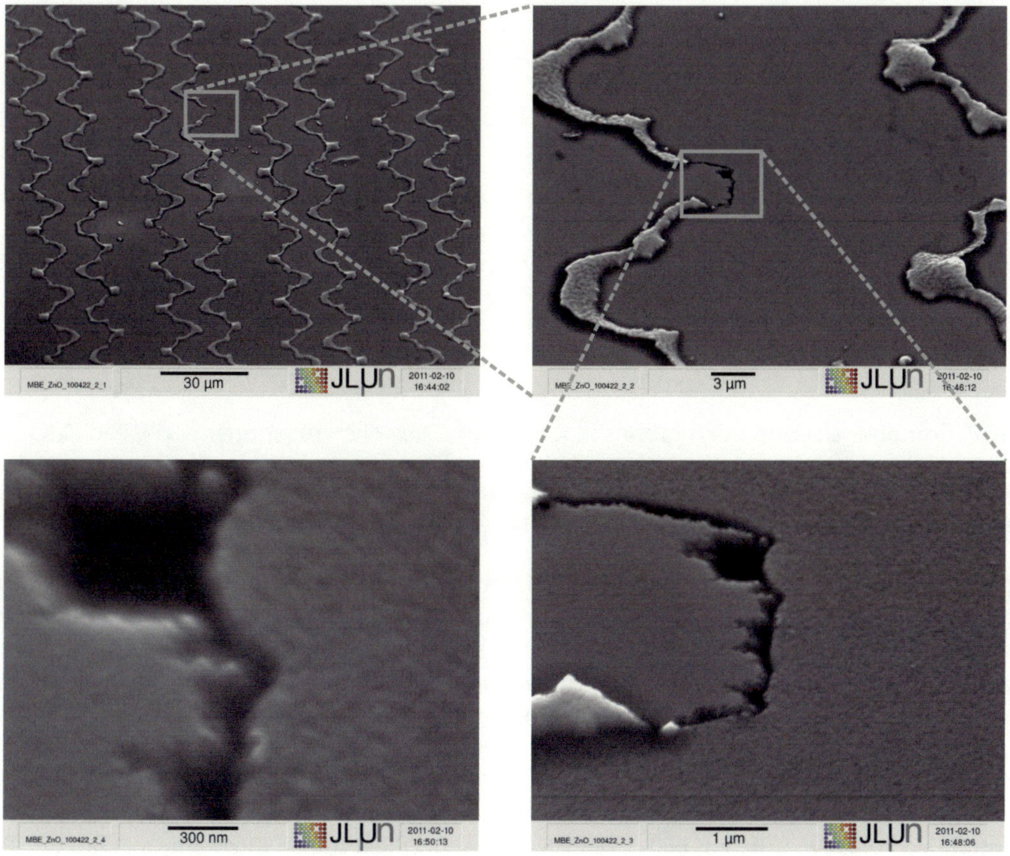

Fig. 7. Scanning electron micrographs (zoom-in series) of a sample with MBE-grown ZnO (material A) and Ga-doped ZnO (material B), processed using positive photoresist and wet-chemical etching. The wavy structures extending over the sample surface are remnants of the sputter coated ZnO layer that had been deposited on the resist sidewalls and hence were not removed in the lift-off step.

limit imposed by the requirement that lift-off still has to be possible. For a succesful lift-off, the solvent bath had to be heated to 50 °C, and ultrasonic excitation had to be applied to the bath.

Figure 7 shows a series of scanning electron micrographs of a sample processed in the way described. At high magnification, the point-wise nature of the contact along the interface is clearly visible. The remnants of the Ga-doped ZnO due to incomplete lift-off (so-called "garden fences", resulting from material deposited on the resist sidewalls) had no influence on the electrical transport properties.

A point-wise contact along the interface similar to that in fig. 7 can also be achieved using negative instead of positive photoresist. The substrate in this experiment was glass, and material A was a thin layer of sputter deposited ZnO. The negative resist ma-N 1420 (again from micro resist technology, Berlin) was spincoated to a thickness of approximately 2 µm and exposed using the same UV broadband source mentioned above, with doses of 560 mJ cm^{-2} at

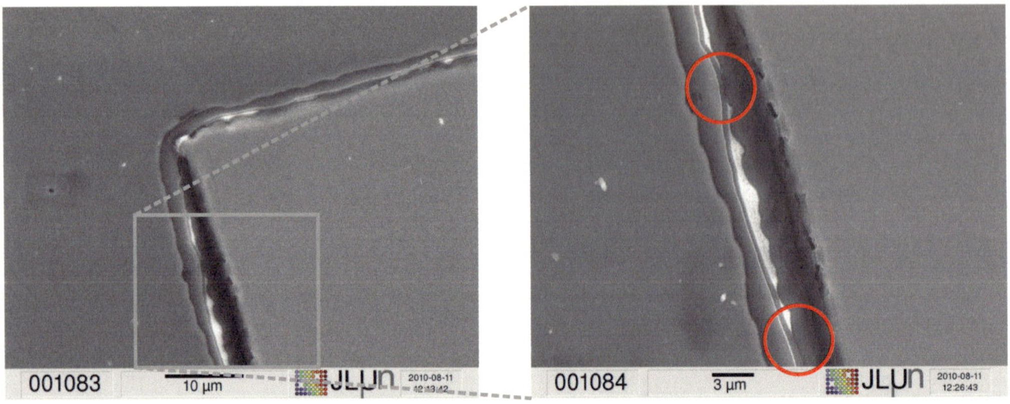

Fig. 8. Scanning electron micrographs (zoom-in) of a sample with sputter deposited ZnO (material A) and sputter deposited Cu_2O (material B), processed using negative photoresist and wet-chemical etching. The point-wise contacts (circled in the right image) had a density of approximately one per 10 μm of interface length.

365 nm wavelength and 1002 mJ cm^{-2} at 405 nm wavelength. The most important parameter is the development time, since this time determines the width of the undercut profile in the developed resist. A development time of 65 s in the developer ma-D 533S (from the same supplier) resulted in an undercut of approximately 0.6 μm (mic, 2011). Again, hardbake has to be subject to tight temperature control in order to preserve the undercut profile. The ZnO was wet-chemically etched as in the experiments with positive photoresist detailed above. Copper-I-oxide (Cu_2O) was sputter coated on the surface. The hardbaked negative resist could not be removed with acetone, so the lift-off was carried out using the specially formulated remover mR-rem 660 (from the same supplier) under ultrasonic excitation. The scanning electron microscopic characterization of the resulting interface, see fig. 8, shows the point-wise nature of the contact, with a contact density of about one per 10 μm of interface length.

Using extreme undercut profiles allows one to separate materials A and B by a gap. Such ***interfaces with a gap*** can be obtained using the same preparation of material A as described in fig. 8, but extending the development time to 120 s. This created a rather extreme undercut of approximately 2.1 μm (mic, 2011). Exposure doses in this case were 504 mJ cm^{-2} (365 nm) and 918 mJ cm^{-2} (405 nm). Material B again was a 150 nm thick layer of Ga-doped ZnO deposited by RF sputter coating. Lift-off was easily achieved with such a large undercut, and fig. 9 shows the result of the complete process in a series of scanning electron micrographs. The isotropic character of the wet-chemical etching process in combination with the undercut profile of the resist sidewalls prevents the creation of interfaces with continuous contact.

Continuous almost ideal interfaces can be obtained by using ion beam etching (IBE) instead of wet-chemical etching as the dry etching avoids an undercut. In IBE, chemically inert ions (Ar^+ in our case) are accelerated by a voltage of several hundred volts (in our case 700 V, in a Kaufman source, at a current density of 190 μA cm^{-2}) and directed onto the sample, resulting in a material removal caused by physical effects ("bombardment") with a certain degree of anisotropy. The etching rate of ZnO under these conditions was approximately 30 nm min^{-1}.

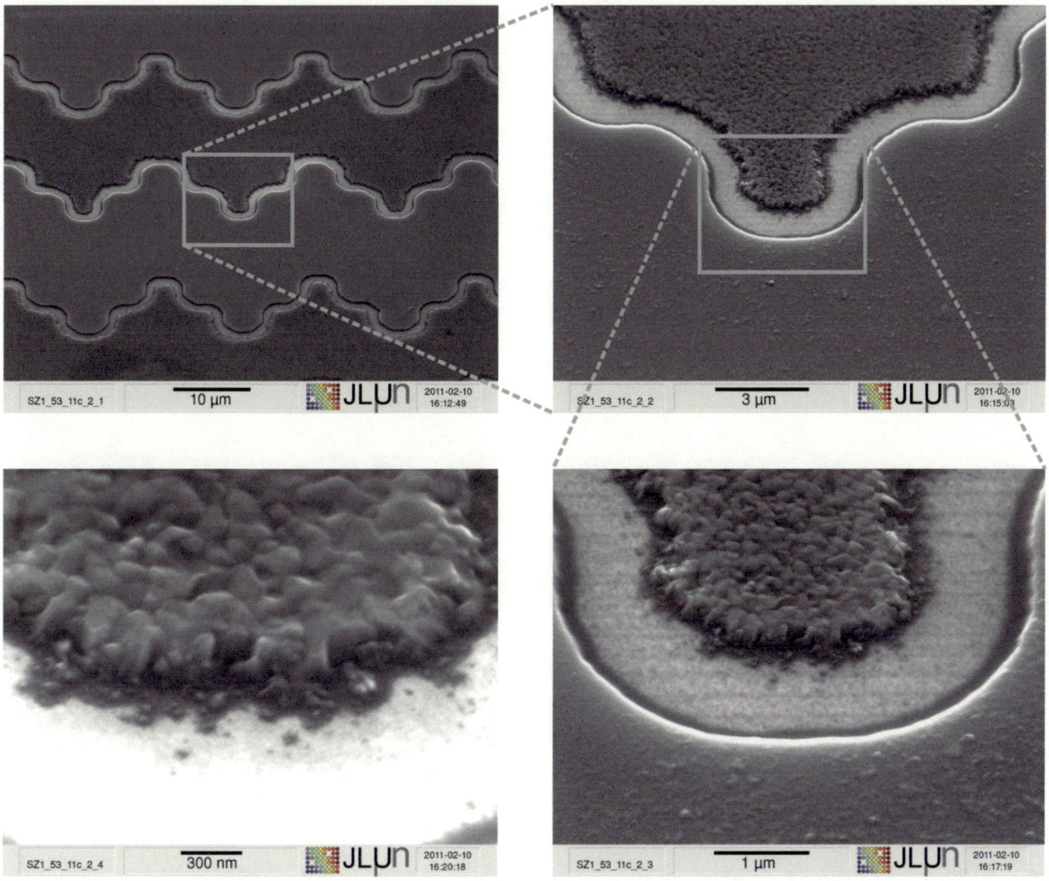

Fig. 9. Scanning electron micrographs (zoom-in series) of a sample with MBE-grown ZnO (material A) and Ga-doped ZnO (material B), processed using negative photoresist and wet-chemical etching. The undercut of the negative resist was tuned to around 2 µm by extending the development time to 120 s, resulting in a well-defined gap along the entire length of the interface.

Alternating etching cycles of no more than 3 minutes and cooling cycles had to be employed to avoid excessive heating of the sample by the ion beam.

The result of a process flow with CVD-grown ZnO (material A), positive photoresist, ion beam etching, and sputter deposited ZnO (material B) is shown in fig. 10. Positive photoresist was used here since in general it provides less undercut than negative resist, which obviously is advantageous for a continuous interface. Lift-off could be achieved with heated acetone under strong ultrasonic excitation. As can be seen in fig. 10, the two materials are in contact along the entire interface, i. e., the combination of photoresist processing and thin film etching method approaches the perfect interface.

Figure 11 summarizes variations of the interface morphology at the submicrometer scale possible with self-aligned pattern transfer based on a single photolithography step.

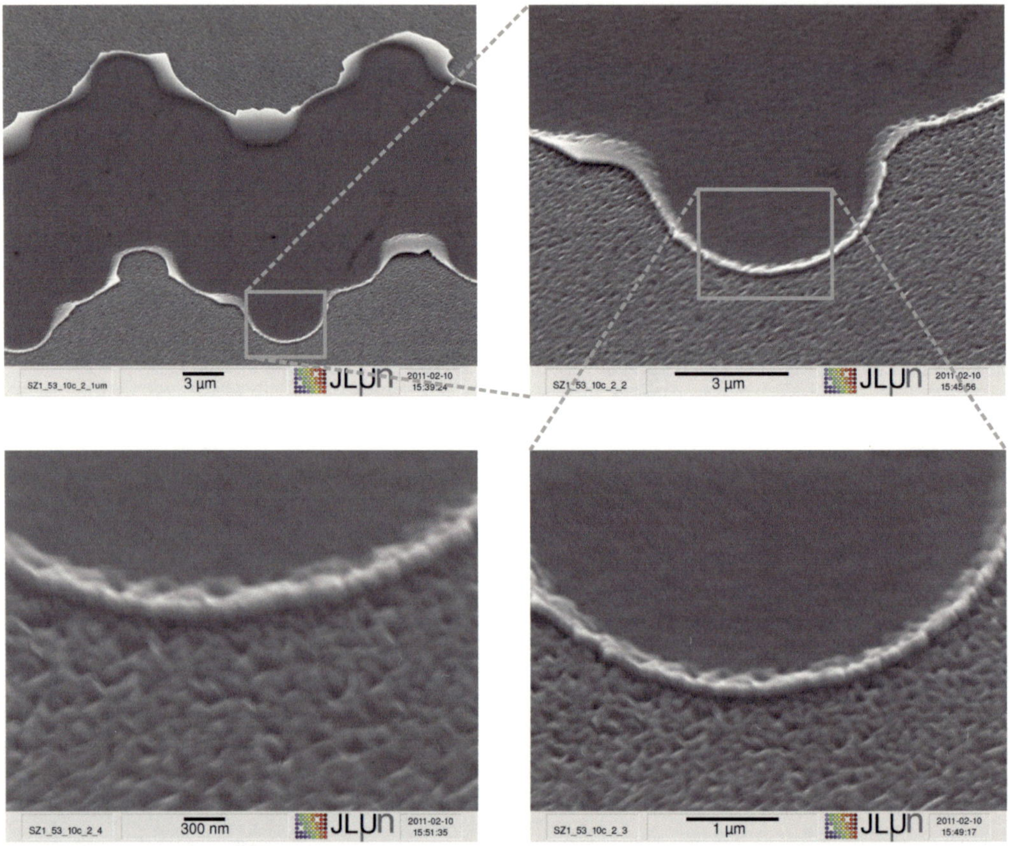

Fig. 10. Scanning electron micrographs (zoom-in series) of a sample with CVD-grown ZnO (material A) and Ga-doped ZnO (material B), processed using positive photoresist and ion beam etching. Some remnants of material B are visible, indicating imperfect lift-off. The two materials are in electrical contact along the entire length of the interface.

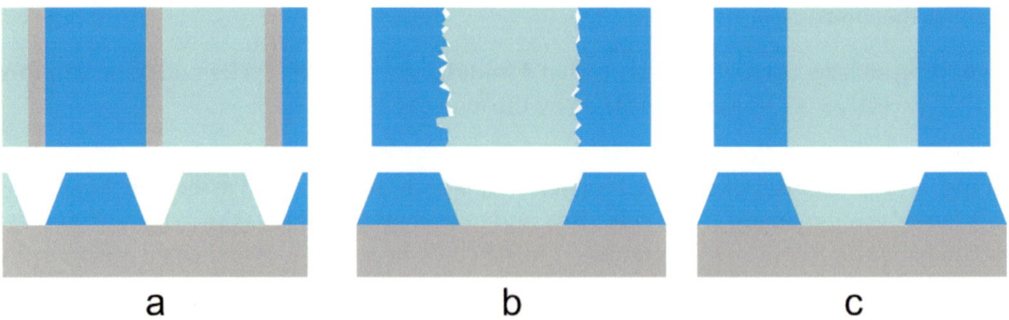

Fig. 11. Interface morphologies realized by a single photolithography step and self-aligned pattern transfer, by variation of the photoresist (undercut profile) and the etching process (anisotropy). From left to right: (a) insulating gap along the length of the interface, (b) point-wise contacts along the length of the interface, and (c) continuous electrical contact over the entire length of the interface.

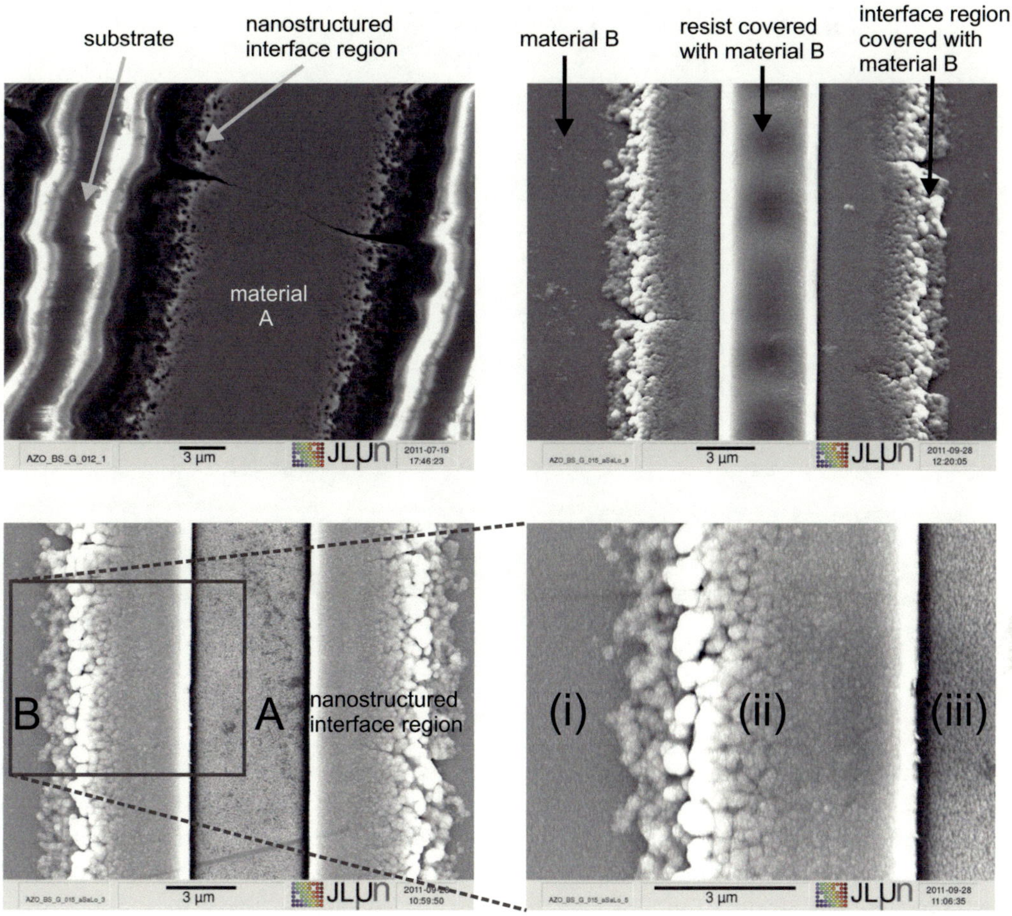

Fig. 12. Scanning electron micrographs of a wet chemically etched layer of ZnO:Al. The four images show samples after different steps of the processing: After wet-chemical etching (upper left), after second lithography and sputter deposition (upper right), after liftoff (lower left), and a magnification of the resulting interface (lower right). The roughness of the edge is predominantly caused by the crystallinity of the material and is on the order of a few tens to about one hundred nanometers.

4.2 Examples of controlled structuring of interfaces on the sub-micrometer scale

In combination with the self-aligned patterning process it is even possible to achieve controlled structure definition of interface regions on the sub-micrometer scale. Here, the possibilities are manifold and we will briefly discuss two examples.

The first example makes use of the properties of wet etching in sputter deposited oxide layers. The grain structure of the oxide thin films is columnar with characteristic diameters of grain columns in the range of 50 to 100 nm. Wet-chemical etching proceeds faster along the grain boundaries than through the nanocrystalline grains, resulting in an irregular array of freestanding nanocrystallites, often columnar in shape, along the edge of the etched thin film, as shown in the SEM image of fig. 12 (top left) and the AFM image in fig. 6 (a). The

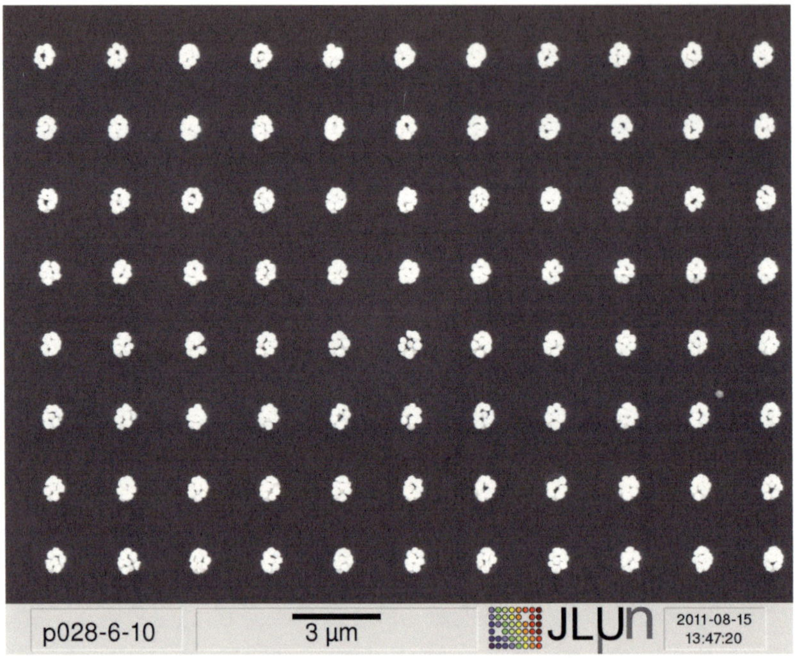

Fig. 13. Gold nanoparticles arranged in circular indentations (arranged on a square grid with 3 µm pitch) in a thin film by the meniscus force method.

width of this region is determined by the characteristic radius of the columnar grains and the ratio of the etch rates along the grain boundaries and through the bulk of the grains. The region can be created in a single photolithography step. Let us assume that the interfaces of a sputter deposited thin film of material A are nanostructured this way, then using the structured film as a starting point of a second lithography step followed by the self-aligned deposition of material B, allows one to embed the nanostructures of material A in the interface region between A and B, yielding a defined interface region. The SEM images of fig. 12 depict different stages of the processing. The magnification of the interface shows three distinct regions: Region (i) is material B coming from the second sputter deposition. Region (ii) is the nanostructured interface, where both materials are on top of each other. In region (iii) material A can be seen. This is the region which was protected by the resist (compare upper right image of fig. 12) and is now uncovered after the lift-off.

The second example may use structures with a well defined gap between A and B (as obtained in fig. 9) as the starting point. Employing a second lithography step to cover the remnants of materials A and B enables one to fill the gap between the two materials (or any other indentation in the photoresist) with nanoparticles. One possibility of achieving the filling in a self-organized way is employing the meniscus force method (Cui et al., 2004; Yin et al., 2001), where during the evaporation of a drop of suspension of nanoparticles, the latter are pushed into the indentations in the sample surface due to the surface tension of the drop. Exemplarily, fig. 13 shows gold nanoparticles self-organized on a substrate with circular indentations. Again, after the gap is filled with nanoparticles a self-aligned deposition of a material C may be carried out to obtain a closed surface.

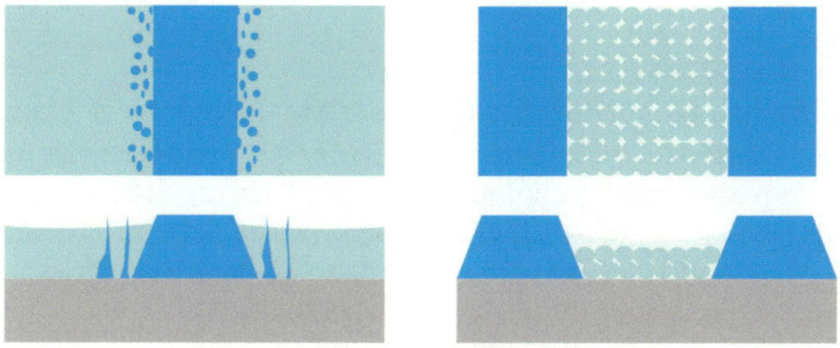

Fig. 14. Some variations of the interface morphology at the sub-micrometer scale that have been (left) or at the micrometer scale that could be (right) realized with self-aligned pattern transfer. Left: nanocolumnar structures of material A between the areas of bulk materials A and B, respectively. Right: gap between strips of material A filled with nanoparticles before the deposition of material B.

These two examples summarized schematically in fig. 14 demonstrate the potential of integrating the self-aligned pattern transfer method into extended and more complex process schemes.

5. Conclusions and outlook

We have shown how a variety of morphologies of the interface between two conducting oxides can be created by a single photolithography process and subsequent self-aligned pattern transfer. The choice of the process parameters allows one to adjust the morphology of the interfaces ranging from almost ideal interfaces via interfaces with point-wise contacts to interfaces with a gap of controlled width, even below the resolution limit of the photolithographic process step. This is one of the characteristic features of self-aligned pattern transfer.

Incorporating the self-aligned pattern transfer into more complex processing schemes allows one to even achieve controlled interface structuring on the nanometer scale. We have demonstrated that the method of self-aligned pattern transfer offers versatile ways of lateral structuring of thin-film materials.

6. References

Alvarez-Quintana, J., Martínez, E., Pérez-Tijerina, E., Pérez-García, S. A. & Rodríguez-Viejo, J. (2010). Temperature dependent thermal conductivity of polycrystalline ZnO films, *Journal of Applied Physics* 107(6): 063713.
URL: *http://dx.doi.org/doi/10.1063/1.3330755*

Bachmann, M., Czerner, M., Edalati-Boostan, S. & Heiliger, C. (2011). Ab initio calculations of phonon transport in ZnO and ZnS.
URL: *http://arxiv.org/abs/1111.2540*

Bies, W. E., Radtke, R. J. & Ehrenreich, H. (2000). Phonon dispersion effects and the thermal conductivity reduction in GaAs/AlAs superlattices, *Journal of Applied Physics*

88(3): 1498–1503.
URL: *http://dx.doi.org/doi/10.1063/1.373845*

Capinski, W. S., Maris, H. J., Ruf, T., Cardona, M., Ploog, K. & Katzer, D. S. (1999). Thermal-conductivity measurements of GaAs/AlAs superlattices using a picosecond optical pump-and-probe technique, *Phys. Rev. B* 59(12): 8105–8113.
URL: *http://link.aps.org/doi/10.1103/PhysRevB.59.8105*

Chen, G. (1998). Thermal conductivity and ballistic-phonon transport in the cross-plane direction of superlattices, *Phys. Rev. B* 57: 14958–14973.
URL: *http://link.aps.org/doi/10.1103/PhysRevB.57.14958*

Cui, Y., Björk, M. T., Liddle, J. A., Sönnichsen, C., Boussert, B. & Alivisatos, A. P. (2004). Integration of colloidal nanocrystals into lithographically patterned devices, *Nano Letters* 4(6): 1093–1098.
URL: *http://dx.doi.org/10.1021/nl049488i*

Daly, B. C., Maris, H. J., Tanaka, Y. & Tamura, S. (2003). Molecular dynamics calculation of the in-plane thermal conductivity of GaAs/AlAs superlattices, *Phys. Rev. B* 67: 033308.
URL: *http://link.aps.org/doi/10.1103/PhysRevB.67.033308*

Dresselhaus, M., Chen, G., Tang, M., Yang, R., Lee, H., Wang, D., Ren, Z., Fleurial, J.-P. & Gogna, P. (2007). New directions for low-dimensional thermoelectric materials, *Advanced Materials* 19(8): 1043–1053.
URL: *http://dx.doi.org/10.1002/adma.200600527*

Elm, M. T., Henning, T., Klar, P. J. & Szyszka, B. (2008). Effects of artificially structured micrometer holes on the transport behavior of Al-doped ZnO layers, *Applied Physics Letters* 93(23): 232101.
URL: *http://dx.doi.org/doi/10.1063/1.3040312*

Fasol, G., Tanaka, M., Sakaki, H. & Horikoshi, Y. (1988). Interface roughness and the dispersion of confined LO phonons in GaAs/AlAs quantum wells, *Phys. Rev. B* 38: 6056–6065.
URL: *http://link.aps.org/doi/10.1103/PhysRevB.38.6056*

Franz, R. & Wiedemann, G. (1853). Ueber die Wärme-Leitungsfähigkeit der Metalle, *Annalen der Physik* 165(8): 497–531.
URL: *http://dx.doi.org/10.1002/andp.18531650802*

Hicks, L. D. & Dresselhaus, M. S. (1993a). Effect of quantum-well structures on the thermoelectric figure of merit, *Phys. Rev. B* 47(19): 12727–12731.
URL: *http://link.aps.org/doi/10.1103/PhysRevB.47.12727*

Hicks, L. D. & Dresselhaus, M. S. (1993b). Thermoelectric figure of merit of a one-dimensional conductor, *Phys. Rev. B* 47(24): 16631–16634.
URL: *http://link.aps.org/doi/10.1103/PhysRevB.47.16631*

Hillmer, H., Forchel, A., Sauer, R. & Tu, C. W. (1990). Interface-roughness-controlled exciton mobilities in GaAs/$Al_{0.37}Ga_{0.63}As$ quantum wells, *Phys. Rev. B* 42: 3220–3223.
URL: *http://link.aps.org/doi/10.1103/PhysRevB.42.3220*

Homm, G., Petznick, S., Gather, F., Henning, T., Heiliger, C., Meyer, B. K. & Klar, P. J. (2011). Effect of Interface Regions on the Thermoelectric Properties of Alternating ZnO/ZnO:Al Stripe Structures, *Journal of Electronic Materials* 40: 801–806.
URL: *http://dx.doi.org/10.1007/s11664-011-1574-4*

Huang, Z. X., Tang, Z. A., Yu, J. & Bai, S. (2011). Thermal conductivity of nanoscale polycrystalline ZnO thin films, *Physica B: Condensed Matter* 406(4): 818 – 823. URL: *http://www.sciencedirect.com/science/article/pii/S0921452610011671*

Igamberdiev, K. T., Yuldashev, S. U., Kurbanov, S. S., Kang, T. W., Khabibullaev, P. K., Rakhimova, S. M., Pelenovich, V. O. & Shashkov, A. G. (2010). Thermal properties of semiconductor zinc oxide nanostructures, *Journal of Engineering Physics and Thermophysics* 83(4): 863 – 868.
URL: *http://dx.doi.org/10.1007/s10891-010-0407-2*

Lorenz, L. (1872). Bestimmung der Wärmegrade in absolutem Maaße, *Annalen der Physik* 223(11): 429–452.
URL: *http://dx.doi.org/10.1002/andp.18722231107*

Madou, M. J. (2002). *Fundamentals of Microfabrication - The Science of Miniaturization*, CRCPress.

mic (2011). *ma-N 400 and ma-N 1400 - Negative Tone Photoresists*.
URL: *http://www.microresist.de/products/negative_photoresists/pdf/po_pi_1400_400_man_08041003_en_ls.pdf*

Müller, W., Bertram, D., Grahn, H. T., von Klitzing, K. & Ploog, K. (1994). Competition between thermally induced resonant tunneling and phonon-assisted tunneling in semiconductor superlattices, *Phys. Rev. B* 50: 10998–11001.
URL: *http://link.aps.org/doi/10.1103/PhysRevB.50.10998*

Schwartz, A. (1998). The potential engineering of grain boundaries through thermomechanical processing, *JOM Journal of the Minerals, Metals and Materials Society* 50: 50–55.
URL: *http://dx.doi.org/10.1007/s11837-998-0250-5*

Sootsman, J. R., Chung, D. Y. & Kanatzidis, M. G. (2009). Alte und neue Konzepte für thermoelektrische Materialien, *Angewandte Chemie* 121: 8768–8792.
URL: *http://dx.doi.org/10.1002/ange.200900598*

Venkatasubramanian, R. (2000). Lattice thermal conductivity reduction and phonon localization like behavior in superlattice structures, *Phys. Rev. B* 61(4): 3091–3097.
URL: *http://link.aps.org/doi/10.1103/PhysRevB.61.3091*

Venkatasubramanian, R., Siivola, E., Colpitts, T. & O'Quinn, B. (2001). Thin-film thermoelectric devices with high room-temperature figures of merit, *Nature* 413(6856): 597–602.
URL: *http://dx.doi.org/10.1038/35098012*

Völklein, F. & Zetterer, T. (2006). *Praxiswissen Mikrosystemtechnik*, Vieweg Praxiswissen.

Watanabe, T. (1985). Structural effects on grain boundary segregation, hardening and fracture, *Journal de Physique Colloques* 46(C4): C4–555–C4–566.
URL: *http://hal.archives-ouvertes.fr/jpa-00224714/en/*

Watanabe, T. (1993). Grain boundary design and control for high temperature materials, *Materials Science and Engineering: A* 166(1-2): 11–28.
URL: *http://www.sciencedirect.com/science/article/pii/092150939390306Y*

Watanabe, T. & Tsurekawa, S. (1999). The control of brittleness and development of desirable mechanical properties in polycrystalline systems by grain boundary engineering, *Acta Materialia* 47(15-16): 4171–4185.
URL: *http://www.sciencedirect.com/science/article/pii/S135964549900275X*

Yang, B. & Chen, G. (2003). Partially coherent phonon heat conduction in superlattices, *Phys. Rev. B* 67: 195311.
URL: *http://link.aps.org/doi/10.1103/PhysRevB.67.195311*

Yao, T. (1987). Thermal properties of AlAs/GaAs superlattices, *Applied Physics Letters* 51(22): 1798–1800.
URL: *http://dx.doi.org/doi/10.1063/1.98526*

Yin, Y., Lu, Y., Gates, B. & Xia, Y. (2001). Template-assisted self-assembly: A practical route to complex aggregates of monodispersed colloids with well-defined sizes, shapes, and structures, *Journal of the American Chemical Society* 123(36): 8718–8729.
URL: *http://dx.doi.org/10.1021/ja011048v*

4

Design, Manufacturing and Testing of Polymer Composite Multi-Leaf Spring for Light Passenger Automobiles - A Review

Senthilkumar Mouleeswaran
Assistant Professor (Senior Grade),
Department of Mechanical Engineering
PSG College of Technology
Coimbatore
India

1. Introduction

Weight reduction has been the main/primary focus of automobile manufactures. Suspension leaf spring, a potential item for weight reduction in automobiles, accounts for 10-25 percent of unsprung weight, (which is considered to be the mass not supported by leaf spring). Application of composite materials reduces the weight of leaf spring without any reduction on the load carrying capacity and stiffness in automobile suspension system(Daugherty,1981;Breadmore,1986;Morris,1986). A double tapered beam for automotive suspension leaf spring has been designed and optimized(Yu & Kim,1988). Composite mono leaf spring(Rajendran & Vijayarangan,2001) has also been analyzed and optimized.

Leaf spring should absorb vertical vibrations due to road irregularities by means of variations in the spring deflection so that potential energy is stored in the spring as strain energy and then released slowly. So, increasing energy storage capability of a leaf spring ensures a more compliant suspension system. A material with maximum strength and minimum modulus of elasticity in longitudinal direction is the most suitable material(Corvi,1990) for a leaf spring. Important characteristics of composites(Springer & Kollar,2003) that make them excellent for leaf spring instead of steel are higher strength-to-weight ratio, superior fatigue strength, excellent corrosion resistance, smoother ride, higher natural frequency, etc. Fatigue failure is the predominant mode of in-service failure of many automobile components, especially the springs used in automobile suspension systems. Fatigue behaviour of Glass Fiber Reinforced Plastic epoxy (GFRP) composite materials has been studied(Hawang & Han,1986). A composite mono-leaf spring has been designed and their end joints are anlysed and tested for a light weight vehicle (Shivasankar & Vijayarangan, 2006). Experimental and numerical analysis are carried out on a single leaf constant cross section composite leaf spring (Jadhao & Dalu, 2011). Theoretical equation for predicting fatigue life, formulated using fatigue modulus and its degrading rate, is simplified by strain failure criterion for practical application. A prediction method for fatigue strength of composite structures at an arbitrary combination of frequency, stress ratio and temperature has been presented (Yasushi,1997).

In the present work, a 7-leaf steel spring used in a passenger car is replaced with a composite multi leaf spring made of glass/epoxy composites. Dimensions and number of leaves of steel leaf spring (SLS) and composite leaf spring (CLS) are considered to be same. Primary objective is to compare their load carrying capacity, stiffness and weight savings of CLS. Ride comfort of both SLS and CLS is found and compared. Also, fatigue life of SLS and CLS is also predicted. This chapter of the book explores the work done on design optimisation, finite element analysis, analytical & experimental studies and life data analysis of steel and composite leaf springs (Senthilkumar & Vijayarangan,2007).

2. Steel Leaf Spring (SLS)

2.1 Design and finite element analysis

Design parameters of 7-leaf steel spring that exists in a light passenger car (available in India) are given in Table 1. The spring is assumed to be a double cantilever beam, even though the leaf spring is simply supported at the ends. Also, this spring is geometrically and materially symmetrical so that only one half is considered with cantilever beam boundary conditions for the analysis to save the calculation time. Axle seat is assumed to be fixed and loading is applied at free eye end.

Parameters	Values
Total length (eye-to-eye), mm	1150
Arc height at axle seat(Camber), mm	175
Spring rate, N/mm	20
Number of full length leaves	2
Number of graduated leaves	5
Width of the leaves, mm	34
Thickness of the leaves, mm	5.5
Full bump loading, N	3250
Spring weight, kg	13.5

Table 1. Parameters of steel leaf spring.

A stress analysis was performed using two-dimensional, plane strain finite element model (FEM). A plane strain solution is considered because of the high ratio of width to thickness of a leaf. Model is restrained to the right half part only because the spring is symmetric. The contact between leaves is emulated by interface elements and all the calculations are done using ANSYS (version 7.1) (Eliahu Zahavi,1992). Nodes are created based on the values of co-ordinates calculated and each pair of coincident nodes is joined by the interface elements that simulate action between neighboring leaves. Element selected for this analysis is SOLID42(Eliahu Zahavi,1992) which behaves as the spring having plasticity, creep, swelling, stress stiffening, large deflection, and large strain capabilities. Element is defined by four nodes having two degrees of freedom at each node: translations in the nodal x and y directions. Interface elements CONTA174 that is defined by eight nodes and TARGE170 are used to represent contact and sliding between adjacent surfaces of leaves. The contact elements themselves overlay the solid elements describing the boundary of a deformable body and are potentially in contact with the target surface, defined by TARGE170. This target surface is discretized by a set of target segment elements (TARGE170) and is paired with its associated contact surface via a shared real constant set. An average coefficient of friction 0.03 is taken between surfaces(SAE manual). Also, analytical solution is carried out using spring design SAE manual.

2.2 Static testing

The static testing on existing steel leaf spring was carried out using an electro-hydraulic test rig which is depicted in Fig.1. The rig has the ability to apply a maximum static load of 10 kN. It has a display unit to show both load and corresponding deflection. The loading was gradually from no load to full bump load of 3250 N. The strain gauges were employed to measure strain and to calculate stress. The experimental data corresponding to steel leaf spring is given in Table 2. Maximum normal stress, $\sigma 11$ from FEM is compared to the experimental solution under full bump loading (error, 8.63%). There is a good correlation for stiffness in experimental, analytical and FEM methods (Table 2).

Parameters	Experiment	Analytical	FEM
Load, N	3250	3250	3250
Maximum stress, MPa	680.05	982.05	744.32
Maximum deflection, mm	155	133.03	134.67
Maximum stiffness, N/mm	20.96	24.43	24.13

Table 2. Stress analysis of steel leaf spring using experimental, analytical and FEM.

3. Composite Leaf Spring (CLS) (Senthilkumar & Vijayarangan,2007)

Applicability of CLS in automobiles is evaluated by considering the types of vehicles and different loading on them. Theoretical details of composite mono-leaf spring are reported (Ryan,1985; Richrad et al., 1990). In some designs, width is fixed and in each section the thickness is varied hyperbolically so that thickness is minimum at two edges and is maximum in the middle (Nickel,1986). Another design, in which width and thickness are fixed from eyes to middle of spring and towards the axle seat width decreases hyperbolically and thickness increases linearly, has been presented (Yu & Kim, 1988). In this design, curvature of spring and fiber misalignment in the width and thickness direction are neglected. A double tapered CLS has been designed and tested with optimizing its size for minimum weight(Rajendran & Vijayarangan,2002). A composite mono-leaf spring has also been designed and optimized with joint design(Mahmood & Davood,2003). The mono-leaf spring is not easily replaceable on its catastrophic failure. Hence, in this work, a composite multi leaf spring is designed and tested for its load carrying capacity, stiffness and fatigue life prediction using a more realistic situation.

3.1 Material selection

Material selected should be capable of storing more strain energy in leaf spring. Specific elastic strain energy can be written as:

$$S = \frac{1}{2}\frac{\sigma_t^2}{\rho E} \quad (1)$$

where σ_t is allowable stress, E is modulus of elasticity and ρ is density. Based on specific strain energy of steel spring and some composite materials(Yu&Kim,1988), E-glass/epoxy is selected as spring material having the mechanical properties given in Table 3. This material is assumed to be linearly elastic and orthotropic.

Parameters	Values
Modulus of elasticity, GPa	E_{11}, 38.6
Modulus of elasticity, GPa	E_{22}, 8.27
Modulus of shear, GPa	G_{12}, 4.14
Poisson ratio,	v_{xy}, 0.26
Tensile strength, MPa	σ_{t11}, 1062
Tensile strength, MPa	σ_{t22}, 31
Compressive strength, MPa	σ_{c11}, 610
Compressive strength, MPa	σ_{c22}, 118
Shear strength, MPa	τ_{12}, 71

Table 3. Mechanical properties of E-glass/epoxy(Springer & Kollar,2003).

3.2 Layup selection

Stored elastic strain energy in a leaf spring varies directly with the square of maximum allowable stress and inversely with the modulus of elasticity both in the longitudinal and transverse directions according to Eq. (1). Composite materials like E-glass/epoxy in the direction of fibers have good characteristics for storing strain energy. So, lay up is selected to be unidirectional along the longitudinal direction of spring. This also helped in fabricating process using filament-winding machine.

3.3 Design and finite element analysis of composite leaf spring

With the extensive use of laminated composite materials in almost all engineering fields, the optimal design of laminated composites has been an extensive subject of research in recent years. The dimensions of the composite leaf spring are taken as that of the conventional steel leaf spring. Each leaf of the composite leaf spring consists of 20 plies of thickness 0.275 mm each. The number of leaves is also the same for composite leaf spring. The design parameters selected are listed in Table 4.

A 3-D model of the leaf spring is used for the analysis in ANSYS 7.1, since the properties of the composite leaf spring vary with the directions of the fiber. The loading conditions are assumed to be static.The element chosen is SOLID46, which is a layered version of the 8-node structural solid element to model layered thick shells or solids. The element allows up to 250 different material layers. To establish contact between the leaves, the interface elements CONTACT174 and TARGET170 are chosen.

Individual leaves are fabricated using a filament-winding machine. A fiber volume fraction of 0.6 is used. All individual leaves are assembled together using a center bolt and four side clamps. Also metal spring eyes are fixed at both the ends. CLS is tested with an electro-hydraulic leaf spring test rig (Fig. 1). Four CLSs were manufactured and tested. The spring, which provided the lowest stiffness and highest stress values, has been considered for comparative purpose because it satisfies the fail-safe condition. The reason for the stiffness and stress variations may be due to variation in volume fraction obtained in the fabrication process or due to lack of complete curing. A reasonably good weight reduction (68.15%) is achieved by using CLS (4.3 kg) in place of SLS (13.5 kg).

Parameters	Values
Thickness of each leaf, mm	5.5
Width of the each leaf ,mm	34
Thickness of the fiber, mm	0.2
Width of the fiber, mm	34
Thickness of the resin, mm	0.075
Width of the resin, mm	34
Thickness of single layer, mm	0.275
Number of layers	20

Table 4. Design parameters of composite leaf spring.

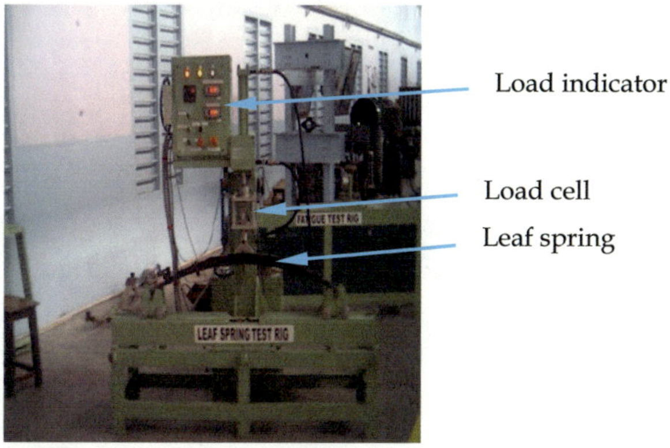

Fig. 1. Electro-hydraulic leaf spring test rig.

For a light passenger vehicle with a camber height of 175 mm, static load to flatten the leaf spring is theoretically estimated to be 3250 N. Therefore, a static vertical force of 3250 N is applied to determine the load-deflection curves (Fig. 2).The load is gradually increased to obtain the deflection of steel spring first untill it becomes comepletely flat. Then, for similar deflection in composite leaf spring, the loads are measured for composite leaf spring. From Fig.2, it is understood that the deflection increases linearly as load increases in both steel and composite leaf springs. For a full bump load of 3250 N, composite leaf spring deflects to 94 mm only while steel leaf spring deflects 175 mm. The FEM results of longitudinal stress and deflection of CLS are shown in Figs.3&4. During full bump load test, experimental stress measurement (Fig. 5) is carried out to verify the results of FEM analysis (Figs 3 & 4). Fig.3 shows the variation of stress in CLS along the length of the spring. Fig.4 shows the deflection of CLS at various points along the length. It is found that CLS developes the maximum stress of about 215 MPa and it deflectes about 60 mm. E-glass/epoxy composite leaf spring has spring constants 34.57-53.59 N/mm. Thus, all the data of spring constants for CLSs are greater than the design value, 20 N/mm. The reason for increased stiffness is lower density of E-glass/epoxy composite combination.

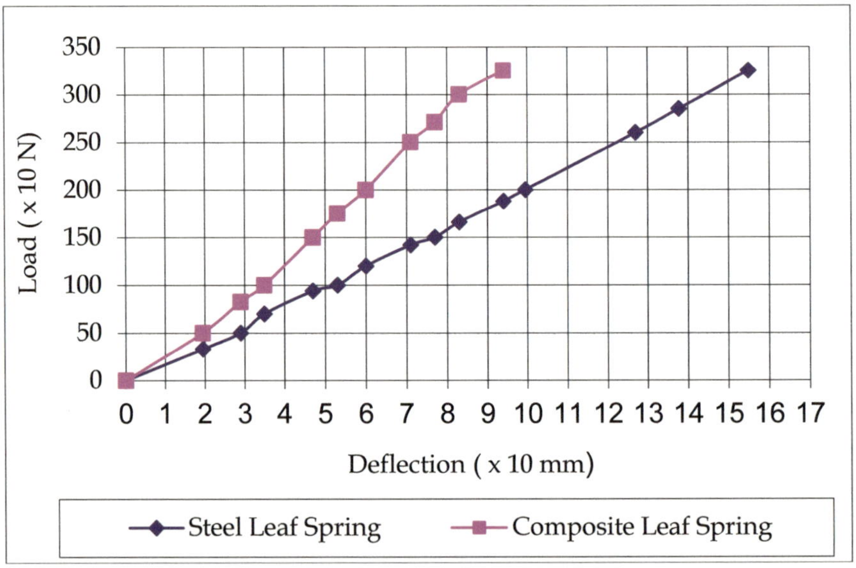

Fig. 2. Load - deflection curves for steel and composite leaf springs.

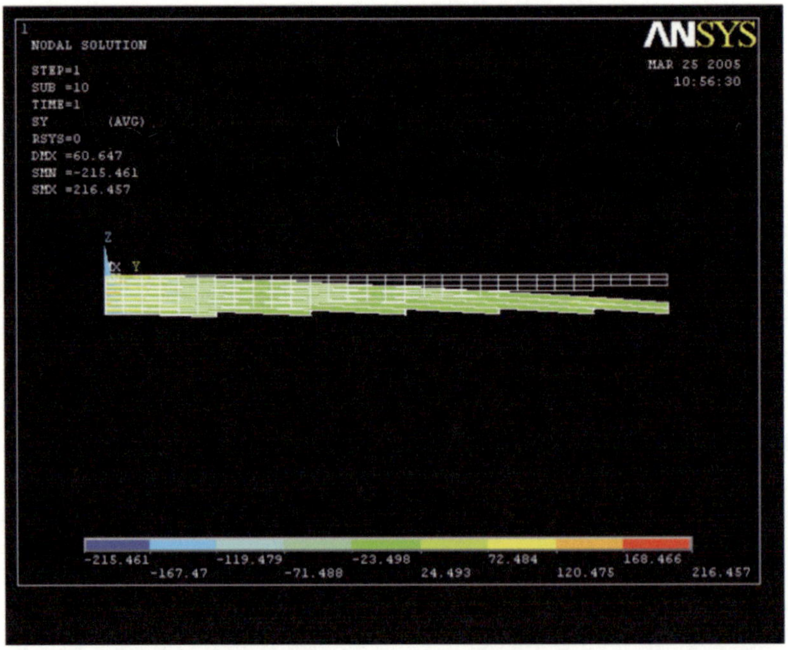

Fig. 3. FEM results of longitudinal stress of composite spring.

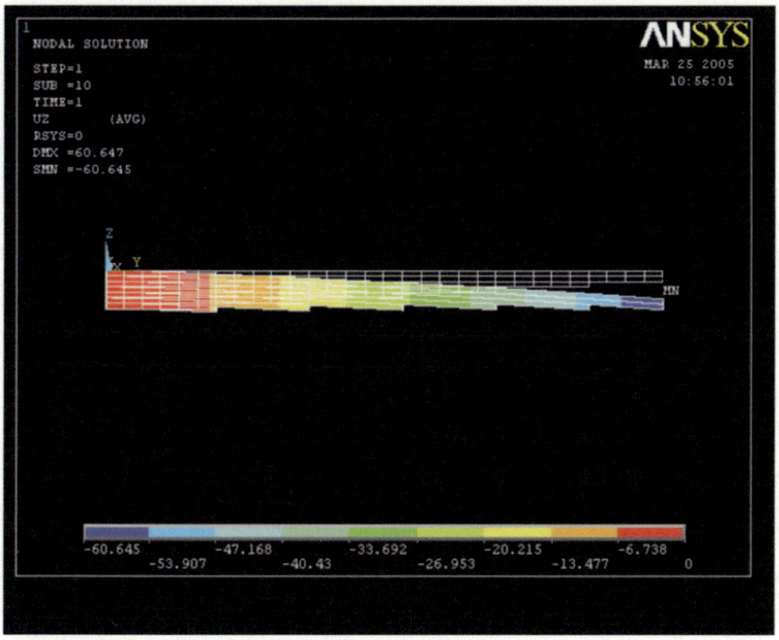

Fig. 4. FEM results of deflection of composite spring.

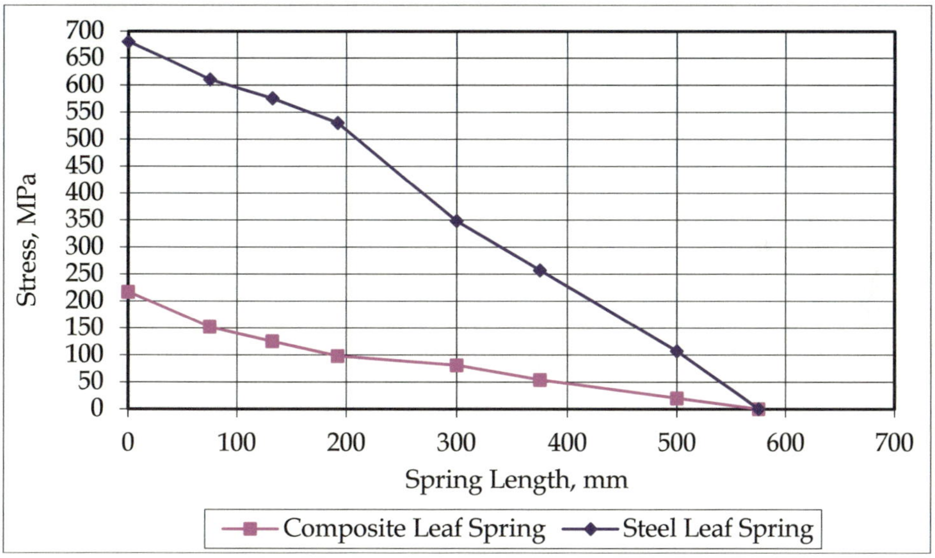

Fig. 5. Variation of experimental stress of steel and composite springs.

Leaf spring is analyzed under transverse loading condition. The longitudinal compressive strength of composite used in this analysis is less than its longitudinal tensile strength. So failure criterion stress is longitudinal compressive stress. Maximum longitudinal compressive stress (Fig. 5) is 222 MPa for CLS. At a same loading, maximum stress developed in SLS is 680 MPa (Table 2, Fig. 5). When compared with stress developed in SLS, less stress (67.35%) is developed in CLS. Compressive strength of fiber glass/epoxy is 610

MPa and yielding stress of steel is 1175 MPa. So, factor of safety obtained in SLS is 1.73, while in CLS it is 2.75. Experimental deflection of CLS under full bump loading is 94 mm (Fig. 2), which is less than the maximum value (175 mm). It shows that CLS is stiffer (64.95 %) than SLS. Table 5 gives the results of analysis of CLS using experimental, analytical and finite element methods. The variation of longitudinal stress of SLS and CLS is also presented in Figs.6&7. Fig. 6 shows the variation of longitudinal stress of steel leaf spring in FEM and experimental analysis. There is about 8% higher stress value obtained in the FEM than experiments. This is due to the fact that the constraints given at the ends of leaf spring are stiffer than the actual stiffness due the cantilever configuration in FEM. There could also be deviation from the material properties used in FEM analysis.

Parameters	Experiment	Analytical	FEM
Load, N	3250	3250	3250
Maximum stress, MPa	222	310.82	215.46
Maximum deflection, mm	94	59.20	60.65
Maximum stiffness, N/mm	34.57	54.89	53.59

Table 5. Stress analysis of composite leaf spring using experimental, analytical and FEM.

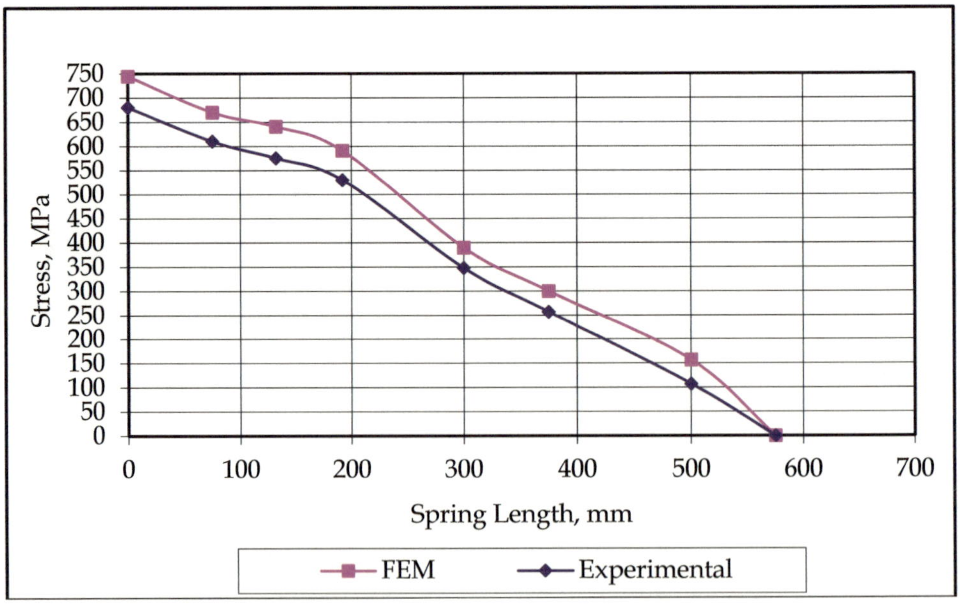

Fig. 6. Variation of longitudinal stress of steel leaf spring.

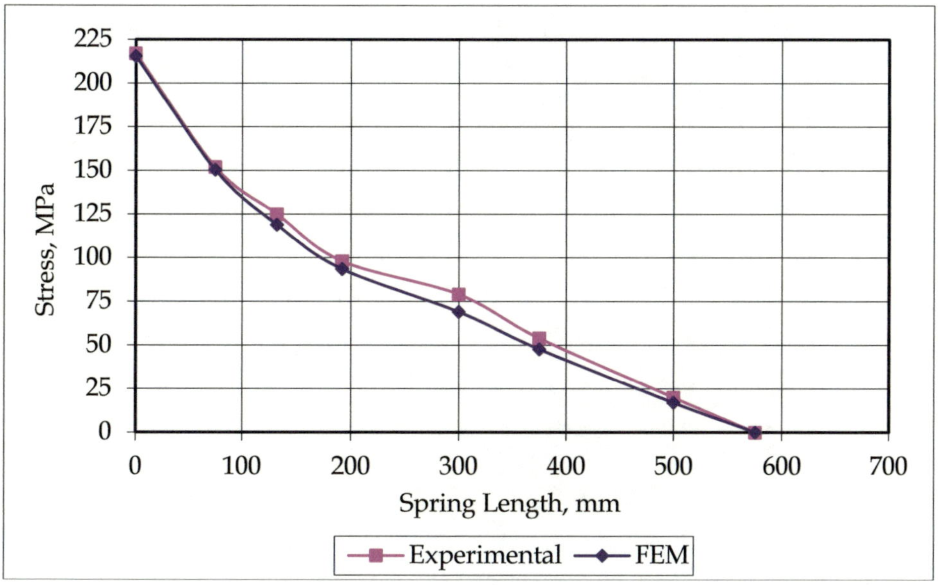

Fig. 7. Variation of longitudinal stress of composite leaf spring.

4. Fatigue analysis (Senthilkumar & Vijayarangan, 2007)

Main factors that contribute to fatigue failures include number of load cycles experienced, range of stress and mean stress experienced in each load cycle and presence of local stress concentrations. Testing of leaf springs using the regular procedure consumes a lot of time. Hence (SAE manual,1990) suggests a procedure for accelerated tests, which give quick results, particularly for SLSs. As per the outlined procedure(SAE manual,1990 & Yasushi,1997), fatigue tests are conducted on SLSs and CLSs. Fatigue life(Yasushi,1997) is expressed as the number of deflection cycles a spring will withstand without failure (Fig. 8).

4.1 Fatigue life of Steel Leaf Spring (SLS)

Fatigue life calculation of SLS is given as follows: stroke available in fatigue testing machine, 0-200 mm; initial deflection of SLS, 100 mm; initial stress (measured by experiment), 420 MPa; final deflection of SLS (camber), 175 mm; maximum stress in the final position (measured by experiment), 805 MPa. Fatigue life cycles predicted for SLS is less than 10,00,000 cycles (Fig. 8) by the procedure outlined in (SAE manual,1990).

4.2 Fatigue life of Composite Leaf Spring (CLS)

A load is applied further from the static load to maximum load with the help of the electro-hydraulic test rig, up to 3250 N, which is already obtained in static analysis. Test rig is set to operate for a deflection of 75 mm. This is the amplitude of loading cycle, which is very high. Frequency of load cycle is fixed at 33 mHz, as only 20 strokes/min is available in the test rig. This leads to high amplitude low frequency fatigue test.

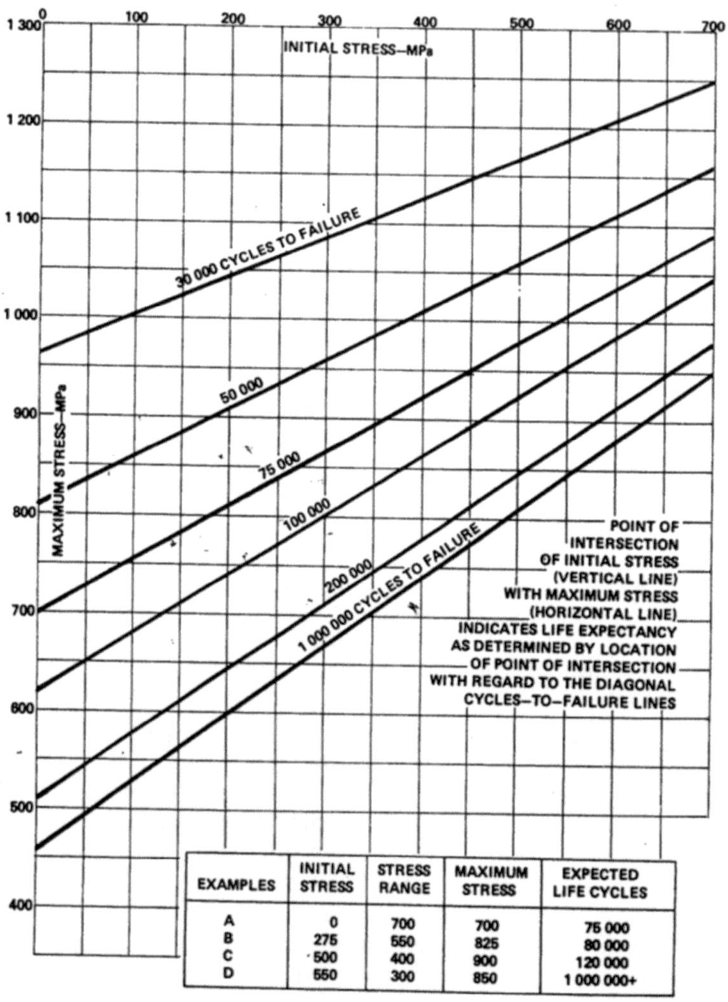

EXAMPLES	INITIAL STRESS	STRESS RANGE	MAXIMUM STRESS	EXPECTED LIFE CYCLES
A	0	700	700	75 000
B	275	550	825	80 000
C	500	400	900	120 000
D	550	300	850	1 000 000+

Fig. 8. Estimation of fatigue life cycles of steel leaf springs (SAE manual,1990).

Maximum and minimum stress values obtained at the first cycle of the CLS are 222 MPa and 133 MPa respectively. As the cycles go on increasing, stress convergence is happening only after 25000 cycles. These maximum and minimum operating stress values are 240 MPa and 140 MPa respectively. Because of very low stress level, fatigue life of CLS is very high under simulated conditions.

4.2.1 Life data analysis

Life data analysis (Weibull,1961) which is a statistical approach is used to find the reliability of predictions about the life of composite leaf springs by fitting a statistical distribution to life data from representative sample units. For the GFRP leaf spring, the life data is measured in terms of the number of cycles to fail for the four leaf springs and are presented in Table 6.

Leaf spring No.	Cycles to fail	Stress level
1	10,800	0.65
2	6,950	0.65
3	19,240	0.65
4	14,350	0.65

Table 6. Number of cycles to failure for composite leaf spring.

4.2.2 Life time distribution

Weibull life distribution model is selected which has previously been used successfully for the same or a similar failure mechanism. The Weibull distribution is used to find the reliability of the life data and it helps in selecting the particular data that is to be used in life prediction model. The Weibull distribution uses two parameters, namely, 'b' and 'Θ' to estimate the reliability of the life data. 'b' is referred to as shape parameter and 'Θ' is referred to as scale parameter.

The reliability of Weibull distribution is given by,

$$R(t) = 1\text{-}\exp[-x / \theta]^{b} \tag{2}$$

where, X is the life; b is the Weibull slope or 'shape' parameter and Θ is the characteristic life or 'scale' parameter.

The parameters of the Weibull distribution are calculated using probability plotting (Weibull, 1961). The life cycles of leaf spring are arranged in increasing order and the median rank is calculated using the Equation (3) and are shown in Table 7.

$$Medianrank = 100 * (j\text{-}0.3) / (\mathrm{N} + 0.4) \tag{3}$$

where, j is the order number and N is the total quantity of sample.

Order No.	Cycles to fail	Median rank
1	6,950	15.9
2	10,800	38.5
3	14,350	61.4
4	19,240	84.1

Table 7. Median rank of composite leaf springs.

The value of Θ is found to be 14,600 cycles. The reliability of the life data is calculated and shown in Table 8. It is found that the reliability of 3rd GFRP spring is higher than that of other leaf springs and the fatigue life data of 3rd GFRP spring has been considered for fatigue life prediction.

The fatigue test is conducted up to 20000 cycles and it is examined that no crack initiation is visible. The details of test results at 0 and 20000 cycles are as follows: maximum load cycle range, 1850-3250 N; amplitude, 75 mm; frequency, 33 mHz; spring rate, 27.66 N/mm; maximum operating stress, 240 MPa; minimum operating stress, 140MPa and time taken 17 h. The experimental results are available only up to 20000 cycles. With no crack initiation,

Leaf spring no.	Life (cycles)	Median rank	Reliability (%)
1.	10,800	38.5	43
2.	6,950	15.9	39
3.	19,240	84.1	73
4.	14,350	61.4	62.7

Table 8. Reliability of fatigue life data.

there is a necessity to go for analytical model for finding number of cycles to failure from analytical results. An analytical fatigue model to predict the number of fatigue cycles to failure for the components made up of composite material has been developed (Hwang and Han,1986). They have proposed two constants in their relation on the basis of experimental results. It is proved that the analytical formula predicts the fatigue life of component with E-Glass/epoxy composite material.

$$\text{Hwang and Han relation: } N=\{B(1-r)\}^{1/C} \quad (4)$$

where, N is the number of cycles to failure; B= 10.33; C= 0.14012; $r = \sigma_{max}/\sigma_U$; σ_{max} is the maximum stress; σ_U is the ultimate tensile strength and r is the applied stress level. Equation (4) is applied for different stress levels and fatigue life is calculated for the composite leaf spring. The results are obtained based on the analytical results (Table 9) and the resulting S-N graph is shown in Fig.9. From Fig.9, it is observed that the composite leaf spring, which is made up of E-glass/epoxy is withstanding more than 10 Lakh cycles under the stress level of 0.24.

The test was conducted for nearly 17 hours to complete 20000 cycles. The variations in stress level were reduced to very low level after 20000 cycles. There was no crack initiation up to 20000 cycles. The stress level of 0.24 is obtained from experimental analysis. This is very much helpful for the determination of remaining number of cycles to failure using fatigue model (Hwang and Han,1986). According to this fatigue model, the failure of the composite leaf spring takes place only after 10 Lakh cycles. Since the composite leaf spring is expected to crack only after 10 Lakh cycles, it is required to conduct the leaf spring fatigue test up to 10 Lakh cycles for finding type and place of crack initiation and propagation. For completing full fatigue test up to crack initiation with the same frequency, nearly 830 hours of fatigue test is required.

From the design and experimental fatigue analysis of composite multi leaf spring using glass fibre reinforced polymer are carried out using life data analysis, it is found that the composite leaf spring is found to have 67.35% lesser stress, 64.95% higher stiffness and 126.98% higher natural frequency than that of existing steel leaf spring. The conventional multi leaf spring weighs about 13.5 kg whereas the E-glass/Epoxy multi leaf spring weighs only 4.3 kg. Thus the weight reduction of 68.15% is achieved. Besides the reduction of weight, the fatigue life of composite leaf spring is predicted to be higher than that of steel leaf spring. Life data analysis is found to be a tool to predict the fatigue life of composite multi leaf spring. It is found that the life of composite leaf spring is much higher than that of steel leaf spring.

Maximum stress MPa	Applied stress level	Number of cycles to failure
100	0.1	8143500
200	0.2	3515500
300	0.3	1354800
400	0.4	450900
500	0.5	122700
600	0.6	25000
700	0.7	3200
800	0.8	200
900	0.9	-
1000	1.0	-

Table 9. Fatigue life at different stress levels of composite leaf spring.

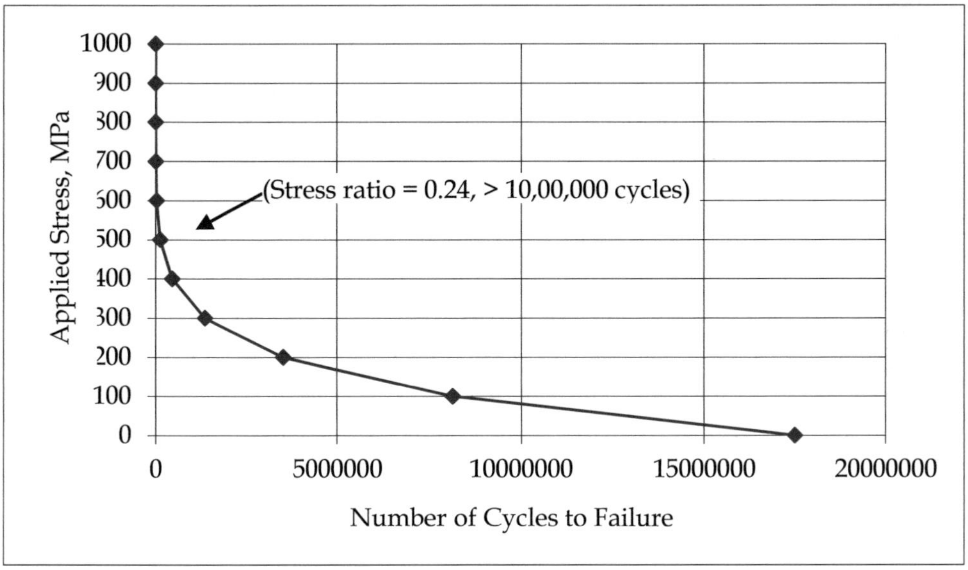

Fig. 9. S-N curve for composite leaf spring.

5. Optimisation

Optimisation of weight reduction in CLS has been carried out(Senthilkumar & Vijayarangan,2007). The finite element analysis of composite multi leaf spring with 7 leaves shows that maximum deflection of the leaf spring and the bending stress induced are well within the allowable limit (factor of safety 2.8). So, it is decided to optimize the number of leaves with minimum of 2 leaves and maximum of 7 leaves using ANSYS 7.1 itself, for the minimum weight. The number of leaves is reduced one by one, without changing any of the dimensions, and found out results for same full bump load. The results are shown in Table 10. From the FEM results in Table 10, it is understood that the leaf spring with 3 leaves violates the constraints of deflection and stress. Therefore, further optimization is terminated at this point. It is evident that only 4 leaves (2 full length leaves and 2 graduated leaves) are sufficient to withstand the applied load. The composite leaf spring with 4 leaves weighs about 3.18 kg only. It gives a weight reduction of 76.4%.

No.of Leaves	**Maximum Deflection (mm)**		**Maximum direct stress along the length of leaf spring (MPa)**	
	ANSYS	Allowable	ANSYS	Allowable
7	61	175	217	610
6	97		324	
5	104		409	
4	150		607	
3	308		1079	

Table 10. Results of optimization of number of leaves.

6. Ride comfort

To provide ride comfort to passenger, leaf spring has to be designed in such a way that its natural frequency is maintained to avoid resonant condition with respect to road frequency. The road irregularities usually have the maximum frequency of 12 Hz(Yu&Kim,1988). Therefore, leaf spring should be designed to have a natural frequency, which is away from 12 Hz to avoid the resonance. Stiffness is more and weight is lower of CLS than that of SLS. Therefore, first natural frequency of CLS (14.3 Hz) will be higher (126.98%) than that of SLS (6.3 Hz). First natural frequency of CLS is nearly 1.2 times the maximum road frequency and therefore resonance will not occur, and it provides improved ride comfort. After optimization, CLS has a fundamental natural frequency of 41.5 Hz, which is 3.46 times the maximum road frequency, which ensures that resonance will not occur.

7. Conclusion

Design and experimental analysis of composite multi leaf spring using glass fibre reinforced polymer are carried out. Compared to steel spring, the composite leaf spring is found to have 67.35% lesser stress, 64.95% higher stiffness and 126.98% higher natural frequency than that of existing steel leaf spring. The conventional multi leaf spring weighs about 13.5 kg whereas the E-glass/Epoxy multi leaf spring weighs only 4.3 kg. Thus the weight reduction of 68.15% is achieved. Besides the reduction of weight, the performance of the leaf spring is also increased. Compared to the steel leaf spring (13.5 kg), the optimised composite leaf spring weighs nearly 76.4% less than the steel spring. Ride comfort and life of CLS are also more when compared to SLS. Therefore, it is concluded that composite multi leaf spring is an effective replacement for the existing steel leaf spring in light passenger vehicles.

8. Acknowledgement

This work is carried out as part of research project supported by Defense Research and Development Organisation (DRDO), India. Author is grateful for the financial support provided by DRDO.

9. References

ANSYS 7.1. Manual, (1997). Ansys Inc, New York

Breadmore, P. & Johnson, C. F. (1986). The potential for composites in structural automotive Applications, *Composites science and Technology*, Vol. 26, No.4, pp. (251- 281), ISSN 0266-3538

Corvi, A. (1990). A preliminary approach to composite beam design using FEM analysis, *Composite structures*, Vol. 16, No.(1-3), pp. (259 - 275), ISSN 0263-8223

Daugherty, R.L. (1981). Composite leaf springs in heavy truck applications, *Proceedings of International conference of composites material*, ISBN 493113601X, Tokyo

Eliahu Zahavi. (1992). *The finite element method in machine design*, Prentice Hall, ISBN 0133182398, New York.

Hawang, W. & Han, K.S. (1986). Fatigue of composites—fatigue modulus concept and life prediction, *Journal of composite material*, Vol. 20, pp. (154 - 165), ISSN 0021-9983

Jadhao, K.K. & Dalu, R.S. (2011). Experimental investigation & numerical analysis of composite leaf spring, *International journal of engineering science and technology*, Vol.3, No.6,pp.(4759-4764), ISSN 0975-5462.

Morris, C. J. (1986). Composites structures for automobiles, *Composite structures*, Vol. 5, No.3, pp. (163 - 176), ISSN 0263-8223

Mahmood, M.S. & Davood, R. (2003). Analysis and optimization of a composite leaf spring, *Composite structures*, Vol.60, pp.(317-325), ISSN 0263-8223

Nickel, H.W. (January 1986). Bushing construction for a fiber reinforced plastic leaf spring, *US pat 4,565,356*

Ryan, W.E. (December 1985). Method of making a molded fiber reinforced plastic leaf spring, *US pat* 4, 560, 525

Richrad, D.S.; Mutzner, J.E. & Eiler, J.F. (January, 1990). Method of forming a composite leaf spring with fabric wear pad, *US pat 4,894,108*

Rajendran, I. & Vijayarangan, S. (2001). Optimal design of a composites leaf spring using genetic algorithm, *Computers and structures*, Vol. 79, No.11, pp. (1121 - 1129), ISSN 0045-7949

Senthilkumar, M. & Vijayarangan, S. (2007). Design optimization and experimental analysis of composite leaf spring for light passenger vehicles, *Advances in vibration engineering*, Vol.6, No.3, pp.(175-183), ISSN 0972-5768

Senthilkumar, M. & Vijayarangan, S. (2007). Static analysis and fatigue life prediction of steel and composite leaf spring for light passenger vehicles, *Journal of scientific & industrial research*, Vol.66, No.2,pp.(128-134), ISSN: 0022-4456

Senthilkumar, M. & Vijayarangan, S. (2007. Analytical and experimental studies on fatigue life prediction of steel and composite multi-leaf spring for light passenger vehicles using life data analysis, *Materials science (Medžiagotyra)*, Vol.13, No.2, pp.(141-146), ISSN 1392-1320

Shiva Shankar, G.S. & Vijayarangan, S. (2006). Mono composite leaf spring for light weight vehicle - design, end joint analysis and testing, *Materials science*, Vol.12, No.3, pp. (220-225), ISSN 1392-1320

Springer, G.S. & Kollar,L.P. (2003). *Mechanics of composite structure*, Cambridge university press, ISBN 0-521-80165-6, New york

Weibull, W. (1961). Fatigue testing and analysis of test results, *Pergamon press*, Paris

Yu, W.J. & Kim, H.C. (1988). Double tapered FRP beam for automotive suspension leaf spring. *Composite structures*, Vol. 9, No.4, pp. (279- 300), ISSN 0263-8223

Yasushi, M. (1997). Prediction of flexural fatigue strength of CFRP composites under arbitrary frequency, stress ratio and temperature, *Journal of composite material*, Vol. 31, pp. (619 - 638), ISSN 0021-9983

(1980). *Manual on Design and Application of Leaf Springs*, HS - *788*(edition: 1), SAE International, ISBN 978-0-89883-383-6, New York

5

Modeling, Simulation and Experimental Studies of Distortions, Residual Stresses and Hydrogen Diffusion During Laser Welding of As-Rolled Steels

T. Böhme[1,*], C. Dornscheidt[2,*], T. Pretorius[3], J. Scharlack[3] and F. Spelleken[3]
[1]*Freudenberg-Schwab Vibration Control GmbH & Co. KG, Velten b Berlin*
[2]*Strip Processing Lines Division, SMS Siemag AG, Hilden*
[3]*Department Research and Development,*
Thyssen Krupp Steel Europe AG, Duisburg
[*]*Formerly with ThyssenKrupp Steel Europe AG*
Germany

1. Introduction

Beyond the assembling of components laser welding procedures are also increasingly applied in steel manufacturing. Reasons for this trend are: high joining velocities, concentrated heat input resulting in extremely thin heating zones, and nearly constant seam geometries along the welding line. So, for example, welding aggregates are installed at the beginning of electrolytic or hot-galvanizing manufacturing lines to join coils of different thickness or grade for continuous production, cf. Figure 1. To minimize shutdowns welding must be extremely reliable, since the joining zone of the 'endless strip' has to resist high thermal and mechanical loads, such subsequent annealing, bending and tension. Failures lead to technical breakdowns, repairs, and loss of production, which should be avoided as much as possible. Moreover, the welding process must be such robust that material with imperfections, such as oxidations or materials ripples, would be also accurately joined. Welding defects, for instances cracks, pores or seam shrinkages, cf. Figure 2, lead to reduction of the cross-sectional area and, therefore, represent critical regions w. r. t. damage and failure. From the engineering point-of-view it is essential to know, whether residual stresses or heat-induced degradations of the materials strength following from joining become critical or not. This question can be answered by e.g. extensive experimental investigations, during which material combinations, geometries (i.e., thickness) and welding parameters such as joining velocity, welding power or laser caustic are varied. Subsequent tensile tests of the different seams yield the critical strength, which is compared with the loading conditions of the production line.

However, such experiments are time-consuming and, therefore, expensive. Moreover, the derived predictions only hold for the used materials, geometries and investigated process parameters; extrapolations of the results beyond these conditions are not possible. Consequently, it is desirable to have a general framework, which allows for the prediction of the thermal and mechanical material response following from an arbitrary choice of

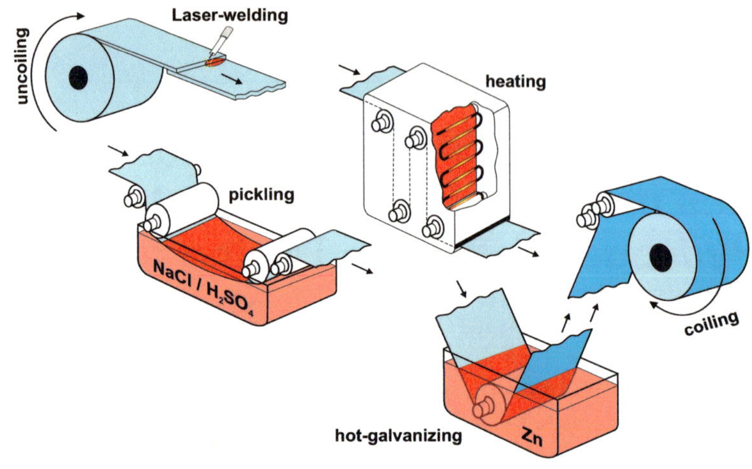

Fig. 1. Illustration of the continuous (hot-)galvanizing process including laser welding at the beginning.

materials, geometries, and welding parameters. Numerical simulations of welding processes increasingly satisfy this demand. They provide information about heat input, resulting thermal distortions, residual stresses or accompanying microstructural changes, and that mostly much faster than corresponding experiments. Thus it is possible to identify critical regions, welding parameters, or materials combinations in advance. In the following

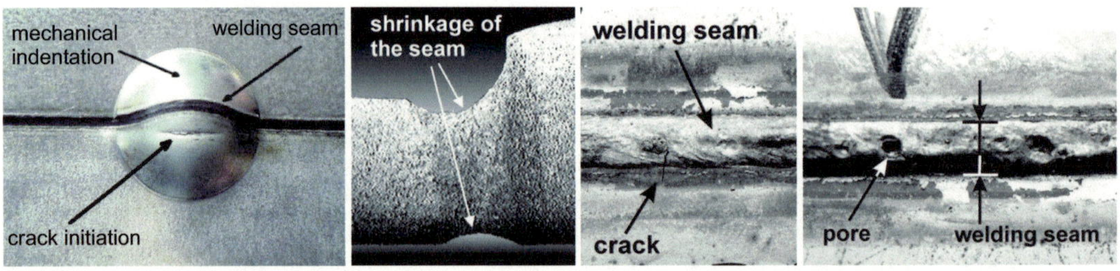

Fig. 2. *Left*: crack initiation within the Heat Affected Zone (HAZ) during the so-called ERICHSEN test. *2nd picture*: shrinkage of the seam at the upper and lower side of two dissimilar sheet geometries. *3rd picture*: Transverse crack due to shrinkage. *Right*: welding pore at the surface of the seam.

sections the laser welding process, with process data typically used during continuous steel manufacturing, will be investigated from the numerical and experimental point-of-view. After an outline to the characteristic continuum-mechanical equations the mathematical framework of Finite Element (FE) simulations is briefly described. Subsequently all required, thermal and mechanical materials data are determined, partially by means of tensile tests with varying temperatures. Different FE simulations are presented using the commercial FE-code ANSYS®, (ANSYS®, 2007), and considering a multiphase TRIP steel (TRIP 700) of 1.8mm thickness. The numerical results are compared with simultaneously performed welding experiments, and, finally, current limitations, open questions and further tasks for (laser-) welding modelling are discussed.

2. A brief review of the physical and mathematical modelling[1]

2.1 Thermodynamic states, balances and processes

The thermodynamic state of an arbitrary material point at the position[2] $\underline{\mathbf{x}}$ at time t of a system (body) can be completely described by *five field quantities*[3] within the so-called 5-field-theory, viz.

- $\rho(\underline{\mathbf{x}}, t)$: mass density,
- $\underline{\mathbf{u}}(\underline{\mathbf{x}}, t) = \underline{\mathbf{x}}(t) - \underline{\mathbf{X}}$: displacements,
- $e(\underline{\mathbf{x}}, t)$: mass specific internal energy.

$\underline{\mathbf{u}}$ represents the displacements of the material point w.r.t. a chosen reference position $\underline{\mathbf{X}} = \underline{\mathbf{x}}(t_0)$ with (reference) time t_0. The internal energy is defined as the difference between total energy and the energy of motion (here: translation), namely $e(\underline{\mathbf{x}}, t) = e_{\text{tot}}(\underline{\mathbf{x}}, t) - \frac{1}{2}\dot{\underline{\mathbf{x}}}^2$.

The temporal and spatial evolutions of the above five fields are determined by so-called *balance equations* (abbr.: balances) for mass, momentum and energy. These equations represent axioms[4] and read:

$$\frac{\partial \rho}{\partial t} + \underline{\boldsymbol{\nabla}} \cdot (\rho \underline{\mathbf{v}}) = 0\,, \tag{1}$$

$$\frac{\partial \dot{\underline{\mathbf{u}}}}{\partial t} + \underline{\boldsymbol{\nabla}} \cdot (\rho \underline{\mathbf{v}}\,\underline{\mathbf{v}} - \underline{\underline{\boldsymbol{\sigma}}}^T) = \underline{\mathbf{f}}\,, \tag{2}$$

$$\frac{\partial e}{\partial t} + \underline{\boldsymbol{\nabla}} \cdot (\rho e\, \underline{\mathbf{v}} + \underline{\mathbf{q}}) = \underline{\underline{\boldsymbol{\sigma}}}^T \cdot\cdot\, \underline{\boldsymbol{\nabla}}\, \underline{\mathbf{v}} + r\,. \tag{3}$$

Please note the identity between the velocity $\underline{\mathbf{v}}$ and the time derivative of $\underline{\mathbf{u}}$. Furthermore the quantities $\underline{\mathbf{f}}$, $\underline{\underline{\boldsymbol{\sigma}}}^T$, $\underline{\mathbf{q}}$ and r stand for the sum of externally applied body forces, the transposed CAUCHY stress tensor, the heat flux, and an arbitrary energy production term (e.g. due to latent heat during phase transitions). Eqs. (1-3) are equations of motion for the five unknown fields ρ, $\underline{\mathbf{u}}$ and e. They are universal, namely material-independent. To solve these equations, the *constitutive quantities*, viz. heat flux and stress tensor, must be replaced by constitutive equations (cf. subsequent paragraph) $\underline{\mathbf{q}} = \tilde{\underline{\mathbf{q}}}(T, \underline{\boldsymbol{\nabla}}T, ...)$ and $\underline{\underline{\boldsymbol{\sigma}}} = \tilde{\underline{\underline{\boldsymbol{\sigma}}}}(T, \underline{\mathbf{u}}, ...)$. Moreover, up to now no temperature T occur in the balances (1-3). For this reason a caloric state equation, $e = \tilde{e}(T)$, must be introduced, which allows for replacing the internal energy e by temperature T.

---

[1] All considerations refer to Cartesian coordinates. Italic letters stand for scalar values, vectors and tensors of 2nd (or higher) order are characterized by bold and underlined, double underlined (or double underlined with an index of order) letters. Scalar products of vectors and tensors are given by a dot or two dots, respectively.

[2] An EULER-configuration is used in the following considerations, i.e., all equations and quantities refer to the current position; further details can be found in (Gross & Seelig, 2006; Wriggers, 2000).

[3] Field quantities depend on time and position in a continuous and 'sufficiently smooth' (i.e., differentiable) manner, see also (Müller, 1984).

[4] Indeed, the balance for momentum or energy, Eqs. (2, 3), are an alternative formulations of NEWTON's 1st axiom or the 1st law of thermodynamics, respectively, (Müller, 1999; Müller & Müller, 2009).

If all constitutive equations are known and included, the balances represent a closed, coupled and mostly nonlinear System of Partial Differential Equations (PDE), which could be solved (in the majority of cases) by numerical methods. A *thermodynamic process* is defined by the solution of this PDE system, i.e., the knowledge of ρ, $\underline{\mathbf{u}}$, and T for all positions $\underline{\mathbf{x}}$ and times t. In particular the coupling between mass, momentum and energy - or with other words between materials structure, mechanical and thermal behaviour - lead to the notation of a 'multi-physical process' for welding, cf. Figure 3.

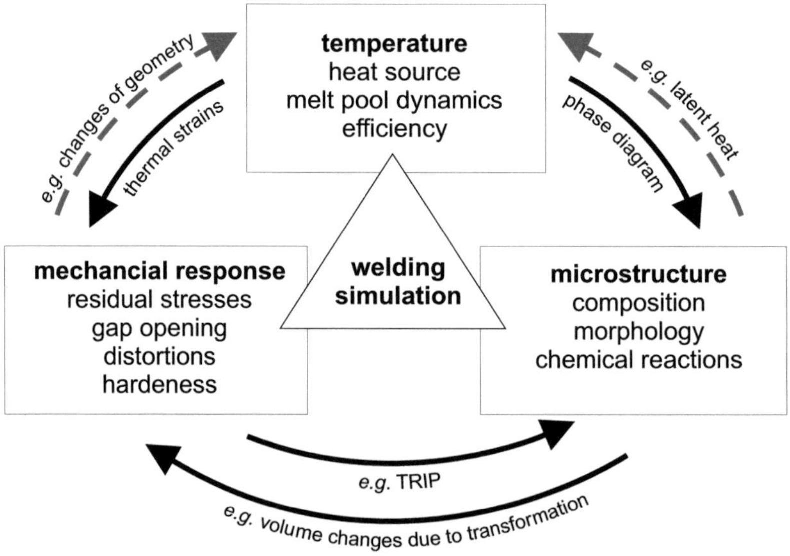

Fig. 3. Interaction of different physical disciplines during welding (dashed indicated couplings are neglected).

2.2 Constitutive equations for simulations

Typical constitutive equations, which are provided by commercial FE program packages, such as ANSYS®, are for instances the FOURIER's law of heat conduction:

$$\underline{\mathbf{q}} = -\underline{\underline{\boldsymbol{\kappa}}} \cdot \underline{\boldsymbol{\nabla}} T, \tag{4}$$

the caloric equation of state:

$$\frac{\mathrm{d}e}{\mathrm{d}t} = c_{\mathrm{p}} \frac{\mathrm{d}T}{\mathrm{d}t}, \quad \text{(pressure } p = \text{const)}, \tag{5}$$

or HOOKE's law for isotropic, elastic solids:

$$\underline{\underline{\boldsymbol{\sigma}}} = \underline{\underline{\boldsymbol{\sigma}}}^{\mathrm{el}} = 2\mu \underline{\underline{\boldsymbol{\varepsilon}}} + \lambda (\mathrm{Tr}\, \underline{\underline{\boldsymbol{\varepsilon}}}) \underline{\underline{\mathbf{I}}} - (3\lambda + 2\mu) \underline{\underline{\boldsymbol{\varepsilon}}}^{\mathrm{th}}, \tag{6}$$

with with linearized and thermal strains as follows: $\underline{\underline{\boldsymbol{\varepsilon}}} = \frac{1}{2}(\underline{\boldsymbol{\nabla}}\,\underline{\mathbf{u}} + (\underline{\boldsymbol{\nabla}}\,\underline{\mathbf{u}})^T)$ or $\varepsilon = (x - X)/X$ (for one dimension) and $\underline{\underline{\boldsymbol{\varepsilon}}}^{\mathrm{th}} = \varepsilon^{\mathrm{th}} \underline{\underline{\mathbf{I}}} = \alpha(T - T_0)\underline{\underline{\mathbf{I}}}$. Furthermore the material dependent quantities $\underline{\underline{\boldsymbol{\kappa}}}$, c_{p}, λ, μ and α identify the heat conductivity, the specific heat capacity, the LAME's constants, and the thermal expansion coefficient.

A constitutive equation of the form $\underline{\underline{\sigma}} = \tilde{\underline{\underline{\sigma}}}(\underline{\mathbf{u}})$ or $\underline{\underline{\sigma}} = \hat{\underline{\underline{\sigma}}}(\underline{\underline{\varepsilon}})$ for elasto-plastic material behaviour[5], correspondingly to Eq. (6), cannot be derived. The main reasons are **(I)**: there is no mathematical uniqueness for cyclic loading/ unloading, cf. Figure 4 (left), i.e., the appropriate stress for a given strain value depends on the loading history, and **(II)**: the multi-axial stress state and the corresponding strain cannot be directly derived by comparison with the one-dimensional stress-strain curve. To overcome these two items the following strategies are typically applied: **(i)** consideration of increments, which allows taking the loading history into account and **(ii)** introduction of a scalar 'equivalent stress' measure representing the multi-axial state. One of the most established equivalent measures are the VON MISES stress,

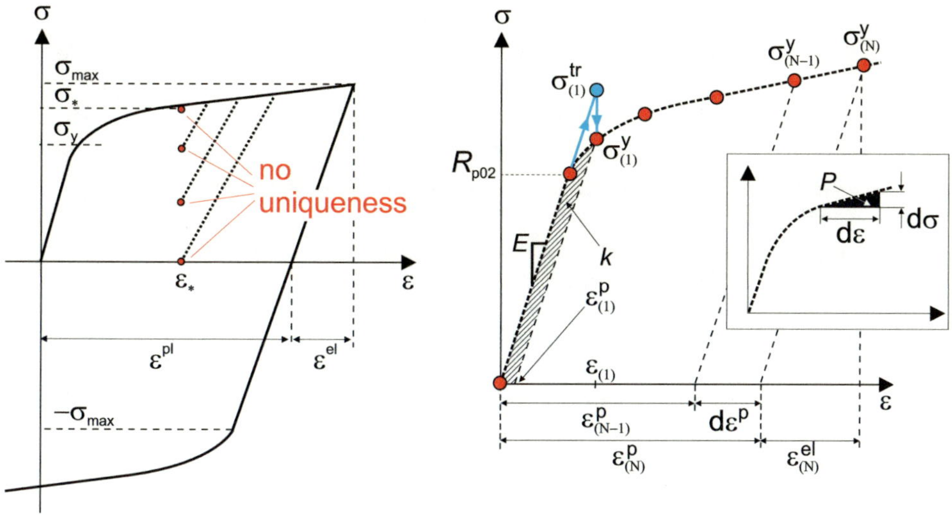

Fig. 4. *Left*: Elasto-plastic (un-)loading including a non-uniqueness regime. *Right*: Illustration of the 1D-stress-strain-curve interpolation, the yield stress calculation and DRUCKER's postulate.

or in the inverse formulation, the VON MISES strain[6]:

$$\sigma_{\mathrm{V}} = \sqrt{\frac{3}{2}\mathrm{dev}\,\underline{\underline{\sigma}} : \mathrm{dev}\,\underline{\underline{\sigma}}} \quad \text{and} \quad \varepsilon_{\mathrm{V}} = \sqrt{\frac{2}{3}\underline{\underline{\varepsilon}} : \underline{\underline{\varepsilon}}}\,, \tag{7}$$

where $\mathrm{dev}\,\underline{\underline{\sigma}} = \underline{\underline{\sigma}} - p\underline{\underline{\mathbf{I}}}$ denotes the deviator (i.e., the traceless part) of a tensor, since hydrostatic stresses do not contribute to plastic deformations. By means of a so-called yield criteria, containing the so-called yield-function Φ, it is possible to decide, whether the present stresses lead to yielding or not:

$$\Phi(\underline{\underline{\sigma}}) = \sigma_{\mathrm{V}}^2(\underline{\underline{\sigma}}) - \sigma_{\mathrm{y}}^2 = 0\,. \tag{8}$$

Here σ_{V} stands for the yield stress, which represents a cylindrical surface in the space spanned by the principle stresses, for details see e.g. (Gross & Seelig, 2006), and which is equal to R_{p02} for 'virgin' (no strain-hardening) material. If σ_{V}, calculated by Eq. (7), is at the surface

[5] The following considerations are restricted to time-independent plasticity with isotropic hardening; further models can be found in (Ansys theory ref., 2007).

[6] These quantities are typically used for metals and show realistic results.

or outside of the cylinder yielding will occur; for values inside the cylinder the material will remain elastic.

The yield stress σ_y increases with rising plastic deformation, which is called strain-hardening of the material. The mathematical description of this effect follows e.g. by means of a strain-hardening parameter k with $\sigma_y = \tilde{\sigma}_y(k)$ and, therefore, $\Phi(\underline{\underline{\sigma}}) \rightarrow \Phi(\underline{\underline{\sigma}}, k)$. For isotropic hardening (Figure 4, right) this parameter is directly linked to the consumed plastic work P of the material via the relation $\dot{k} = \dot{P} = \underline{\underline{\sigma}} \cdot\cdot\, \underline{\underline{\dot{\varepsilon}}}^{\text{pl}}$, (Sloan, 1987). Thus the yield condition of Eq. (8) can be used to derive a consistency condition, which - as will be seen subsequently - plays an important role for the determination of the plastic strains. It reads:

$$\dot{\Phi} = \frac{\partial \Phi}{\partial \underline{\underline{\sigma}}} \underline{\underline{\dot{\sigma}}} + \frac{\partial \Phi}{\partial k} \dot{k} = 0 \,. \tag{9}$$

The derivative $\partial\Phi/\partial\underline{\underline{\sigma}}$ follows from Eq. (7) and the expression $\partial\Phi/\partial k$ results from the graphical exploitation of the material specific, temperature-depending, one-dimensional stress-strain curves, cf. Figure 4 (right), provided by the user of the FE program package. The remaining variables $\underline{\underline{\dot{\sigma}}}$ and $\dot{k}$ can be calculated by HOOKE's law in combination with an additive decomposition of the total strains into an elastic and plastic part, $\underline{\underline{\varepsilon}} = \underline{\underline{\varepsilon}}^{\text{el}} + \underline{\underline{\varepsilon}}^{\text{pl}}$ ($\underline{\underline{\mathbf{C}}}_4$ denotes the 4th order stiffness matrix):

$$\underline{\underline{\dot{\sigma}}} = \underline{\underline{\mathbf{C}}}_4 \cdot\cdot (\underline{\underline{\dot{\varepsilon}}} - \underline{\underline{\dot{\varepsilon}}}^{\text{pl}}) \quad \text{and} \quad \dot{k} = \underline{\underline{\mathbf{C}}}_4 \cdot\cdot (\underline{\underline{\dot{\varepsilon}}} - \underline{\underline{\dot{\varepsilon}}}^{\text{pl}}) \cdot\cdot \underline{\underline{\dot{\varepsilon}}}^{\text{pl}} \,. \tag{10}$$

By knowledge of the total strain the only remaining unknown variable is $\underline{\underline{\varepsilon}}^{\text{pl}}$. In other words, the decomposition of $\underline{\underline{\varepsilon}}$ into an elastic and plastic part must be determined, see Figure 4. For this reason DRUCKER's postulate[7] can be used, which yields, inter alia, an associated flow rule as follows:

$$\underline{\underline{\dot{\varepsilon}}}^{\text{pl}} = \dot{\Lambda} \frac{\partial \Phi}{\partial (\text{dev}\underline{\underline{\sigma}})} \quad \text{or} \quad \text{d}\underline{\underline{\varepsilon}}^{\text{pl}} = \text{d}\Lambda \frac{\partial \Phi}{\partial (\text{dev}\underline{\underline{\sigma}})} \,. \tag{11}$$

Obviously the plastic strains are parallel to the derivative of the yield function w.r.t. the deviator of the stress tensor. The coefficient $\text{d}\Lambda$ characterizes the absolute value of $\text{d}\underline{\underline{\varepsilon}}^{\text{pl}}$ and must be iteratively determined by means of the consistency condition, cf. (Ansys theory ref., 2007).

Finally a schematic chronology for the computational calculation of the elastic-plastic stresses by means of the above equations is given. Here the total strains (following e.g. by thermal expansion due to heat supply) are assumed to be known:

(1) Consideration of the total strains as the sum of temporally successive strains, $\underline{\underline{\varepsilon}} = \sum_n \underline{\underline{\varepsilon}}(n)$, representing the deformation history.

---

[7] The postulate states, that for stable yielding the consumed plastic work is always positive (or zero), viz.

$$\text{d}P = \text{d}\underline{\underline{\sigma}} \cdot\cdot \text{d}\underline{\underline{\varepsilon}}^{\text{pl}} \geq 0 \,,$$

cf. Figure 4 (right) or explanations in various textbooks, for example in (Stouffer & Dame, 1996). By decomposition of $\text{d}\underline{\underline{\sigma}} = \text{d}\underline{\underline{\sigma}}^{\perp} + \text{d}\underline{\underline{\sigma}}^{||}$ into a normal and tangential part w. r. t. the yielding surface one can geometrically proof, that $\text{d}\underline{\underline{\varepsilon}}^{\text{pl}}$ is parallel to $\partial\Phi/\partial\underline{\underline{\sigma}}$ or $\partial\Phi/\partial(\text{dev}\underline{\underline{\sigma}})$, respectively. As a consequence of this fact Eq. (11) directly follows.

(2) Calculation of a trial stress by applying HOOKE's law for the n-th step as follows[8]:

$$\underline{\underline{\sigma}}^{\mathrm{tr}}(n) = \underline{\underline{\mathbf{C}}}_4 \cdot\cdot \left[\underline{\underline{\varepsilon}}(n) - \underline{\underline{\varepsilon}}^{\mathrm{pl}}(n-1)\right] , \tag{12}$$

which - in turn - can be used to compute the equivalent stress $\sigma_{\mathrm{V}}^{\mathrm{tr}}(n)$ according to Eq. (7).

(3) Determination of the yield stress σ_{y} by means of the known, material-specific stress-strain-curve and $\underline{\underline{\varepsilon}}^{\mathrm{pl}}(n-1)$. If $\underline{\underline{\varepsilon}}^{\mathrm{pl}}(n-1) = 0$, then σ_{y} will be equal to R_{p02}, otherwise its determination follows from the graphical exploitation according to Figure 4 (right).

(4) If the equivalent stress - calculated in the second item - is outside of the yielding surface, then plastic deformations will occur. During this process the plastic strains will reduce the corresponding stresses such that the yield criteria of Eq. (8) will finally hold. Consequently, $\underline{\underline{\sigma}}^{\mathrm{tr}}$ must be projected onto the yielding curve, which is done by adjusting $\mathrm{d}\underline{\underline{\varepsilon}}^{\mathrm{pl}}$, i.e., by varying the factor $\mathrm{d}\Lambda$ within the consistency condition of Eq. (9) and subsequent insertion into Eq. (11).

(5) Update of plastic and elastic strains:

$$\underline{\underline{\varepsilon}}^{\mathrm{pl}}(n) = \underline{\underline{\varepsilon}}^{\mathrm{pl}}(n-1) + \mathrm{d}\underline{\underline{\varepsilon}}^{\mathrm{pl}} \quad \text{and} \quad \underline{\underline{\varepsilon}}^{\mathrm{el}}(n) = \underline{\underline{\varepsilon}}(n) - \underline{\underline{\varepsilon}}^{\mathrm{pl}}(n) . \tag{13}$$

(6) Calculation of current stresses by:

$$\underline{\underline{\sigma}}(n) = \underline{\underline{\mathbf{C}}}_4 \cdot\cdot \underline{\underline{\varepsilon}}^{\mathrm{el}}(n) . \tag{14}$$

(7) Continuation with items (2) - (6) for the next steps $(n+1)$, $(n+2)$, etc. .

2.3 The principle of finite element analysis

To solve the PDE system following from the balances and the inserted constitutive equations different numerical methods can be applied, e.g. the Finite Element Method (FEM), for details see in various publications and textbooks, e.g. (Dean & Hidekazu, 2006; Kang & Im, 2007; Voss, 2001; Wriggers, 2000). For this reason, today a multitude of commercial programs, such as ANSYS® (ANSYS®, 2007), are available. They provide to the materials engineer a powerful mathematical framework, which conveniently select and solve the problem-specific equations under almost arbitrary conditions. However, a sovereign handling of these programs presumes knowledge about the underlying principles and framework. Therefore a brief explanation about the FEM will be given in the following paragraph. Starting point is the arbitrary PDE of *balance type* for a general scalar quantity w:

$$R(\underline{\mathbf{x}}, t) = \frac{\partial w(\underline{\mathbf{x}}, t)}{\partial t} + \underline{\boldsymbol{\nabla}} \cdot \underline{\tilde{\mathbf{f}}}\left[w(\underline{\mathbf{x}}, t)\right] - s(\underline{\mathbf{x}}, t) = 0 . \tag{15}$$

Here s denotes an arbitrary production or supply and $\underline{\tilde{\mathbf{f}}}$ identifies a vector function (see e.g. the mass balance).

The basic idea of FEM is the consideration of the so-called spatial *weak formulation* of the above PDE using a sufficiently smooth[9] test function $\eta(\underline{\mathbf{x}}) \in \mathbb{H}^1$. This function must vanish at the

[8] The plastic strains $\underline{\underline{\varepsilon}}^{\mathrm{pl}}(n-1)$ vanish, if elastic material behaviour occurred during the time step $(n-1)$.

[9] In the mathematical sense this condition ensures differentiability. The set of all suitable functions is denoted by $\mathbb{H}^1$, the so-called HILBERT space or the space of square-integrable functions.

boundary, $\partial\Omega$, of the domain, Ω, of the body, i.e., $\eta = 0$ for all $\underline{\mathbf{x}} \in \partial\Omega$ (the so-called zero conditon). Multiplying Eq. (15) with η and subsequent integration on $\Omega = \{\underline{\mathbf{x}}|\underline{\mathbf{x}} \in \mathbb{R}^3\}$ yields:

$$\int_\Omega R\eta \, \mathrm{d}\underline{\mathbf{x}} = 0 \,. \tag{16}$$

Thus, the quantity w must be chosen such, that the integral (or functional) of Eq. (16) becomes minimal (i.e., zero). This task is typical for so-called variation problems, in which extrema of 'functions of functions' are calculated. The variable R is often called the residuum and represents the deviation from the 'correct solution'.

For the most engineering problems the formulations of Eq. (15/16) are equivalent. The advantage of Eq. (16) follows by partial integration, viz.

$$\int_\Omega \left[\eta\partial_t w + \underline{\nabla}\eta \cdot \underline{\tilde{\mathbf{f}}}(w) - s\eta\right] \mathrm{d}\underline{\mathbf{x}} = 0 \,. \tag{17}$$

in which the boundary terms vanish due to the above mentioned zero-condition. Two items are obvious in Eq. (17): **(a)** the order of derivation of w w.r.t. $\underline{\mathbf{x}}$ has reduced and **(b)** the derivation changed to the test function.

So the required order of differentiability for the function w has reduced by one. Therefore the solution of Eq. (17) for arbitrary test functions η is called the weak solution of the origin PDE.

The unknown field quantity w, the test function, and all required derivatives of these quantities can be decomposed into contributions of the n elements following from the discretization of the system. By means of weighted basis functions (also denoted as ansatz functions) the following approximation is derived (GALERKIN method):

$$w(\underline{\mathbf{x}}) = \sum_i^n \alpha_i N_i(\underline{\mathbf{x}}) \quad , \quad \underline{\mathbf{w}}'(\underline{\mathbf{x}}) = \sum_i^n \alpha_i \underline{\mathbf{N}}_i'(\underline{\mathbf{x}}) \,, \tag{18}$$

$$\eta(\underline{\mathbf{x}}) = \sum_i^n \beta_i N_i(\underline{\mathbf{x}}) \quad , \quad \underline{\eta}'(\underline{\mathbf{x}}) = \sum_i^n \beta_i \underline{\mathbf{N}}_i'(\underline{\mathbf{x}}) \,, \tag{19}$$

with the unknown coefficients α_i (also denoted as node variables), the arbitrary factors β_i and the ansatz functions N_i. Usually the ansatz functions are approximated by power series of first, second or higher order, which - consequently - lead to the notation of linear, square, or higher order elements. Note that the computational effort considerably increases with rising order, since the number of nodes required for the calculation of the ansatz functions increases, namely n+1 (linear), 2n+1 (square), etc..

Inserting Eq. (18, 19) into Eq. (17) yields the following expression:

$$\sum_j \beta_j \left[\frac{\mathrm{d}}{\mathrm{d}t} \sum_i \alpha_i \int N_i N_j \, \mathrm{d}\underline{\mathbf{x}} + \sum_i \alpha_i \int \underline{\tilde{\mathbf{f}}}(N_i, \underline{\mathbf{N}}_i', \ldots) \cdot \underline{\mathbf{N}}_j' \, \mathrm{d}\underline{\mathbf{x}} - \int s N_j \, \mathrm{d}\underline{\mathbf{x}}\right] = 0 \tag{20}$$

or - by successive consideration of ($\beta_1 = 1, \beta_2 = 0, \ldots, \beta_n = 0$); ($\beta_1 = 0, \beta_2 = 1, \ldots, \beta_n = 0$); etc. - an Ordinary Differential Equation (ODE) system of dimension n:

$$\frac{\mathrm{d}}{\mathrm{d}t}\left[\underline{\underline{\mathbf{M}}} \cdot \underline{\alpha}\right] + \underline{\underline{\mathbf{K}}} \cdot \underline{\alpha} - \underline{\mathbf{s}} = 0 \,, \tag{21}$$

which, in turn, can be solved by forward/backward time integration methods (e.g. EULER prodedure). Finally it is worth-mentioning that the calculation of $\underline{\underline{\mathbf{M}}}$, $\underline{\underline{\mathbf{K}}}$ (the so-called mass and stiffness matrix) and $\underline{\mathbf{s}}$ requires a numerical integration. For this reason the integrals are replaced by weighted sums as follows:

$$\int N_i(\underline{\mathbf{x}})\,\mathrm{d}\underline{\mathbf{x}} \cong \sum_{p=1}^{n_p} N_i(\underline{\mathbf{x}}_p) W_p \,. \tag{22}$$

Here the functions N_i are only calculated at representative points (the so-called GAUSS points[10]) by using the weight factors W_p. Position and typical number of the points $\underline{\mathbf{x}}_p$ as well as the according weight factors can be found in the literature, e.g. (Wriggers, 2000).

2.4 An example for the FEM formalism - heat transfer equation

In order to illustrate the above, slightly abtract, explanations the corresponding equations will be derived for the heat transfer problem within this paragraph. For this purpose the balance of internal energy in Eq. (3) including FOURIER's law of heat conduction in Eq. (4) and the caloric equation of state in Eq. (5) is considered and terms of velocity gradients (dissipation) are neglected. The resulting equation is known as the instationary heat transfer equation and reads:

$$\rho c_{\mathrm{p}} \frac{\mathrm{d}T}{\mathrm{d}t} - \underline{\boldsymbol{\nabla}} \cdot \underline{\underline{\boldsymbol{\kappa}}} \cdot \underline{\boldsymbol{\nabla}} T = r \quad \text{with} \quad \mathrm{d}_t(\cdot) = \partial_t(\cdot) + \underline{\mathbf{v}} \cdot \underline{\boldsymbol{\nabla}}(\cdot) \,. \tag{23}$$

The weak formulation follows by applying the product rule and the divergence theorem analogously to Eq. (17):

$$\int_\Omega \rho c_{\mathrm{p}} (\mathrm{d}_t T) \eta \,\mathrm{d}\underline{\mathbf{x}} + \int_\Omega \underline{\underline{\boldsymbol{\kappa}}} \cdot \underline{\boldsymbol{\nabla}} T \cdot \underline{\boldsymbol{\nabla}} \eta \,\mathrm{d}\underline{\mathbf{x}} = \int_{\partial\Omega} \underbrace{(\underline{\underline{\boldsymbol{\kappa}}} \cdot \underline{\boldsymbol{\nabla}} T)\,\eta \cdot \underline{\mathbf{n}}}_{=\underline{\mathbf{q}}_n} \mathrm{d}A + \int_\Omega r\eta \,\mathrm{d}\underline{\mathbf{x}} \,. \tag{24}$$

Here the symbol $\underline{\mathbf{q}}_n$ represents the heat flux through the surface of the volume and is usually replaced by the ansatz $\underline{\mathbf{q}}_n = \alpha(T - T_0) \cdot \underline{\mathbf{n}}$, ($\alpha$: heat transfer coefficient and T_0: surrounding temperature). By performing afore mentioned procedure of discretization one finally finds, cp. also (Kang & Im, 2007):

$$\underline{\underline{\mathbf{M}}} \cdot \frac{\mathrm{d}\underline{\mathbf{T}}}{\mathrm{d}t} + \underline{\underline{\mathbf{K}}} \cdot \underline{\mathbf{T}} = \underline{\mathbf{Q}} + \underline{\mathbf{R}} \,, \tag{25}$$

the following components of matrices and vectors ($i, j = 1, \ldots, n$):

$$M_{ij} = \int_\Omega \rho c_{\mathrm{p}} N_i N_j \,\mathrm{d}\underline{\mathbf{x}} \quad , \quad K_{ij} = \int_\Omega \underline{\underline{\boldsymbol{\kappa}}} \cdot \underline{\mathbf{N}}_i' \cdot \underline{\mathbf{N}}_j' \,\mathrm{d}\underline{\mathbf{x}} \,, \tag{26}$$

$$Q_i = \int_{\partial\Omega} \alpha N_i (T - T_0) \underline{\mathbf{n}} \cdot \underline{\mathbf{n}} \,\mathrm{d}A \quad , \quad R_i = \int_\Omega r N_i \,\mathrm{d}\underline{\mathbf{x}} \,. \tag{27}$$

in which $\underline{\mathbf{T}} = (T_1, \ldots, T_n)$ stands for the vector including the unknown temperatures of the single elements. The ODE system of Eq. (25) can be numerically solved for appropriate initial conditions.

[10] The underlying method of numerical integration is often called the GAUSS integration or GAUSS quadrature.

2.5 Geometry, meshing, heat sources, boundary and initial conditions for laser welding

First, the question arises: What means 'simulation of welding'? Basically, this means the calculation of materials response - howsoever - after moving a heat source over two kinds of metals, which are (initially) separated and jointed after the heat supply. This response can be manifold depending on focus and complexity of modelling, e.g.:

- The development of a spatially and temporally changing temperature field including melting zone (welding pool), fluid dynamics (MARAGONI effect) and solidification.
- Thermal expansion, distortion, and the formation of residual stresses within the material during and after the joining process.
- Phase transformation and TRIP (Transformation Induced Plasticity) effects.
- Hydrogen diffusion (activation of trapped hydrogen, increasing hydrogen in regions of high tensile stresses due to diffusion).
- Material income due to filler metals.
- Changes of the geometry, e.g. seam bulging or gap closing.

Obviously, the material response is notedly complex and, therefore, only aspects are considered. Consequently, it is essential to, firstly, clarify the demands on the simulation and the leading effects, which should be modelled.

Starting point of each welding simulation is the (spatially changing) heat input, which is - in case of laser welding - extremely local (in magnitude of the laser diameter, $d = 0.5 \ldots 1.5$ mm). Due to the high power of the laser-beam ($P_{laser} = 38$ KW), considerable temperatures accompanying with high temperature gradients and heat fluxes are usually observed. In contrast, several millimetres adjacent to the seam the temperature is already moderate and partially even below 100 °C. Therefore a spatial discretization (mesh) must be chosen, which is extremely fine in the vicinity of the seam and which coarsens for distant regions. Moreover, to save computational times we restrict the consideration to a Representative Sheet Element (RSE), instead of the complete coil/strip-width ($b_c \approx 800 \ldots 1650$ mm). This restriction only holds, if boundary effects in the middle of the x- and y-direction of the RSE can be neglected or - with other words - if the calculated fields (mass density, temperature and displacements) are geometrically similar for the RSE and the real coil geometry. Figure 5 (left) shows the explained strip, the RSE and the chosen measures for the simulation. Figure 5 (right) illustrates the global FE mesh of the considered RSE together with two detailed views. As one can see, the mesh is much finer in the vicinity of the welding seam. The gap between both sheets is chosen as $s = 0.2$ mm, and the effect of bulging and gap closing is realized by the `Ekill-Ealive` technique provided by the ANSYS® program package. For this purpose the gap as well as the bulging elements are a priori modeled and meshed. Subsequently these elements are deactivated, which means that its contributions within the mass and stiffness matrix are multiplied with a 'numerical zero' (e.g. 10^{-6}). Thus the elements do not contribute to the solution during simulation. Deactivated elements, for which the moving heat source has passed and which fall below the melting temperature, are successively activated, i.e., the above mentioned artificial multiplication is canceled. Hence the corresponding elements contribute to the solution in the proceeding simulation (Ansys theory ref., 2007). This procedure allows to model materials input as well as joining.

The translation of the heat source is defined by the process parameter of welding velocity and direction, whereas the characteristics of heat input (heat source dimension, power

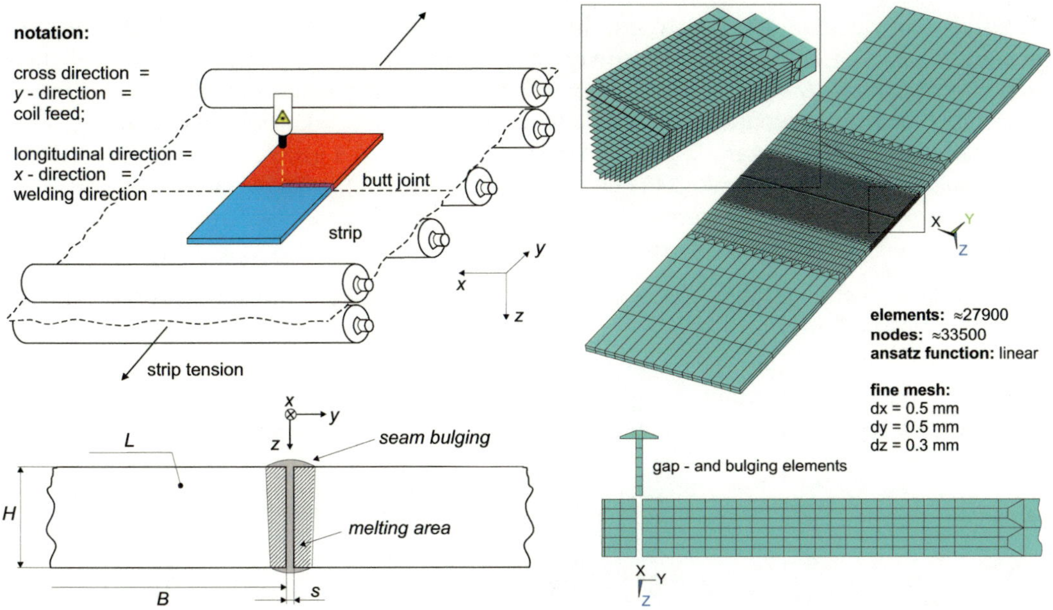

Fig. 5. *Left*: Illustration of the RSE and the chosen geometry data of the RSE: $H = 1.8$ mm, $L = 50$ mm, $B = 75$ mm, $s = 0.2$ mm. *Right*: Illustration of the simulation geometry and the used FE discretization including various meshing data.

distribution, total heat, supply over surface or keyhole) is determined by the method of welding (laser, MIG/MAG, TIG, etc.) and the welding power.

The manner of supplying the heat into the metal as well as the resulting melt pool directly determines the developing temperature distribution. Consequently the realistic modeling of the heat supply is a crucial task of welding simulations. The best strategy in this context would be a coupled consideration of **(a)** the melting process due to heat input, **(b)** the laser-welding-specific vaporization process leading to a vapor capillary in the melt, in which the laser beam can repeatedly be reflected, which - in turn - allows for a deep heat input, and **(c)** the melt pool dynamics, which affects the geometry and dimension of the melting area.

However a complete numerical analysis of the multiphase fluid dynamics problem with subsequent simulation of the solid material w.r.t. resulting temperatures, stresses and microstructure (phases, grains, etc.) is an ongoing challenge. Therefore the literature[11] mainly contains two kinds of investigations:

(1) Considerations of fluid dynamics aspects: This item incorporates e.g. modeling of drop formation and impingement during TIG or MIG/MAG welding (Fan & Kovacevic, 1999; Zhou & Tsai, 2008); investigations of the influence of different chemical elements (such as S) on melt flow and the weld pool geometry (Lee et al., 1998; Mills et al., 1998; Wang et al., 2001; Wang & Tsai, 2001); simulation of the keyhole dynamics during laser welding (Ki et al., 2002; 2002a; Pfeiffer & Schulz, 2009); or the analysis of asymmetric melt pool geometries due to dissimilar materials (Phanikumar et al., 2004). Consequences on residual stresses, distortions or microstructure in the solid state are not - as far as known - considered.

[11] The chosen literature references are not complete; they are examples of the recent years.

(2) Solid-state calculations, which assume an ad-hoc or experimentally verified melt pool geometry: Thus, a temperature or power distribution is *a priori* defined at the melt pool or at the region of heat input. Then the resulting global temperature distribution, stresses, distortions, or microstructures can be calculated (Dean et al., 2003; Hemmer & Grong, 1999; Hemmer et al., 2000; Spina et al., 2007). Moreover, also complex geometries (Tian et al., 2008), multi-pass welding (Dean et al., 2008; Dean & Hidekazu, 2008), cracking caused by shrinkage (Dong & Wei, 2006), or the process-chain bending-welding (Khiabani & Sadrnejad, 2009) have been simulated with sufficient accuracy.

In the following the strategy of the second item is employed[12]. To this end the ANSYS® specific program library LASIM is used, (Junk & Groth, 2004). Here the geometry parameter $x^{1/2/3}_{\mathrm{r/f}}$, $y^{1/2/3}$, and $z^{2/3}$ as well as the total power input must by defined, cf. Figure 6. The (parabolic) power distribution is applied to the nodes of the three illustrated planes such, that the sum reproduces the total power. Alternatively, various authors use a GAUSSian

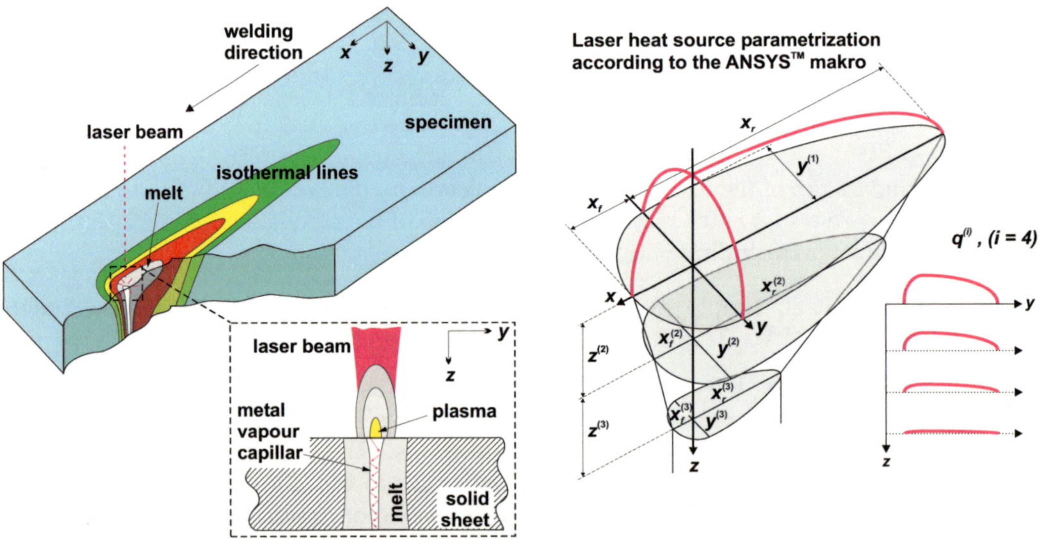

Fig. 6. Melt pool geometry during laser welding and the laser specific LASIM distribution.

distribution, according to the cross-sectional power distribution of the laser beam, e.g. (De et al., 2003), see also Figure 7, to model the welding heat source. It reads:

$$q(x,y,i) = Q(i)\frac{1}{2\pi\sigma^2}\exp\left[-\frac{x^2+y^2}{2\sigma^2}\right] . \tag{28}$$

The variables x and y represent the components of the distance vector between the center of the laser beam and a point of the planar environment. The value of σ stands for the standard deviation of the normal distribution q/Q_i and characterizes the variance of the distribution. The variable i identifies the different planes and allows for a volumetric heat input (depth effect) due to melt pool dynamics. In particular holds for $Q(i)$ in Eq. (28):

$$Q(i) = Qf_i \quad \text{and} \quad \sum_i f_i = 1 . \tag{29}$$

[12] There are also attempts to couple fluid dynamics and solid state calculations (see also the Discussion at the end of the work).

A typical choice, for example, can be: $i = 3$, $f_1 = 0.75$, $f_2 = 0.2$, $f_3 = 0.05$, so that the first three planes of the z-directions are used for heat input. Furthermore it is worth mentioning,

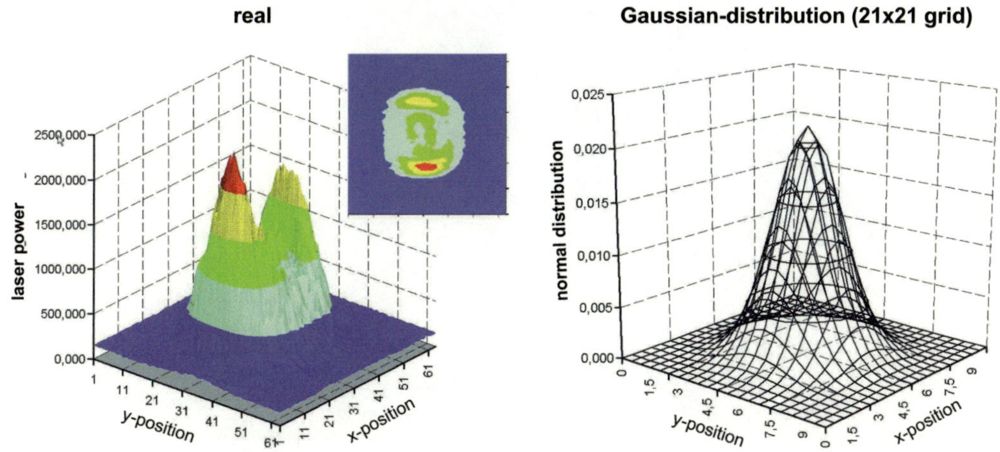

Fig. 7. Cross sectional power distribution of the laser beam: real case (*left*) and modeling case (*right*, GAUSSian distribution).

that the real heat absorption into the sheets considerably differs to the value adjusted at the laser beam source. The majority of power is reflected, passes the material, or dissipates by heating the environment. The remaining power consumed by welding can be described by means of the efficiency η:

$$Q = \eta \, P_0 \quad , \quad (P_0\text{: origin power of the laser source})\,. \tag{30}$$

The efficiency η depends on the material thickness d, the gap width s, the welding velocity v_s, or the laser type (e.g. CO_2- or solid-state-laser). In the present case, i.e., for butt welding with CO_2-laser, experiments show the following empiric relation, (Kotowski, 1998):

$$\eta_{CO2} = 0.0053\, s\, v_s + 3.3854\, d\, v_s + 9.2073\, d + 0.0581\, s + 1.1094\, v_s + 0.1866 \tag{31}$$

with d in [mm], s in [mm] und v_s in [m/min]. Consequently an efficiency of $\eta = 0.43$ is derived for $s = 0.2$ mm and $v_s = 5$ m/min.

In order to numerically solve the required balance equations, thermal and mechanical boundary conditions must be specified. So, the strip is clamped by copper shoes few millimetres adjacent to the seam (3...3.5 mm) from the upper and lower side, cf. Figure 8 (magenta region). The upper side of the strip is exposed to an inert gas (Helium) flow in the vicinity of the seam; thus *forced convection* takes place. Furthermore the lower side of the strip is in *thermal contact* with another steel surface, which leads to heat transfer and to enhanced cooling of the strip. The remaining (free) surfaces of the strip (RSE), e.g. the lower side of the joining area, release heat by *radiation and free convection*, cf. Figure 8. All kinds of heat transfer over the surfaces can be described by means of (different) heat transfer coefficients, α. The corresponding equations and values are summarized in the **Appendix** at the end of this work. Please also note, that the welded parts of the butt joint can conduct heat, and the corresponding surfaces cannot independently move (in contrast to the origin state), since they are mechanically jointed, cf. Figure 8. Finally, initial conditions have to be specified, from which the time integration starts; here we used: $T(\underline{\mathbf{x}}, t = 0) = T_0$ and $\underline{\mathbf{u}}(\underline{\mathbf{x}}, t = 0) = \underline{\mathbf{0}}$.

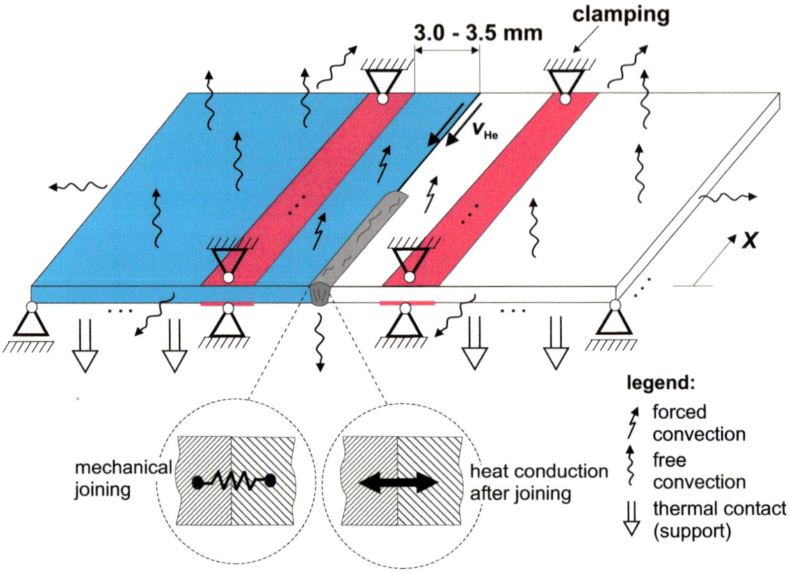

Fig. 8. Illustration of the used boundary conditions due to the conditions of the manufacturing line.

3. Materials data

The following considerations exclusively refer to the investigated TRIP 700 steel; the composition is depicted in Table 1.

	C	Al	Mn	Si	P	B	Cr+Mo	Nb+Ti
max.	0.24	1.60	2.00	0.30	0.04	0.005	0.60	0.20

Table 1. Composition of the investigated TRIP 700 steel.

3.1 Caloric data

To solve the heat conduction equation the following temperature-depending materials data are required additionally to the data accomplished in the **Appendix**: heat conductivity (here for the isotropic case: $\underline{\underline{\kappa}} = \kappa\underline{\mathbf{I}}$), mass density ρ, and specific heat capacity c_{p} (or enthalpy density h instead of ρ and c_{p}). These quantities are determined - in a first approximation - by means of the assumption[13] of infinitely slow heating and quenching. Figure 9 (left) shows the different equilibrium phases, which constitute at various temperatures. The right picture of Figure 9 illustrates the transition between solid and liquid phase in the vicinity of the melting point ($T_{\mathrm{solidus}} = 1457\,°\mathrm{C}$ and $T_{\mathrm{liquidus}} = 1457\,°\mathrm{C}$). The thermal materials data used (later) in the simulations are illustrated in Figure 10. For comparison experimental values, which were

[13] This assumption leads to constant transformation temperatures. Any shifts of the phase transformations due to varying heating/quenching rates $\dot{T}$ are excluded. For the consideration of materials data depending on T and $\dot{T}$ transformation models must be applied, (Avrami, 1939; 1940; Koistinen & Marburger, 1959; Leblond & Devaux, 1984). Such models take the temperature history into account and allow for the calculation of e.g. different phase volume fractions and the resulting materials data, (Kang & Im, 2007).

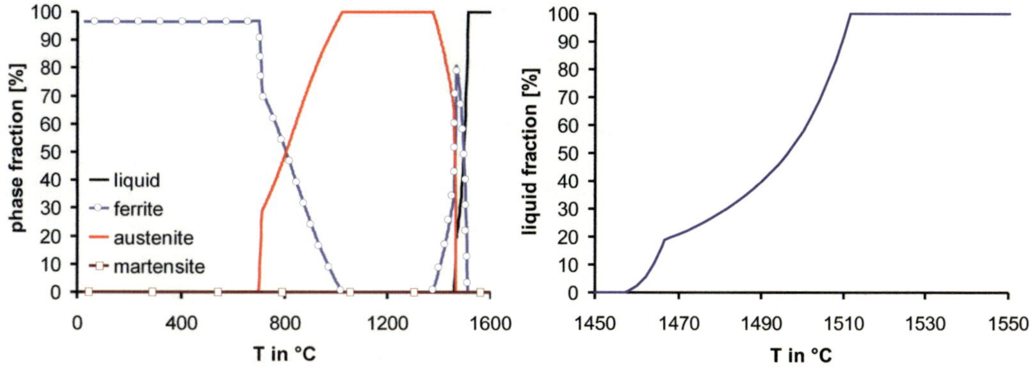

Fig. 9. Different equilibrium phases in TRIP 700, calculated with JMatPro, (JMatPro, 2011).

found in literature[14], (Barth & Groß, 2004), are also displayed. Finally it is worth mentioning, that the conductivity is artificially multiplied by the factor 3 for temperatures $T > \frac{1}{2}(T_{\text{solidus}} + T_{\text{liquidus}})$. Thus the effect of enhanced conductivity due to melt pool dynamics is taken into account.

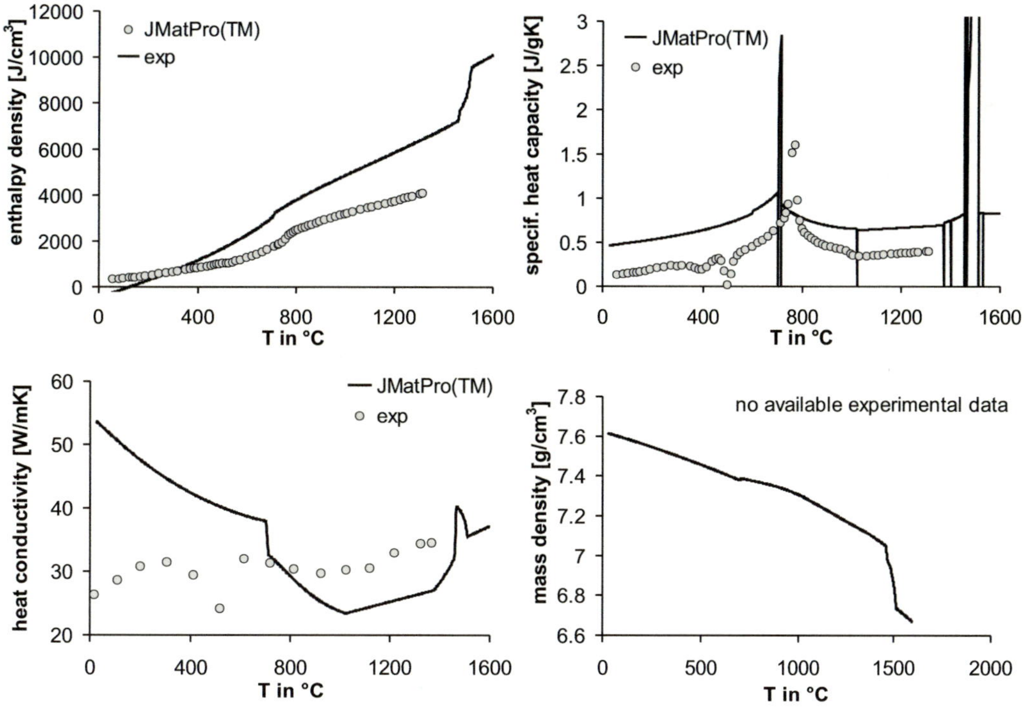

Fig. 10. Calculated and experimental, thermodynamic materials data as functions of the temperature.

[14] The conductivity is indirectly measured by using the temperature conductivity a and the relation $\kappa = \rho\, c_p\, a$.

3.2 Experimental determination of mechanical materials data

Beyond the caloric data, the solution of the balance of momentum requires temperature-depending data for YOUNG's modulus , POISSON ratio (for HOOKE's law), and the work-hardening behavior of the material. In commercial FE-program packages, such as ANSYS®, the work-hardening characteristic of the material is defined via data-sets, representing the (1D-)yielding curve at different temperatures. For values between or beyond the temperatures used for the definition of the curves an interpolation is performed or the curve of the highest temperature is applied.

Mechanical data for as-rolled steels are rarely available and, therefore, experiments must be performed. For this reason tensile tests, realized at the Gleeble 3500®, are analyzed for the temperatures 20 °C, 100 °C, 300 °C, 600 °C, 1100 °C and 1200 °C, cf. Figure 11. Here specimen heating follows from the electrical resistance and JOULE's (heat) effect, viz:

$$dW_{\text{Joule}} = U\,I\,dt \quad , \quad \text{(JOULE's effect)} , \tag{32}$$

$$I = \frac{A\,U}{\rho_R(T)\,l} \quad , \quad \text{(conductor with cross-area } A\text{, length } l) , \tag{33}$$

$$\rho_R(T) \approx \rho_0 \frac{T}{T_0} \quad , \quad \text{(specific resistance)} , \tag{34}$$

with $\rho_0 = 0.1 \ldots 0.2 \cdot 10^{-6}\ \Omega\text{m}$.

It should be pointed out, that the specimen is cooled in the boundary region due to the (cold) clampings; consequently a temperature gradient is formed in loading direction. To minimize temperature gradients within the measured distance a reduced reference length of $L_0 = 25$ mm is used, contrary to the DIN EN 100002-1, (DIN, 2001). The temperature control is carried out via a Thermo-Couple-Element (TCE) at the middle of the specimen-length and -width. The temperatures used for tensile tests are chosen such, that no phase transformation occurs during testing. For this reason, first, the (unloaded) specimen is slowly heated to determine the transformation temperatures. Figure 12 displays the heating procedure (upper left) as well as the resulting width variation of the specimen (upper right). Here the 'jumps' represent the phase transformations, for which the following temperatures hold: The change of width

Ac1 $\approx$ *770 °C (start of austenite formation) ,* *Ar1* $\approx$ *600 °C (start of ferrite formation),*
Ac3 $\approx$ *940 °C (total austenite) ,* *Ar3* $\approx$ *520 °C (total ferrite).*

for different temperatures, cf. Figure 12 (upper right), can be directly used to calculate the averaged thermal expansion coefficient via the relation: $\alpha = \Delta b/[b_0(T - T_0)]$, cf. Figure 12 (lower right). Here only the heating procedure is exploited and - according to the high temperature behaviour of steel, c.f., (Richter, 1973) - a linear extrapolation up to the melting point is performed.

The stress-strain curves measured at different temperatures are illustrated in Figure 13. For each temperature two tensile measurements were exploited to obtain the temperature-depending 1D-stress-strain response of the material. However, the pictures exemplarily display one curve and the averaged discrete values (dots) used - together with YOUNG's modulus, cf. Figure 15 (right) - to construct the final curves for simulation, cf. Figure 13 (lower right). Figure 14 explains the strategy of the tensile tests exemplarily for two temperatures. The specimen is firstly heated (cp. Figure 12, lower right) such, that any displacements are permitted to avoid any internal stresses due to thermal expansion. After reaching the target temperature the specimen is strained up to failure by controlling the displacements. For heating as well as straining moderate rates, i.e., $\mathrm{d}T/\mathrm{d}t = 3$ K/s and $\mathrm{d}l/\mathrm{d}t = 20$ µm/s, are used. Thus, the assumption of phase equilibrium holds at all times, and any viscose effects are minimized.

In order to determine the temperature-depending YOUNG's modules, E, an additional test was performed. Due to clamping relaxations and insufficient measuring deviations the consideration of the primary loading curve for E is questionable. This flaw is also mentioned within the DIN EN 100002, (DIN, 2001), and, therefore the authors recommend the determination of E by using the unloading regime. Here the point of unloading must be chosen beyond yielding but 'appreciable' before necking. In the present investigations two unloading procedures are performed for each tensile test, in which the averaged slope represents YOUNG's modulus. Here the curves in Figure 13 can be exploited to define appropriate stresses for 'inserting' the unloading procedure. The subsequent, separate test is arranged such, that the specimen are stressed (force-controlled) up to the predefined starting points for unloading and, then, unloaded up to 1/10 of the applied maximal stress. After

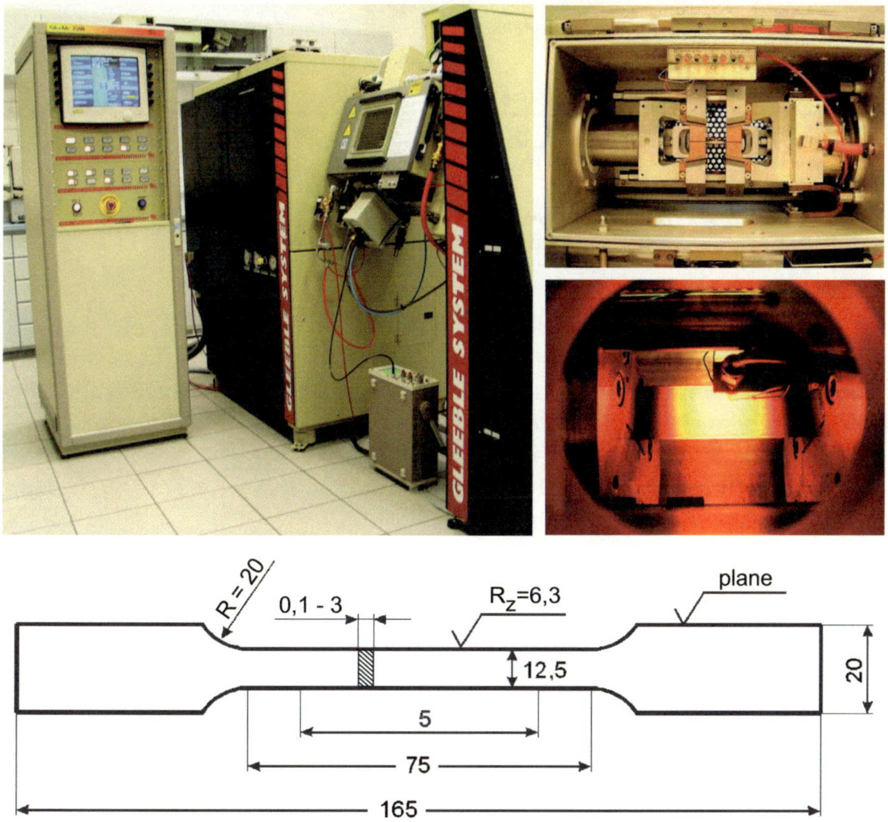

Fig. 11. Gleeble3500® used for the temperature-depending tensile test and geometry of the specimen.

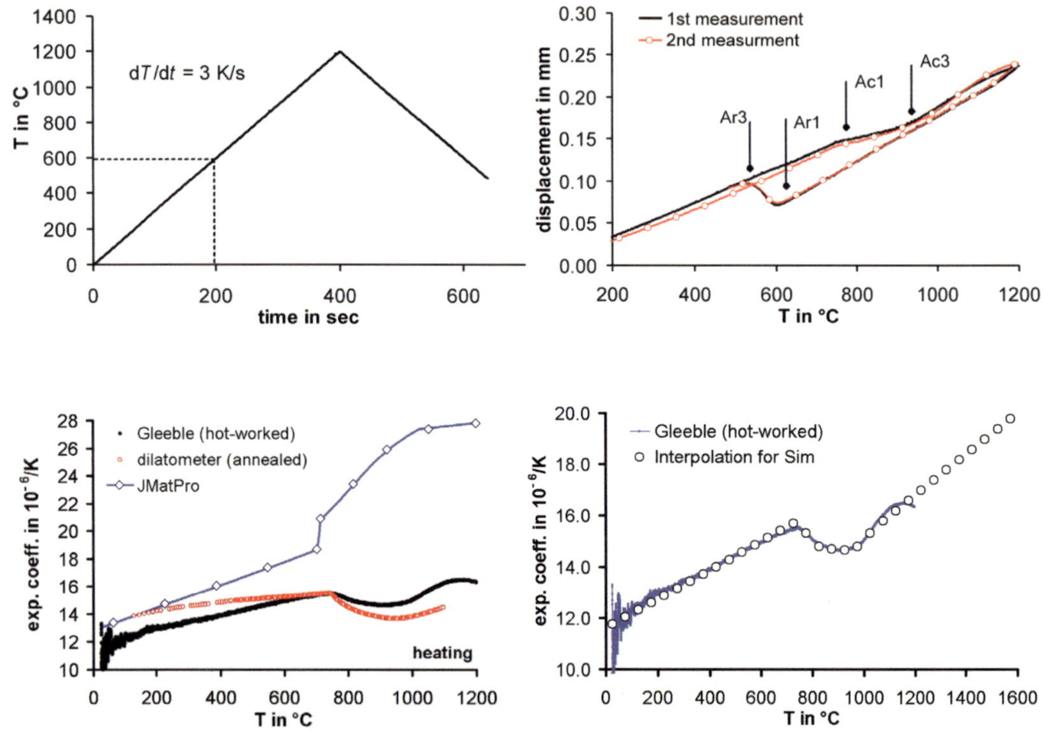

Fig. 12. *First row*: Determination of regions of transformation. *Second row*: Experimentally fitted thermal expansion coefficients (including the comparison with data of annealed material and JMATPRO).

the second unloading the test is finished by fast loading up to failure, cf. Figure 15 (left). The resulting cyclic curves are illustrated in Figure 16 for the temperatures $T = 20$ °C and 100 °C. The right pictures demonstrate the determination of E by averaging the slope of the two unloading regimes. In summary, the values of Table 2 were found, which are displayed in Figure 15 (right). Note, that the fitted function has not physical meaning. It simply reproduces the experimental values and the side condition: $E > 0$ for $T \to \infty$. In many cases,

T [°C]	20	100	300	600	1100	1200
E [GPa]	246.0	177.6	164.8	139.0	103.1	83.7

Table 2. YOUNG's modulus for different temperatures obtained from experiments (TRIP 700, as-rolled).

alternatively to experimentally obtained stress-strain curves, temperature-depending discrete values for E, R_{p02}, R_m and A_g (uniform strain) are available. Then an optimization procedure can be performed by using, for instance, a RAMBERG-OSGOOD power law as follow:

$$\varepsilon = \varepsilon^{\text{elast}} + \varepsilon^{\text{plast}} = \frac{\sigma}{E} + \left(\frac{\sigma}{K}\right)^{1/n} . \tag{35}$$

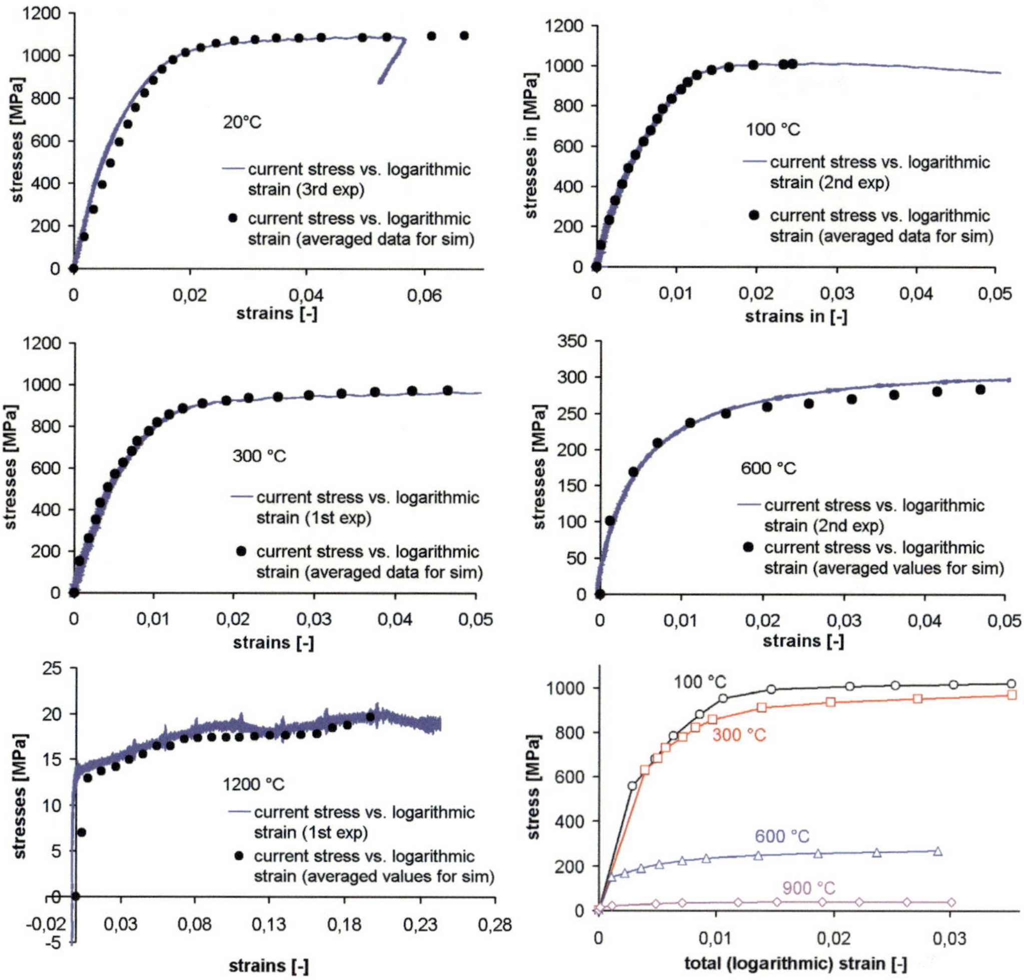

Fig. 13. Measured stress-strain curves for various temperatures including the implemented curves used in the ANSYS® simulation (exemplarily, lower right).

The two arbitrary parameters K and n must be optimized such, that the deviation between the resulting function and the experimental values:

$$\varepsilon^{(1)} = 0.002 + \frac{R_{p02}}{E} \quad , \quad \varepsilon^{(2)} = A_g + \frac{R_m}{E} \tag{36}$$

are sufficiently minimal. For this reason the root-mean-square deviation:

$$\sum_{i=1}^{2} \left[\varepsilon^{(i)} - \frac{\sigma}{E} - \left(\frac{\sigma}{K} \right)^{1/n} \right]^2 \longrightarrow \min \tag{37}$$

can be considered. The minimum of Eq. (37) results e.g. by using a gradient-based algorithm, which is typically available in commercial program packages, such as Matlab® (Mathworks Inc.) or Excel (Microsoft® Corp.). Figure 17 shows the result of the explained procedure using

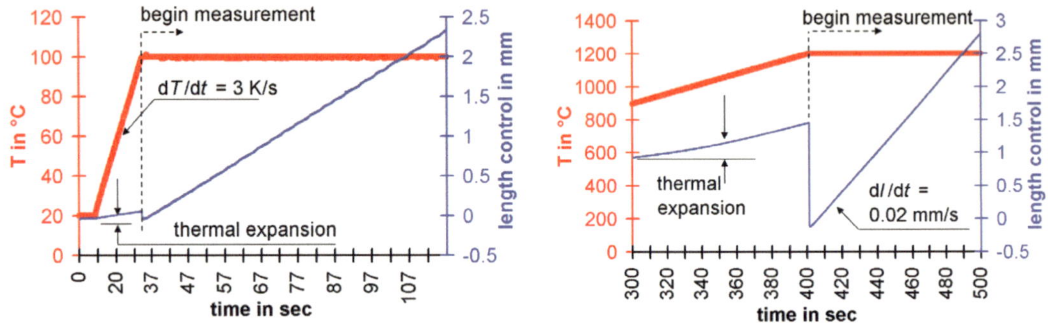

Fig. 14. Heat and loading procedure for two, exemplary, tensile tests (T = 100 °C and 1200 °C).

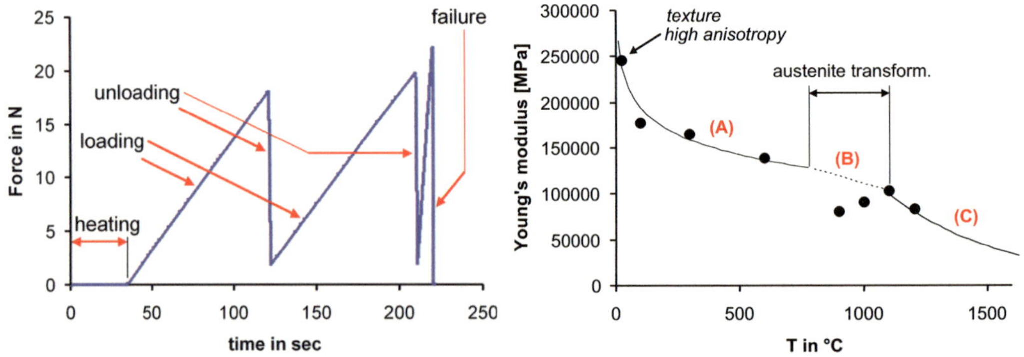

Fig. 15. *Left*: Illustration of (Un-)Loading by means of the force-time characteristics ($T = 100$ °C). *Right*: YOUNG's modulus vs. temperature with the following stepwise interpolations: (A) $E = -31938 \ln(T) + 340924$; (B) $E = 77.9\,T + 188811$; (C) $E = 10^7 \exp[-0.0021\,T]$.

the according values of $T = 20$ °C, i.e., $E = 246$ GPa, $R_{p02} = 760$ MPa, $R_m = 1102.4$ MPa und $A_g = 0.056$. The arbitrary parameter of Eq. (35) obtained by this method are $n = 0.1$ und $K = 1500$ MPa.

A visual comparison of the curves in Figure 13 (upper left) and 17 indicates the adequacy of both methods. The decision of which values will be experimentally determined depends on the experimental equipment and personal preferences.

4. Finite element simulations

4.1 Temperature distribution

The subsequent simulations exclusively refer to the caloric materials data of Section 3.1 as well as to geometry, discretization and side-/initial conditions of Section 2.5. In particular, the heat source macro LASIM provided by ANSYS® is applied, cf. Figure 6 (right), by using the following source geometry data:

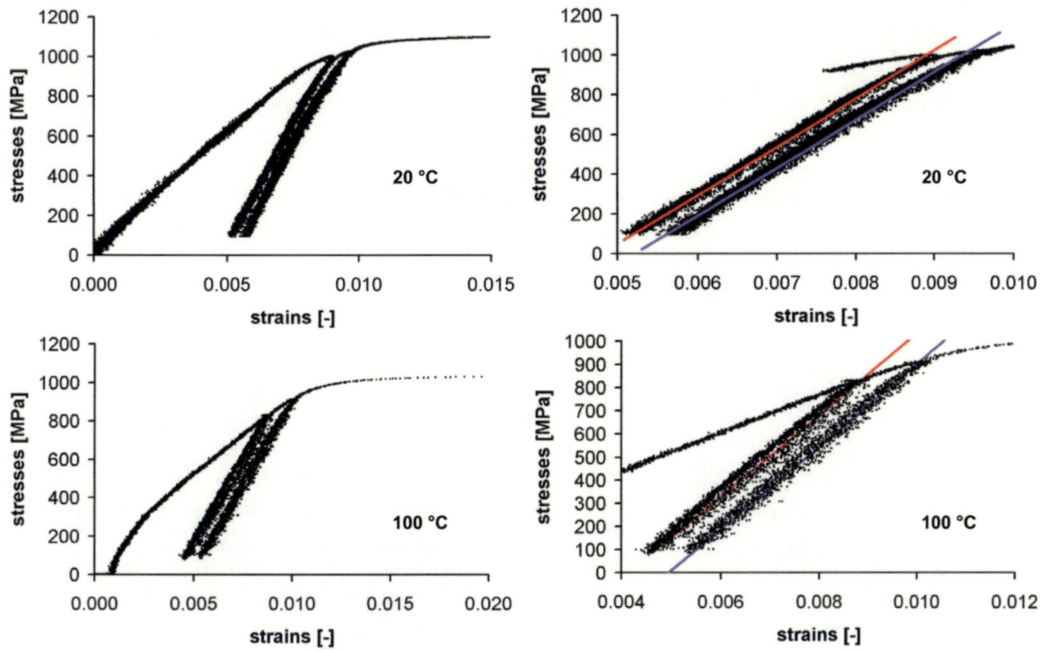

Fig. 16. Cyclic stress-strain history for $T = 20$ °C and $T = 100$ °C including a zoomed view on the unloading regime (right panel).

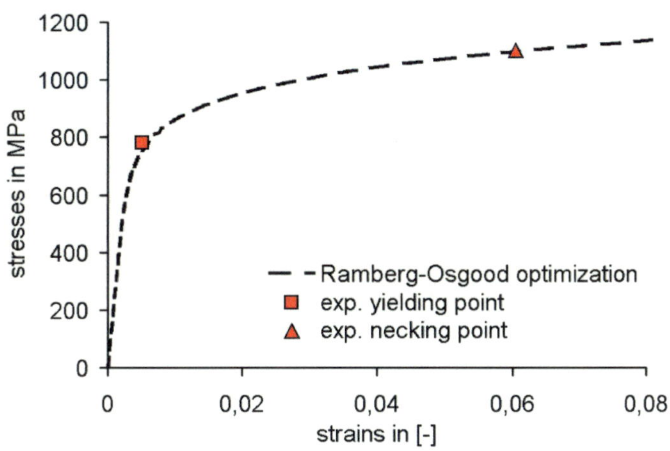

Fig. 17. Stress-strain curves for the as-rolled TRIP 700 steel at temperature $T = 25$ °C following from the optimization procedure and the RAMBERG-OSGOOD-ansatz.

With typical laser welding processes during steel manufacturing, such as continuous galvanizing, in mind, the welding parameters are used as follows:

$$v_s = 6\,\mathrm{m/min} = 100\,\mathrm{mm/s} \quad , \quad P_0 = 5\,\mathrm{kW}\,. \tag{38}$$

Note, that Figure 10 partially contains data of different sources. In order to guarantee consistency with the phase transition temperatures of Figure 9 (and due to the 'moderate'

$$x_1^{\mathrm{f}} = 0.5\,\mathrm{mm}\,, \qquad x_2^{\mathrm{f}} = x_1^{\mathrm{f}}\,, \qquad x_3^{\mathrm{f}} = x_1^{\mathrm{f}}\,,$$
$$x_1^{\mathrm{r}} = 2.5 \times x_1^{\mathrm{f}}\,, \qquad x_2^{\mathrm{r}} = 0.4 \times x_1^{\mathrm{r}}\,, \qquad x_3^{\mathrm{r}} = x_2^{\mathrm{r}}\,,$$
$$y_1 = 1.0\,\mathrm{mm}\,, \qquad y_2 = 0.8 \times y_1\,, \qquad y_3 = y_2\,,$$
$$z_2 = H/2\,, \qquad z_3 = H\,.$$

deviations) data exclusively calculated by JMATPRO are used. Thus the solid-liquid transformation temperatures are: $T_{\mathrm{solidus}} = 1457\ ^{\circ}\mathrm{C}$, $T_{\mathrm{liquidus}} = 1511\ ^{\circ}\mathrm{C}$. Figure 18 displays

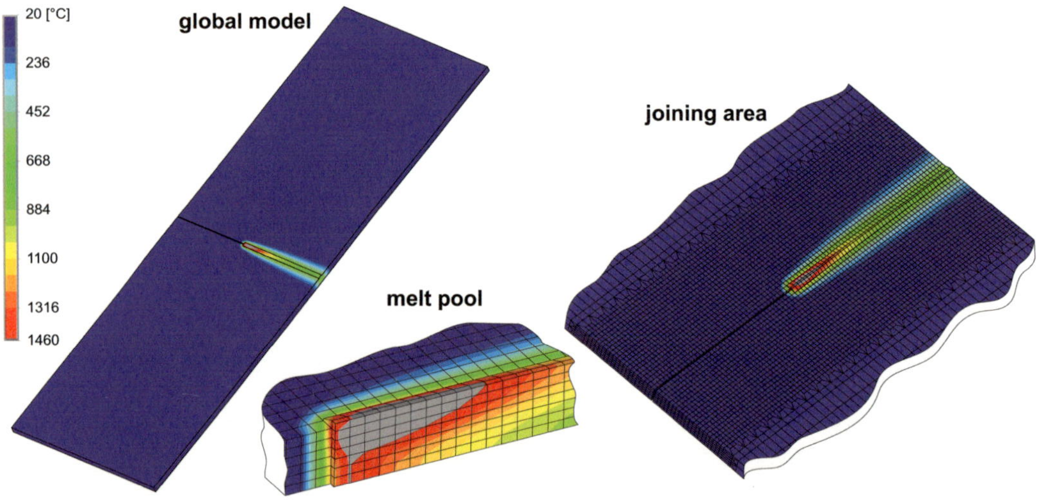

Fig. 18. Different contour plots of the thermal FE simulation including the cross-scetional view of the melt pool.

various contour plots of the thermal simulation after welding time of 0.26 s. As one can clearly see, gap elements will be activated, if one of its nodes achieves the melting temperature T_{liquidus}. Figure 19 (left) compares the calculated melt pool geometry with the according, experimentally obtained, microphotograph. Obviously the a-priori assumed bulging behaviour is overestimated, whereas the calculated melt pool thickness is slightly thinner than the experimental one. However, in summary, there is an overall good agreement between the theoretical and realistic situation. Please note, such agreement is not self-evident, since no adjustments (e.g. of the heat input efficiency η) to experimental findings were used during simulation. Finally, Figure 19 (right) illustrates the calculated temperature vs. time curves of various, exemplary distances from the welding seam at the upper and lower side. The curves are similar to an asymmetric GAUSSian distribution, which is in agreement with purely mathematical formulations of the temperature distributions, (Elmer et al., 2001; Komanduri & Hou, 2000).

4.2 Distortion and residual stresses

The calculated temperature distribution is subsequently used to simulate the thermal deformation and, consequently, the resulting stresses and strains within the welded material. To this end the (thermo-) mechanical materials data, viz. α (cf. Figure 12, lower right), E (cf. Figure 15, right), the stress-strain curves (cf. Figure 13), and POISSON's ratio $\nu = 0.3$ are applied. Furthermore (linear) 8-node hexahedral elements are chosen. Note, that during

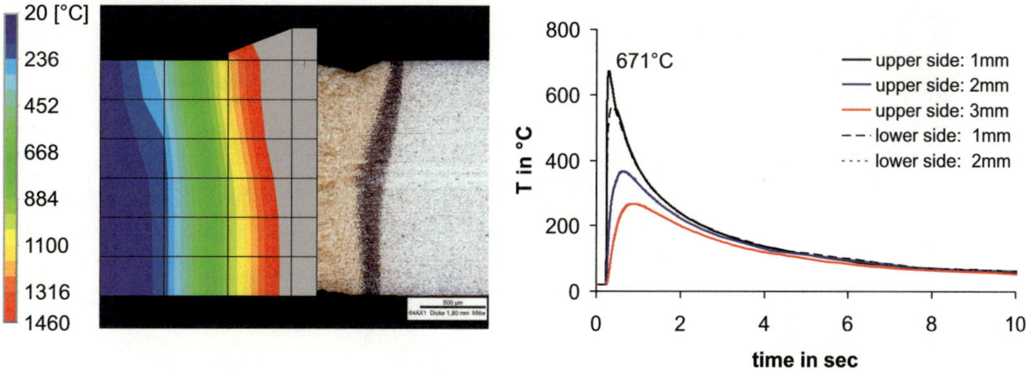

Fig. 19. *Left*: Comparison between calculated and experimentally observed weld pool geometry (sheet edge). *Right*: Calculated temperature vs. time dependence at different distances from the welding seam (upper and lower side, $x = L/2$).

the thermal simulation the initially deactivated gap- and bulging elements are activated as soon as **the melting temperature is reached** due to the heat conduction of the liquid matter. In contrast, the deactivated elements of the mechanical simulation are activated as soon as the reach the solidification temperature **after quenching**, since stresses exclusively develop within the solid state. Moreover, the thermal expansion of the activated elements must be calculated w. r. t. the reference temperature T_{solidus}, whereas the initially activated elements refer to the temperature before welding, i.e., T_0 (surrounding temperature). This fact can be realized by 'material changing' parallel to the element activation.

Furthermore it is worth mentioning, that the application of the `Ekill` / `Ealive` technique requires a sensible adjustment of the ANSYS® calculation option in order to guarantee convergence. In particular option for an adaptive time integration scheme, non-linear geometry, and the SPARSE solver is recommended.

In Figure 20 different contour plots of the VON MISES stress are presented to illustrate the developing stresses during joining as well as the remaining residual stresses after welding and cooling. Clearly to see is, first, the gap opening and, second, the considerable, tensile residual stresses within the welding seam after cooling. On the other hand the seam is nearly stress-free immediately after joining. The boundary regions at the beginning and the end of the welding area contain elevated residual stresses, in particular high compressive stresses in y-direction and tensile stresses in z-direction. This fact is indicated by the components of the stress vector at the material's surface illustrated in Figure 21. A comparison of the residual VON MISES stress with stress-strain curve in Figure 12 (at $T = 20$ °C) indicates a significant remaining stress level after welding, in particular at the middle of length ($x = L/2$). Here the illustrated stresses correspond to the qualitatively plotted distributions found in literature, e.g. (Voss, 2001).

The moving heat source leads to extremely short, positive or negative fluctuating strain amplitude, cf. Figure 22 (first and second row). Here the strain components at the upper side at distances $y = 1$ mm and 10 mm are displayed, which tend to stationary values for increasing time. The final strains resulting from the local thermal expansion are considerable smaller (and show partially opposite sign) at $y = 10$ mm than at 1 mm, which underscores the highly local impact of laser welding processes. This fact is also evident by considering

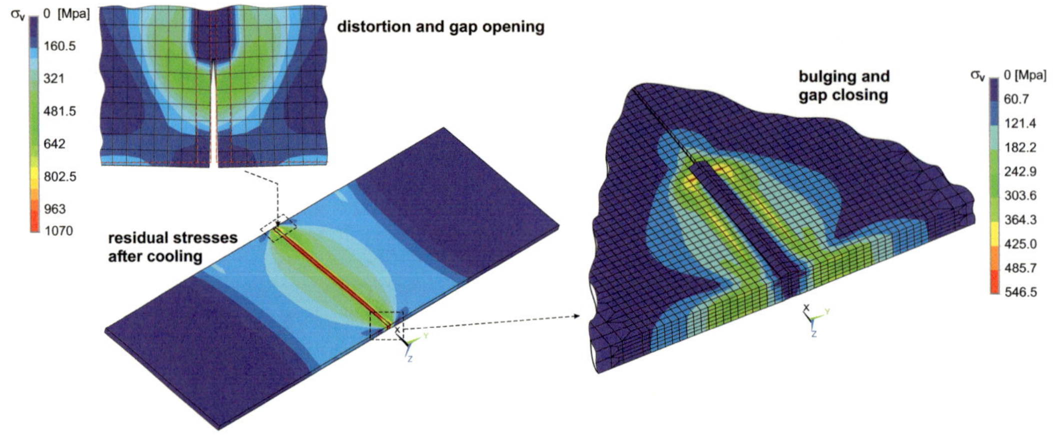

Fig. 20. Gap opening (scaling factor: 3) and VON MISES stress during and after welding of as-rolled TRIP 700.

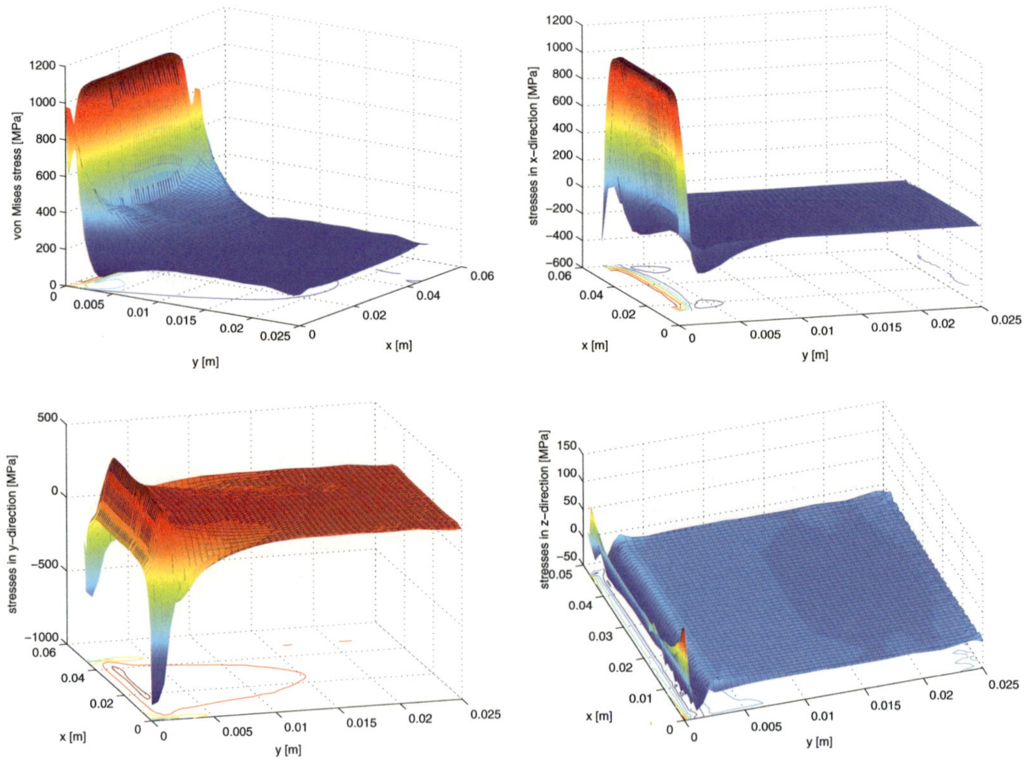

Fig. 21. 2D-distribution of the VON MISES stress and the stress-vector components at the surface after cooling.

Figure 22 (third row), in which most remarkable distortions are found within the first three millimetres beside the seam. Altogether one can say that the demonstrated FE-simulations represent a powerful tool to estimate the mechanical and thermal materials response. In detail

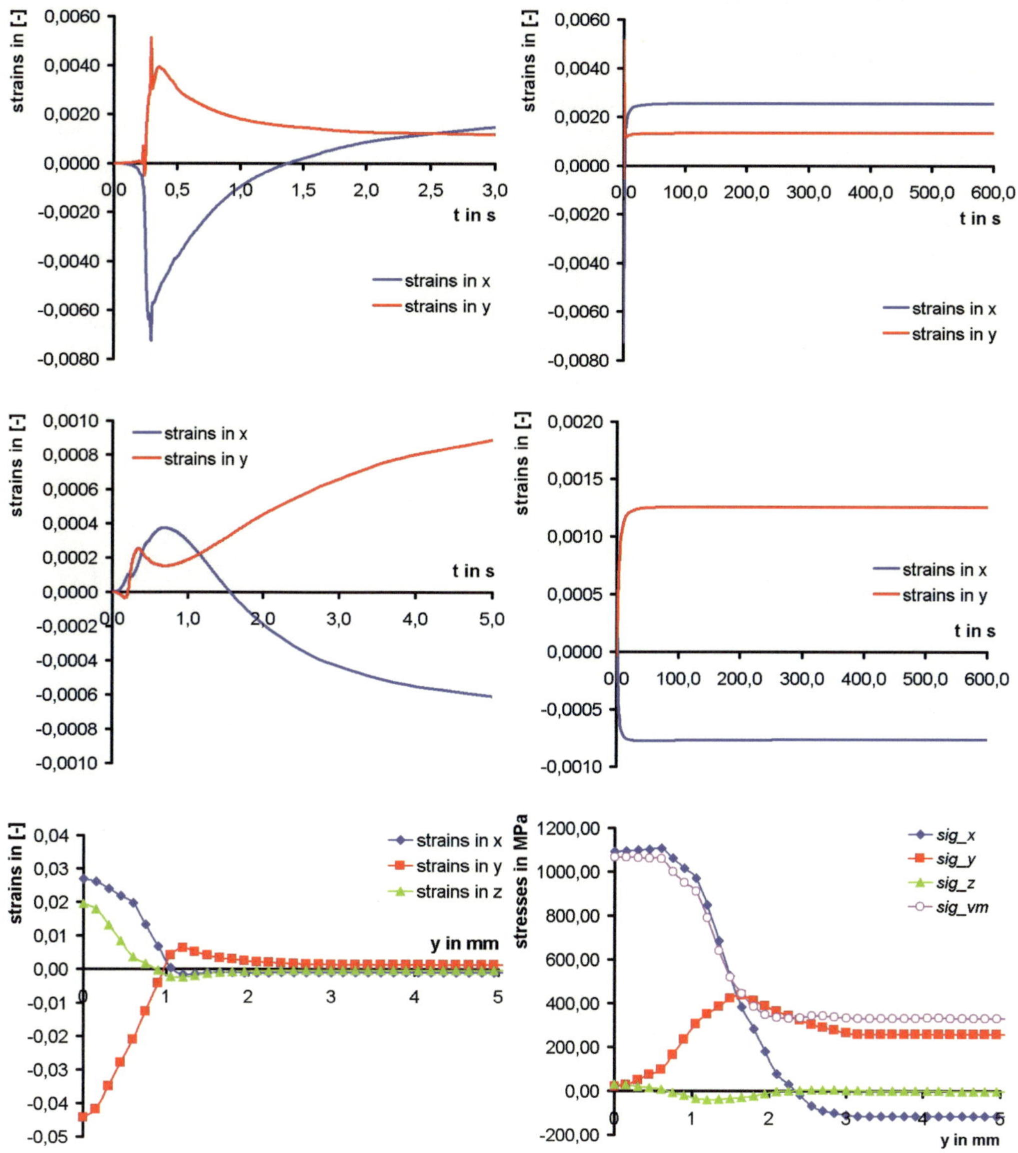

Fig. 22. *First and second row*: Temporal development of the components of the strain vector at the upper surface and at $(x, y) = (L/2, 1\,\text{mm})$ and $(x, y) = (L/2, 10\,\text{mm})$. *Third row*: cross-sectional distribution of the stress components at the upper surface and at $x = L/2$.

it is possible to calculate temperatures, strains, and stresses for arbitrary time and position. Critical regions and values can be identified, which may yield limitations for subsequent process parameters, such as strip tensions during continuous galvanizing, and materials combinations.

5. Experimental, comparative investigations

5.1 Welding experiments

Simultaneously to the numerical simulations experiments have been performed to measure temperatures and distortions at fixed, predefined positions. For this reason two similar sheet geometries[15] are used: $L \times W \times T = 100 \times 150 \times 1.8$ [mm] and $L \times W \times T = 50 \times 75 \times 1.8$ [mm] (geometry of the theoretical investigation). The corresponding specimens are assembled at the upper side (at $X = L/2$) with ThermoCouple Elements (TCE) at different distances to the welding seam as well as Strain Gauges (SGs), rotated by 90 ° to each other, to measure length- and transversal-displacements, cf. Figure 23 and 24. Thus the temperature over time is measured at $Y = 1$ mm, 2 mm, 3 mm and partially 6 mm (for 100×150 [mm] sheets). The displacements can only be integrally quantified along a finite length due to the dimension of the SG. Here changes of the length of a (rosette-type) fine wire, following from changes of the rosette's height, is exploited (in our case the height is 7 . . . 11 mm), cf. Figure 24. The

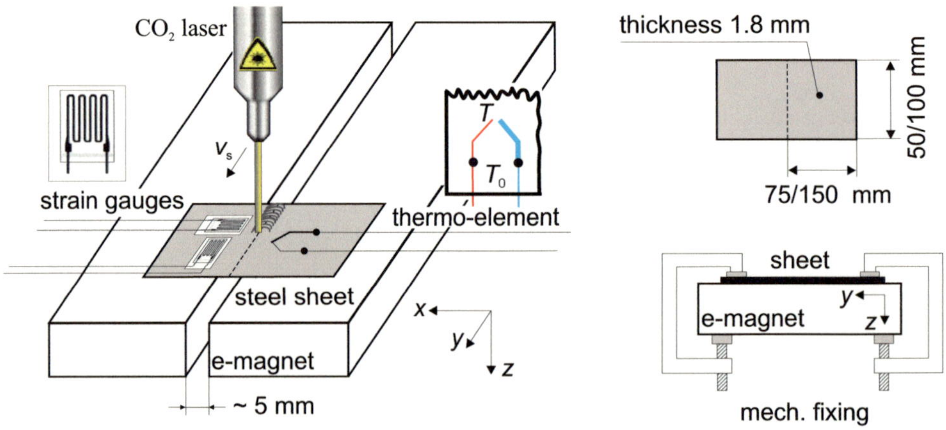

Fig. 23. Schematic view of the experimental setup for the determination of welding distortions and time-temperature curves at fixed distances from the welding seam.

measurement by TCE is based on the so-called SEEBECK effect, i.e. the development of voltage U_m due to different temperatures at the two ends of an electrical conductor (e.g. wire). The TCE results by combining two conductors of different materials in series such, that the ends of both conductors are nearly situated at the position of same temperature. The following relation holds:

$$U_m = k_{AB}(T - T_0) \quad \text{with} \quad k_{AB} = k_A - k_B\,, \tag{39}$$

in which T_0 identifies the reference temperature, measured at the terminals of the voltmeter, and k_{AB} quantifies the proportionality[16] between voltage and temperature difference of the two materials A / B. If T_0 and k_{AB} are known and U_m is measured at the ends of the metallic wires, the unknown temperature T can directly be calculated via Eq. 39. In the present case CrNi-Ni wires of diameter $D = 0.3$ mm with $k_{CrNi} = 22$ V/K and $k_{Ni} = -19.4 \ldots -12$ V/K are used, (Feldmann, 2008). The underlying theory of SG-measurements follows from the

[15] The larger geometry is investigated to ensure the validity of the assumption of RSE, cf. Section 2.5.

[16] For considerable temperature differences the temperature dependence of k_{AB} is not negligible. Then a polynomial ansatz of the form $k_{AB}(T) = a + bT + cT^2 + \ldots$ can be used, which yields a nonlinear relation for T in Eq. (39).

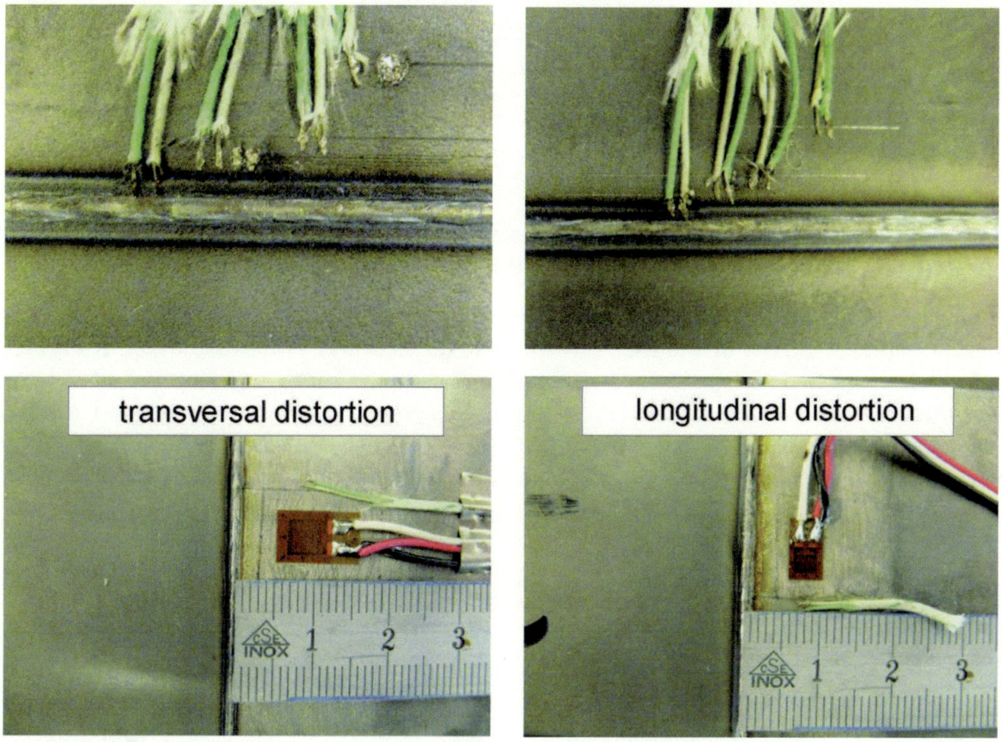

Fig. 24. *First row*: position of the TCEs at the upper side: 1 mm, 2 mm, 3 mm, and 6 mm (for the 150 × 100 [mm] specimen) beside the seam. *Second row*: position of SGs: 7 . . . 11 mm (transversal) or 7 . . . 10 mm (longitudinal) beside the seam.

observation that changes of the length Δl of a cylindrical wire lead to the reduction of its electrical resistance R:

$$\frac{\Delta R}{R} = \hat{k}\frac{\Delta l}{l} \quad \text{with} \quad R = \frac{4l\rho}{\pi D^2} \,. \tag{40}$$

By means of the relation $\Delta R = (\partial R/\partial l)\Delta l + (\partial R/\partial \rho)\Delta\rho + (\partial R/\partial D)\Delta D$ one finds:

$$\frac{\Delta R}{R} = \frac{\Delta l}{l} + \frac{\Delta\rho}{\rho} - 2\frac{\Delta D}{D} \,. \tag{41}$$

By comparing Eq. $(40)_1$ and (41) and by assuming the metal-specific approximation $\Delta\rho = \Delta V = 0$ a value of $\hat{k} = 2$ is derived, which represents 'a standard value' due to the introduced assumptions. Usually $\hat{k}$ values are reported by the provider and manufacturer of the SGs, for e.g. wires made of constantan holds $\hat{k} = 2.05$.

Finally it is worth mentioning, that SGs typically fixed at the (metallic) surface by polymer adhesives, which guarantee the reliable joining up to the critical temperature of 200 °C in the present case. Consequently the SGs must be positioned beyond a critical distance from the welding seam, which is found by the preliminary TCE measurements.

In Table 3 the performed eight experiments are summarized. The joining process is realized by the parameters, listed in Eq. (38). Furthermore a Helium shielding gas is used, and the sheets

no.	quantity of interest	no. of tests	geometry [L × W]
1	temperature	2	50 × 75 mm
2	distortion (transversal)	1	50 × 75 mm
3	distortion (longitudinal)	1	50 × 75 mm
4	temperature	2	100 × 150 mm
5	distortion (transversal)	1	100 × 150 mm
6	distortion (longitudinal)	1	100 × 150 mm

Table 3. List of performed welding experiments.

are magnetically and mechanically fixed as indicated in Figure 23 and 25. The measured time-temperature curves at the TCEs of the test series 1 and 4 are displayed in Figure 26 (first row). Here two measurements are exploited for each series in order to find the mean square deviation Δ via the relation $\Delta = [T_1(t_i) - T_2(t_i)]^2$, cf. Figure 26 (second row). By means of the quantity $\sqrt{\Delta}/2$ the error bars, illustrated in Figure 26 (last row), can be derived. The experimentally obtained welding strains at the upper side and $X = L/2$ are displayed in

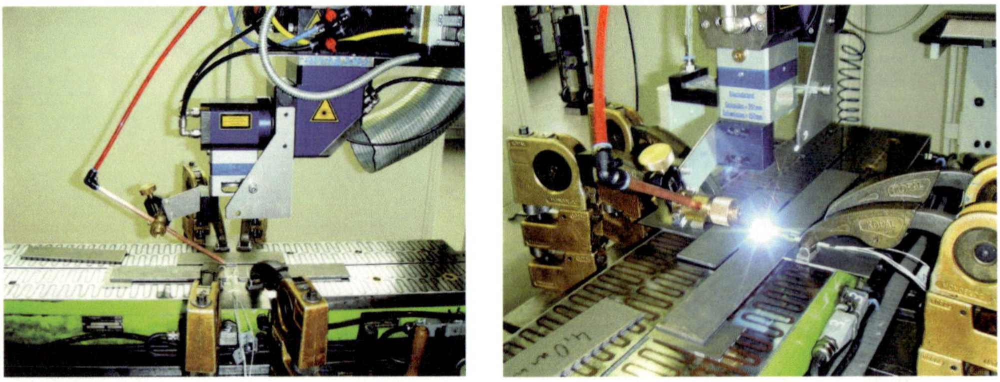

Fig. 25. Illustration of the laser welding equipment (CO2) and the mechanical fixation.

Figure 27. Here the first row shows the longitudinal strains and the second row the transversal strains, 7...10 or 7...11 mm beside the seam, respectively. The geometry of the used strain gauges is $L \times W = 1.6 \times 3.1$ mm (series 2-3) and 3.2×4.6 mm (series 5-6). Obviously, the final residual strains, characterized by the stationary values, are obtained after 40 s. Furthermore the starting point of the measurements slightly varies. Consequently there is a time shift between the different experimental curves. However, this 'shift' has no physical meaning and is cancelled for a better comparison.

As one can easily see, there is a qualitatively and (mainly) quantitatively excellent agreement between the distortion curves of different geometry. This fact is not self-evident but underscores the validity of the RSE assumption. The deviation in the upper left figure results from the preterm release of the mechanical fixation (blue curve). Finally, we did not perform a second comparative measurement to save time and materials.

5.2 Theory vs. experiment

First of all, we compare the temporal temperature evolution following from the numerical simulation and the TCE measurement, cf. Figure 28. In fact, there are slight deviations between the theoretical and experimental peak temperature, however, the overall agreement,

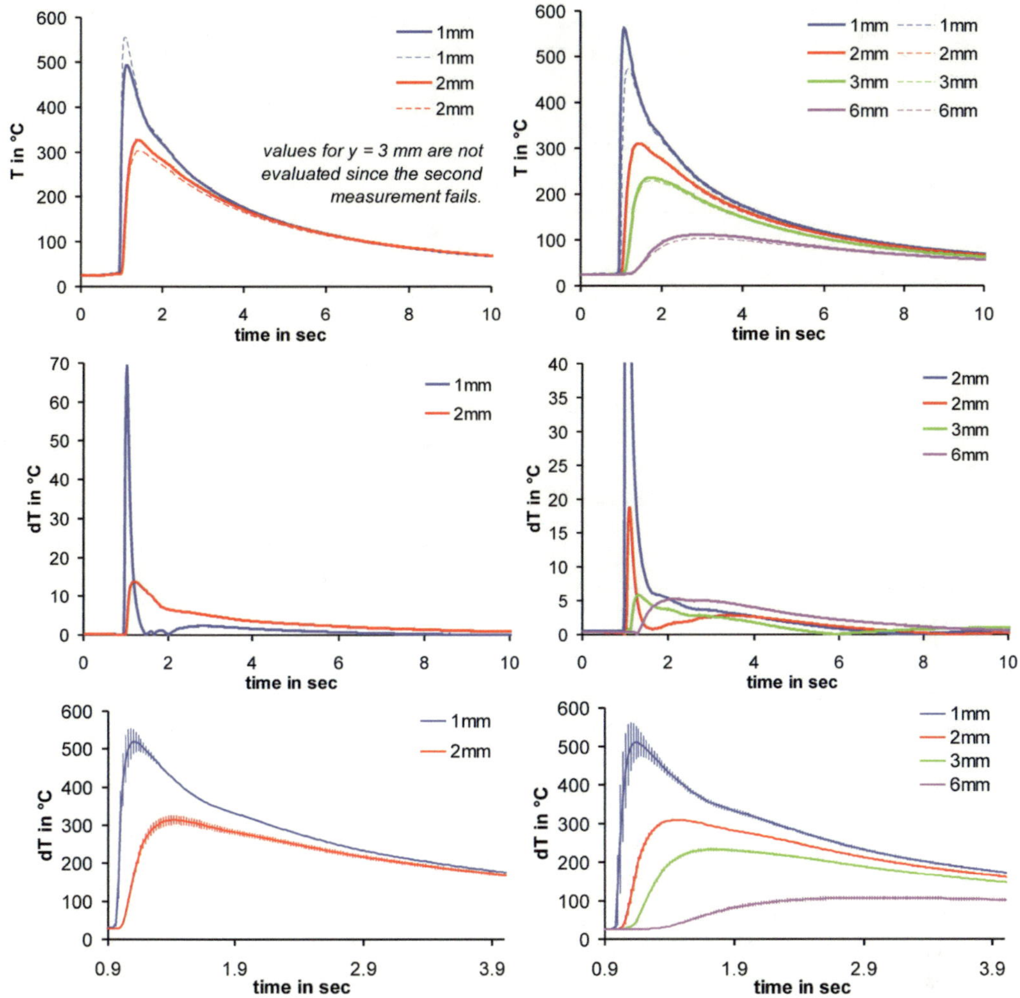

Fig. 26. Temporal temperature evolution at the TCEs for series 1 (left) and 4 (right). *Upper row*: origin data; *Middle row*: mean square deviation ; *Last row*: time-temperature dependence with error bars.

in particular during cooling, is excellent. Two main reasons can be found for the discrepancy of the maximal temperatures (plain vs. dashed lines):

(a) The TCE shows certain initia behaviour, due to the used sample rate of 50 Hz. Thus, values are 'only' registered within the time increment of 0.02 s, whereas the used spatial discretization of the simulation yields a time step of 0.005 s (required time to move the centre of heat source in x-direction one node forward).

(b) Measurements by TCE require additional metallic material, i.e. the TCE wires, at the surface of the specimen, which causes additional cooling and heat transfers. Therefore reduced temperatures are measured at the TCE.

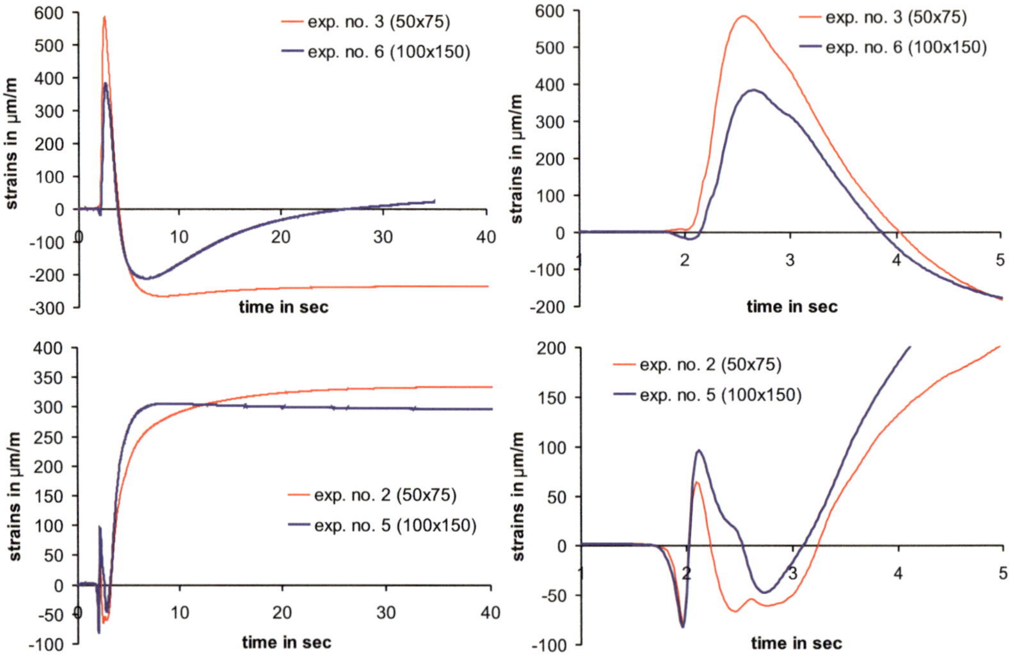

Fig. 27. Experimental results of the SG measurements. *Left*: Time-strain curves up to the stationary state. *Right*: Zoomed view onto the initial process.

Figure 29 compares the experimental and calculated time-strain relations, analogously to the illustration in Figure 22 and 27, at the lateral welding seam distance of 11 mm (sheet geometry $L \times W = 50 \times 75$ [mm], upper side, $X = L/2$). It is easily to see, that the theoretically predicted curves qualitatively follow the experimental ones. Moreover, the calculated stationary strains of the x-direction also render the reality in a quantitative manner. Deviations mainly occur for the stationary values of lateral strains (by factor 3.5 - 4) and for the strains during joining (i.e. the initial fluctuations). In particular, one finds an underestimation for the initial, fluctuating strains in x-direction and an overestimation in y-direction w.r.t. the experiments. To identify the reason of these deviations, one firstly tends to identify the mechanical boundary conditions to be not sufficiently realistic, cf. Figure 8. Especially the upper-side fixation of the sheet is wider and shows an opening for the SG and required cable routings, see Figure 25.

However, this interpretation is in contradiction with the good agreement of the theoretical and experimental values in x-direction. Consequently, unrealistic boundary conditions cannot be the sole origin of the above discrepancy. A second reason - and eventually the main reason - is given by the used, isotropic mechanical materials data. As-rolled steel contains a strong rolling texture and, therefore, a high anisotropy follows. Orientation-depending materials data, e.g. for YOUNG's modulus or expansion coefficient, are required, which require extensive mechanical testing, especially of specimen with different rolling directions. Additionally, one may take an anisotropic yield criterion, such as HILL's yield surface, into account, (Ansys theory ref., 2007).

Finally it should be mentioned, that SG measurements allow for documentation of the temporal strain evolution with certain locality (dimension of the SGs). In contrast, distortion

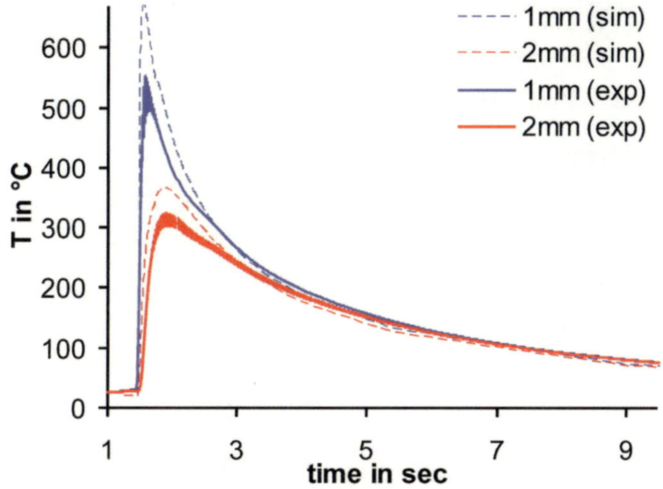

Fig. 28. Experimental vs. theoretical time-temperature curve, distance from the welding seam: 1 mm and 2 mm (upper side).

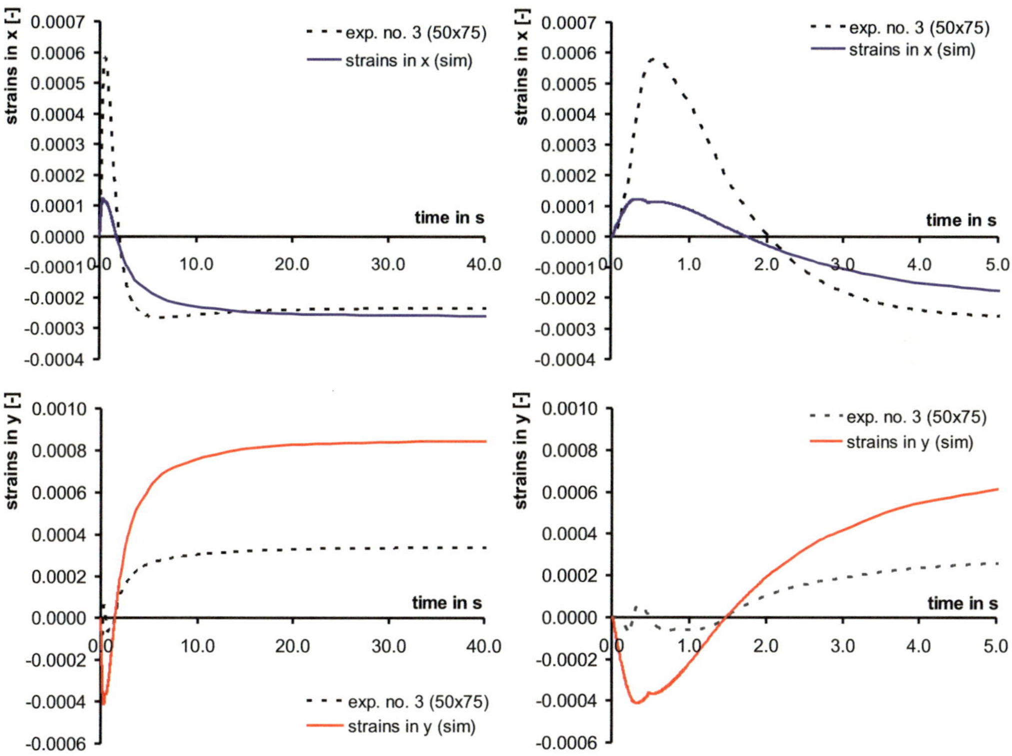

Fig. 29. Comparison between calculated ($y = 10$ mm) and measured ($y = 8 \ldots 12$ mm) mechanical response. *Right*: Zoomed view onto the initial process.

measurements via displacement transducers at the boundaries of the sheet, such as in (Voss, 2001), 'only' yields integral values.

6. Discussion: possibilities and limitations

6.1 Phase transformations

Heat input and welding-induced stress fields yield to changes of the microstructure, cf. Figure 30. This complex phenomenon is mainly triggered by four factors, viz.:

a. initial microstructure (e.g. ratio of retained austenite),

b. heating and cooling rates,

c. peak and final temperature (the latter one is usually the room temperature),

d. distortions and stresses (resulting, e.g. from TRIP effects).

For instance, variations of the content of retained Austenite (A) within the origin microstructure before welding yield different amounts of A within regions of partial austenitization (within the $Ac1$ - $Ac3$ temperature regime). On the other hand different, final Ferrite (F) or Bainit (B) contents result for varying quenching rates, cf. Figure 31[17]: $t_{8/5} = 4$ s vs. $t_{8/5} = 40$ s or B-M vs. F-B-M composition, respectively. Two kinds of microstructure changes can be identified:

1. variations of composition, i.e. of A-, F-, B-, and Martensite (M) contents,

2. changes of the micro-morphology, namely of grain size or phase shape/size.

Changing microstructures influent the thermodynamic and mechanical properties of the material, and may considerably determine resulting welding distortions and residual stresses. Martensite formation in the vicinity of the welding seam due to high cooling rates, for example, lead to increasing volume and to additional, positive strains, (Dean & Hidekazu, 2008). A second example is grain growth within the Heat Affected Zone (HAZ), which reduces the strength of the origin fine-grained material, cf. Figure 30 (region B). From the point-of-view of modelling and simulation physical models and mathematical methods are required, which allow to quantify the changes of item **1.** and **2.** depending on the factors denoted under **a.** - **d.**. However, due to the high complexity of matter, only theoretical investigations of single aspects are found in literature. One of the most established[18] models for estimating phase changes during welding are the so-called JOHNSON-MEHL-AVRAMI-KOLMOGOROV (JMAK) ansatz, (Avrami, 1939; 1940), for diffusion-induced transformations (Bainit $\leftrightarrow$ Ferrite $\leftrightarrow$ Austenite) and the KOISTINEN-MARBURGER[19] (KM) model, (Koistinen & Marburger, 1959), for diffusion-less phase formations. Beyond these the following models are worth-mentioning:

Leblond:
Here a differential equation for the description of nonisothermal, diffusioninduced and -less transformations is derived. Furthermore the model allows for the prediction of the temporal grain size evolution, (Leblond & Devaux, 1984).

---

[17] The graph exclusively holds for the present material and indicated experimental conditions.

[18] One of the reasons is the simple mathematical formulae, resulting in manageable computational efforts.

[19] Surprisingly, this highly cited work is a short comment, in which experimental results are presented and fitted by an obviously universal, linear function.

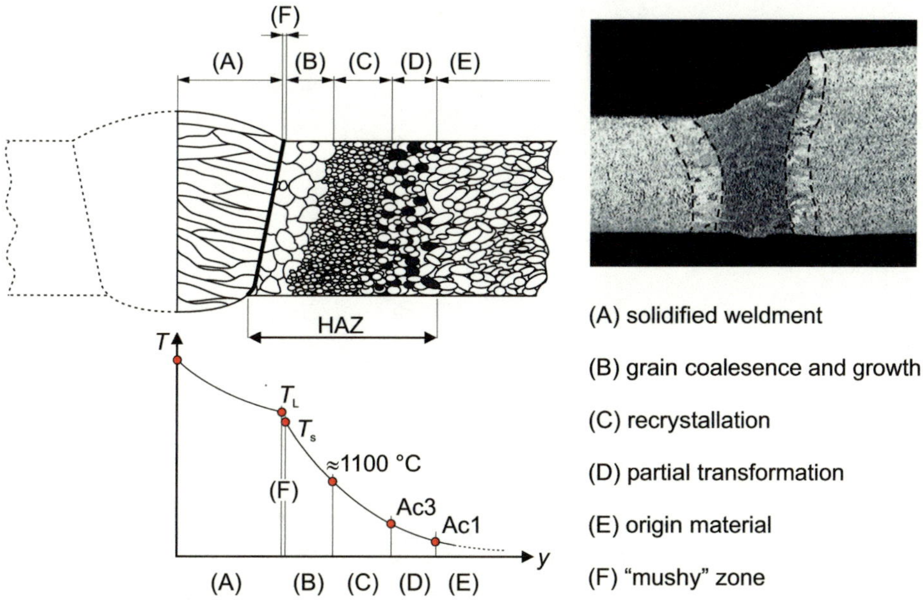

Fig. 30. Schematic illustration of different microstructures within the lateral joining region, following from (Francis et al., 2007); photo by courtesy of O. Voss (TK Tailored Blanks S.A. de S.V., Puebla, Mexico).

Cahn-Hilliard:
So-called phase field models can be used to describe the local evolution of micro-morphologies. This approach is the most general one; but considerable computational efforts are required and materials data as well as initial/boundary conditions are often unknown, (Cahn & Hilliard, 1958; Cahn, 1961; Thiessen et al., 2007).

Zener:
This historical work provides a model to describe the evolution of the so-called growth coordinate (e.g. the phase radius in 3D) for isothermal, diffusion-induced transformations; ratios of different coexisting phases and morphologies are not considered, (Zener, 1949).

STAAZ:[20]
Here the different phases are experimentally determined and the resulting phase regions are plotted within a peak-temperature (y-axis) / cooling time (x-axis) diagram. Thus, correspondingly to the point-wise different heat treatment during welding, the various, calculated temperature histories can be used to identify points, and consequently phases, within the diagram, (Ossenbrink & Michailov, 2007; Ossenbrink, 2008; Voss, 2001).

In what follows we briefly explain the strategy for calculation of welding induced phase transformations by using the JMAK and KM formalism. By assuming spherical (phase-)shapes and the so-called 'isokinetic regime' (isothermal boundary conditions) the model of JMAK applies methods of statistical physics to derive the following relations for

[20] german abbreviation for: Spitzentemperatur-Austenitisierung-Abkühlzeit.

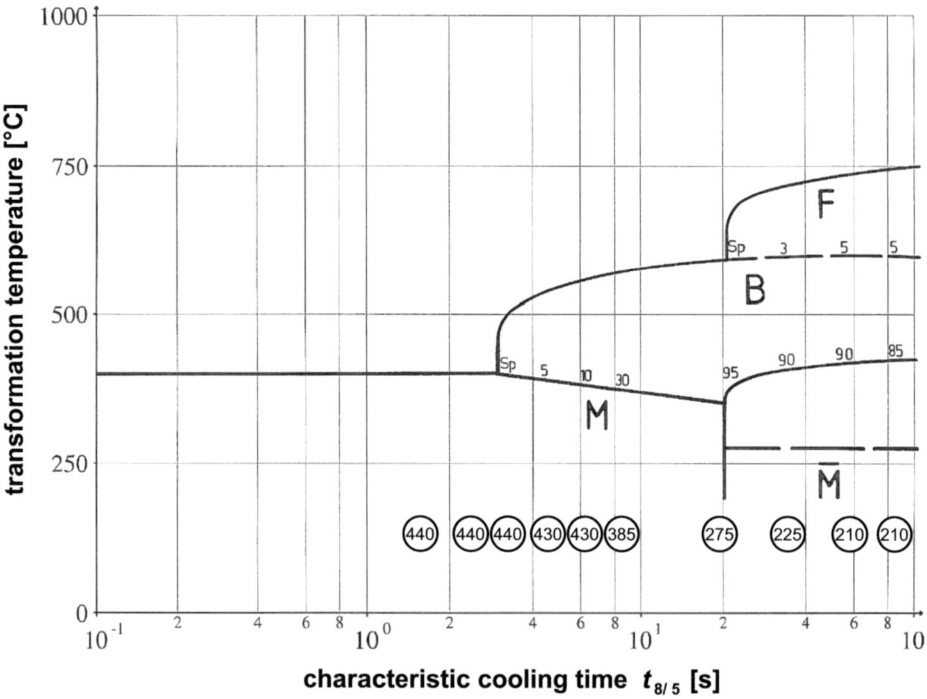

Fig. 31. Continuous-Cooling-Transformation (CCT) diagram for TRIP 700 (measured at Gleeble3500®). The following experimental procedure/material is used: heating (austenitization) up to 1350 °C, holding time: 1 s, $dT/dt = 337.5$ °C/s, austenite grain size: 3.0 (ASTM E112), material: cold rolled strip, thickness: 1.8 mm, $Ms = 400$ °C, $Ac1 = 735$ °C, $Ac3 = 1000$ °C.

the volume fractions of the different phases, (Avrami, 1940):

$$\xi^{\gamma}(t_i) = 1 - \exp\left[-a\, t_i^b\right] \quad \text{with} \quad \gamma = \{\text{A}, \text{B}, \text{C}\}\,. \tag{42}$$

Here t_i represents the i-th time step of the FE calculation and γ stands for the different, developing phases. The material-specific parameters a and b must be fitted to Time-Temperature-Transformation (TTT) diagrams[21], (Kang & Im, 2007). Furthermore the non-isothermal conditions during welding must be taken into account. For this reason, the continuous cooling process is divided into a monotone decreasing step function, and isothermal conditions are assumed for each single plateau. Note that Eq. 42 allows for a separate calculation of the different phases; but, realistically, they coexist and interact each other. In order to correctly combine the calculated volume fractions the so-called additivity rule, (Umemoto et al., 1982; 1983) must be used (detailed explanations can be found e.g. in (Dean et al., 2008; Dean & Hidekazu, 2008; Kang & Im, 2007)). This framework allows to calculate the volume fractions, $\xi^{\text{A/F/B}}$, of each element or node; but the micro-morphology remains undetermined.

[21] To prepare such diagrams the specimen is totally austenitized and subsequently quenched to the target temperature. Then the phase transformation is measured at constant temperature conditions. In contrast CCT-diagrams characterize the phase transformations for constant, predefined cooling rates.

The model of KM, especially in case of M-formation, states, that the volume fraction of the forming phase exclusively depends on degree of undercooling, namely:

$$\xi^{\mathrm{M}}(t_i) = 1 - \sum_{\gamma=\{\mathrm{B,F}\}} \xi^{\gamma}(t_i) - \underbrace{\left\{1 - \exp\left[-c(M_s - T_i)\right]\right\}}_{=\xi^{\mathrm{A}}} , \qquad (43)$$

in which M_s indicates the starting temperature of Martensite formation, $c = \text{konst.} = 0.011$, T_i stands for the present temperature at time t_i, and ξ^{γ} denotes the volume fraction of the γ-th (non-martensitic) phase accordingly to Eq. $(42)_2$. Thus it is possible - starting from the (known) local fractions of A, F, B - to determine the fraction of M formed during the cooling period of the FE-analysis. However, information about the micro-morphology is not derived.

During laser-beam welding extremely high heating and cooling rates occur. Therefore partial austenitization and $t_{8/5}$-times between $0.8 \ldots 1.5$ s result, cf. Figure 28. Consequently the F- and B-areas of Figure 31 are not passed. For this reason it is sufficient to exclusively investigate the M-formation as a first step. Note that the phase evolution takes place on the micro-scale, whereas the used FE-mesh is suitable to simulate the macroscopic welding process. A finer discretization is required, which lead to unmanageable computational times in case of the present full 3D-geometry: A compromise could be the strategy summarized in Table 4:

The final phase composition can be used, for instances, to determine the lateral distribution of hardness (e.g. VICKERS). For this reason the following empirical equations hold, (Voss, 2001):

$$\begin{aligned}
HV_{\mathrm{M}} &= 127 + 949\,\mathrm{C} + 27\,\mathrm{Si} + 11\,\mathrm{Mn} + 8\,\mathrm{Ni} + 16\,\mathrm{Cr} + 21\,\mathrm{Log}(\dot{T}) , \\
HV_{\mathrm{B}} &= 323 + 185\,\mathrm{C} + 330\,\mathrm{Si} + 153\,\mathrm{Mn} + 65\,\mathrm{Ni} + 144\,\mathrm{Cr} + 192\,\mathrm{Mo} + \\
&\quad + (89 + 53\,\mathrm{C} - 55\,\mathrm{Si} - 22\,\mathrm{Mn} - 10\,\mathrm{Ni} - 20\,\mathrm{Cr} - 33\,\mathrm{Mo})\mathrm{Log}(\dot{T}) , \\
HV_{\mathrm{F/P}} &= 42 + 223\,\mathrm{C} + 53\,\mathrm{Si} + 30\,\mathrm{Mn} + 13\,\mathrm{Ni} + 7\,\mathrm{Cr} + 19\,\mathrm{Mo} + \\
&\quad + (10 - 19\,\mathrm{Si} + 4\,\mathrm{Ni} + 8\,\mathrm{Cr} + 130\,\mathrm{V})\mathrm{Log}(\dot{T}) , \\
HV_{\mathrm{tot}} &= HV_{\mathrm{F/P}}\,(\xi^{\mathrm{F}} + \xi^{\mathrm{P}}) + HV_{\mathrm{B}}\,\xi^{\mathrm{B}} + HV_{\mathrm{M}}\,\xi^{\mathrm{M}} . \qquad (44)
\end{aligned}$$

Here $\dot{T}$ must be calculated at temperature of minimal stability for austenite, viz. $T \approx 500$ °C. To investigate the mechanical response of the material following from welding and the induced phase transformations, mechanical materials models are required, in fact before the mechanical FE analysis. Such models allow for the decomposition of the strains by incorporating the phase evolution, $\underline{\underline{\varepsilon}} = \underline{\underline{\varepsilon}}^{\mathrm{el}} + \underline{\underline{\varepsilon}}^{\mathrm{pl}} + \underline{\underline{\varepsilon}}^{\mathrm{th}} + \underline{\underline{\varepsilon}}^{\mathrm{A}\to\mathrm{M}}$. Here $\underline{\underline{\varepsilon}}^{\mathrm{A}\to\mathrm{M}}$ identifies the strain tensor due to the A-M transformation[22], (Dean et al., 2008; Dean & Hidekazu, 2008). It can be quantified by means of crystallographic arguments and the obtained, above volume fractions $\xi^{\mathrm{M/F/B/A}}$. In case of the commercial FE- package ANSYS® phase-depending materials models are not provided by default. To overcome this shortcoming so-called *'user programmable features'* can be used, which enable the user to include own materials models.

In summary one finds, that theoretical investigations of welding induced phase transformations concentrate, in most cases, on the resulting phase volume fractions. It is possible to estimate the distribution of hardness as well as the mechanical material behaviour,

[22] It is also possible to consider TRIP effects by adding an additional term $\underline{\underline{\varepsilon}}^{\mathrm{tr}}$.

(A)	**Defining the region of analysis**
(A.1)	Defining and fine discretizing of an area at $X = L/2$ with normal vector parallel to the welding velocity.
(A.2)	Mapping of the calculated local temperature history of the 3D-mesh to the nodes of the area.
(A.3)	Saving the temporal temperature evolution at each node of the area.
(B)	**Heating period**
(B.1)	Definition of the initial phase composition ($\xi_0^{M}, \xi_0^{RA}, \xi_0^{B}, \xi_0^{F}$) by using experimentally obtained data.
(B.2)	Selection of all nodes with temperatures higher than $Ac3$; definition of $\xi^{A} = 1$ and $\xi^{M,B,F} = 0$ for these nodes.
(B.3)	Selection of nodes with maximal temperature between $Ac1$ and $Ac3$ and linear interpolation as follows: $\xi^{A} = \xi_0^{RA} + T_{max}(1 - \xi_0^{RA})/(Ac3 - Ac1)$ and, correspondingly, $\xi^{F} = 1 - \xi_0^{M} - \xi_0^{B} - \xi_0^{A}$.
(B.4)	The remaining unselected nodes do not change the initial composition.
(C)	**Cooling**
(C.2)	Definition of the the resulting values of b.3 as initial condition.
(C.3)	Calculation of the M- and A-volume fractions by means of Eq. (43).
(C.4)	Finally, the lateral phase distribution, representing the composition for arbitrary X follows ((any boundary effects are neglected).

Table 4. Possible strategy to calculate the Martensite (volume) fraction following from heating and cooling during laser welding.

if - and only if - stress-strain relations (depending on the current phase composition) are available. Furthermore the local micro-morphology (size and shape of phases) can be considered by phase field approaches or partially by the LEBLOND model. However, combining such calculations with thermal and mechanical welding simulations requires considerable computational power, in particular for 3D geometries.

6.2 Hydrogen diffusion and local embrittlement

Hydrogen strongly decreases the mechanical strength (called hydrogen-induced embrittlement, HE) of steels and represents one of the most reasons for failure, (Timmins et al., 1997)[23]. Due to different manufacturing steps, such as annealing or electrolytic galvanizing, little contents of hydrogen are found within the material (0.5 - 3 ppm). Internal H-transport is characterized by interstitial diffusion; additionally, H is increasingly attracted at so-called 'traps', i.e. internal defects, such as precipitates, voids, phase boundaries or dislocations. Trapped hydrogen remains in the atomic state (H) or can recombine to H_2 under certain conditions. The traps act as potential wells, i.e. regions, in which hydrogen have a lower energetic state than at interstitial sites.

As the consequence of trapping, different diffusion coefficients are measured during charging and discharging of steels with H. During charging, first, all traps are filled before purely interstitial, stationary diffusion occurs, whereas during discharging normally only interstitial hydrogen escapes; the trapped hydrogen remains and do not contribute to the diffusion process, (Darken & Smith, 1949). Therefore the use of the 2nd FICKian law, $\partial_t c + \underline{\nabla} \cdot (\underline{\underline{\mathbf{D}}} \cdot$

[23] The authors note that HE provokes annual costs of more than 100 billion U.S.$ within the U.S. industry.

$\underline{\nabla} c) = 0$ ($\underline{\underline{\mathbf{D}}}$: diffusion coefficient, c: H-atoms per unit volume[24]), is limited to some rare cases (e.g. to steels with saturated traps resulting from low density of defects or high H-content). In fact, a source term must be added to the right side of the equation, which was firstly developed by MCNABB and FOSTER, (McNabb & Foster, 1963), and later by ORIANI assuming 'local equilibrium', (Oriani, 1970). Thus, the following equations hold for diffusion of hydrogen under the presence of local stress fields ($c << 1$), (Serebrinsky et al., 2004):

$$\mu(T,c) = \mu_0(T) + RT\ln(c) - \frac{\mathrm{Tr}\,\underline{\underline{\boldsymbol{\sigma}}}}{3} V_{\mathrm{H}}\,, \tag{45}$$

$$\frac{\partial c_{\mathrm{L}}}{\partial t} + \underline{\boldsymbol{\nabla}} \cdot \underline{\boldsymbol{J}} = \frac{\partial c_{\mathrm{T}}}{\partial t}\,, \quad (c = c_{\mathrm{T}} + c_{\mathrm{L}})\,, \tag{46}$$

$$\underline{\boldsymbol{J}} = -\underline{\underline{\mathbf{M}}}(c) \cdot [\underline{\boldsymbol{\nabla}}\mu]\,, \tag{47}$$

$$\underline{\boldsymbol{\nabla}}\mu = \frac{\partial\mu}{\partial T}\underline{\boldsymbol{\nabla}}T + \frac{\partial\mu}{\partial c}\underline{\boldsymbol{\nabla}}c + \frac{\partial\mu}{\partial p}\underline{\boldsymbol{\nabla}}p\,. \tag{48}$$

Here the following notation is used. c_{L} and c_{T}: atomic concentration at interstitial or trap sites; μ and μ_0: molar chemical potential of hydrogen within the 'binary' mixture hydrogen/steel and the temperature-depending, pure substance contribution at reference pressure $p = 0$; $p = \mathrm{Tr}(\underline{\underline{\boldsymbol{\sigma}}}/3)$: hydrostatic stresses (pressure); V_{H}: molar volume of H in solid solution; $\underline{\underline{\mathbf{M}}} = c\,M\,\underline{\underline{\mathbf{I}}}$: mobility of H with $M = D/(\partial\mu/\partial c) = D/(RT)$, ($D$: diffusion coefficient); $R = 8.314\,\mathrm{J/mol\,K}$: universal gas constant and $\underline{\mathbf{J}}$: diffusion flux of H.

For the source term of the right hand side of Eq. (46) holds by statistical argu-ments, (Krom & Bakker, 2000):

$$\frac{\partial c_{\mathrm{T}}}{\partial t} = N_{\mathrm{T}}\frac{\partial\Theta_{\mathrm{T}}}{\partial t} = N_{\mathrm{T}}\left[a\,\Theta_{\mathrm{L}}(1-\Theta_{\mathrm{T}}) - b\,\Theta_{\mathrm{T}}\right]\,, \tag{49}$$

and for equilibrium, where time derivatives will vanish:

$$\frac{\Theta_{\mathrm{T}}}{\Theta_{\mathrm{L}}(1-\Theta_{\mathrm{T}})} = \frac{a}{b} = \exp\left[-\Delta E_{\mathrm{T}}/RT\right] \equiv K_{\mathrm{T}}\,, \tag{50}$$

with two constants, a and b, which characterize the transition $L \to T$ and $T \to L$. Furthermore $N_{\mathrm{T}}, \Theta_{\mathrm{T}}$ denote the (constant) total trap number and the trap occupancy; $N_{\mathrm{L}}, \Theta_{\mathrm{L}}$ stands for the total number of interstitial sites and the corresponding occupancy; and ΔE_{T} identifies the energy difference between a trapped H-particle and an interstitial atom.

By inserting the relations $\Theta_{\mathrm{L}} = c_{\mathrm{L}}/N_{\mathrm{L}}$ and $\Theta_{\mathrm{T}} = c_{\mathrm{T}}/N_{\mathrm{T}}$ into Eq. (50) we obtain:

$$c_{\mathrm{T}} = N_{\mathrm{T}}\left(1 + \frac{N_{\mathrm{L}}}{K_{\mathrm{T}}c_{\mathrm{L}}}\right)^{-1} \tag{51}$$

and finally ($V_{\mathrm{H}} = \mathrm{const}, T = \mathrm{const}$):

$$\frac{\partial c_{\mathrm{L}}}{\partial t} + \underline{\boldsymbol{\nabla}} \cdot \left[-D\,\underline{\boldsymbol{\nabla}}c_{\mathrm{L}} + \frac{DV_{\mathrm{H}}}{RT}(c_{\mathrm{L}} \cdot \underline{\boldsymbol{\nabla}}\,p + p\,\underline{\boldsymbol{\nabla}}c_{\mathrm{L}})\right] = \frac{\partial}{\partial t}\left[N_{\mathrm{T}}\left(1 + \frac{N_{\mathrm{L}}}{K_{\mathrm{T}}c_{\mathrm{L}}}\right)^{-1}\right]\,. \tag{52}$$

[24] By convenience, we will call c the concentration.

Equation (52) represents an evolution equation for the unknown quantity c_L. The required, materials data for $N_L, N_T, \Delta E_T$, which stand for a specific trap configuration, can be found in literature, (Krom & Bakker, 2000; Maroef et al., 2002; Oriani, 1970). In particular the trap density - and thus the total number of traps N_T - increases with degree of cold-working. Furthermore the (residual) hydrostatic stresses, $p = \frac{1}{3}\mathrm{Tr}(\underline{\underline{\sigma}})$, occur in the above equation, which are known as result of the presented welding simulations, see paragraph 4.2 and Figure 32. By means of suitable numerical framework, such as the method of Finite Differences, equation (52) can be solved. Therewith predictions about the hydrogen distribution within the material can be made following from the specific laser welding procedure. In Figure 33 different contourplots are displayed, which show - exemplarily -

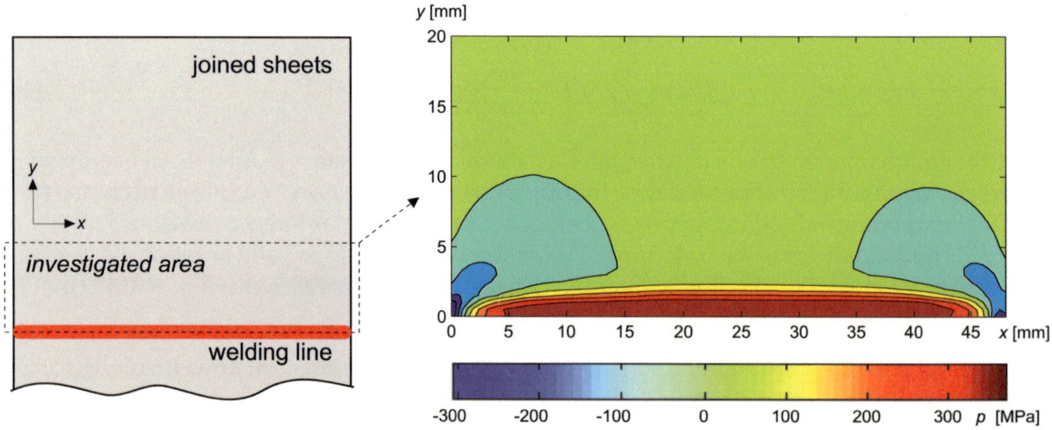

Fig. 32. Illustration of the analyzed region and the calculated, residual hydrostatic stresses used as input data to solve Eq. (52) numerically.

the temporal evolution of hydrogen after laser welding as investigated in Section 4. Here we neglect - as a first approximation - the right hand side of Eq. 52, so that the diffusible hydrogen amount is exclusively investigated (no un-/trapping, $D = 5.8 \cdot 10^{-4}$ mm^2/s). Moreover, a uniform 96 × 40 (length × width) grid is used in combination with NEUMANN boundary conditions ($\underline{\mathrm{J}}_{\partial\Omega} = 0$). A finite difference scheme as well as an EULER time-step scheme is implemented in order to numerically solve the above diffusion equation. Furthermore the spatial distribution of the hydrostatic stresses, calculated by the preliminary welding simulations, is applied to predict hydrogen transport due to local mechanical stress fields.

Obviously, residual stresses lead to the redistribution of hydrogen. In particular, H is increased found in regions of tensile stresses (positive strains), whereas domains of compressive stresses (negative strains) show reduced hydrogen concentrations. The final concentration profile (equilibrium) is determined by the maximal hydrostatic stresses and is found after ~15 h. Furthermore the long-time simulations (12 h, 30 d) indicate, that the homgeneous hydrogen concentration far from welding also decreases due to the H-accumulation in the welding line ($c_{\mathrm{homog}}(t = 30\mathrm{d}) \approx 0.46$). On the other hand the hydrogen concentration near the joining region already increases after eight minutes from 0.5 ppm to 0.52 ppm. However, the present simulations only allow for a rough reproduction of H-diffusion during and after welding. Particularly, the heat input reactivates trapped hydrogen in the vicinity of the joining region, which may drastically increase the diffusible amount. Considerable concentration gradients could result, which are additionally promoted

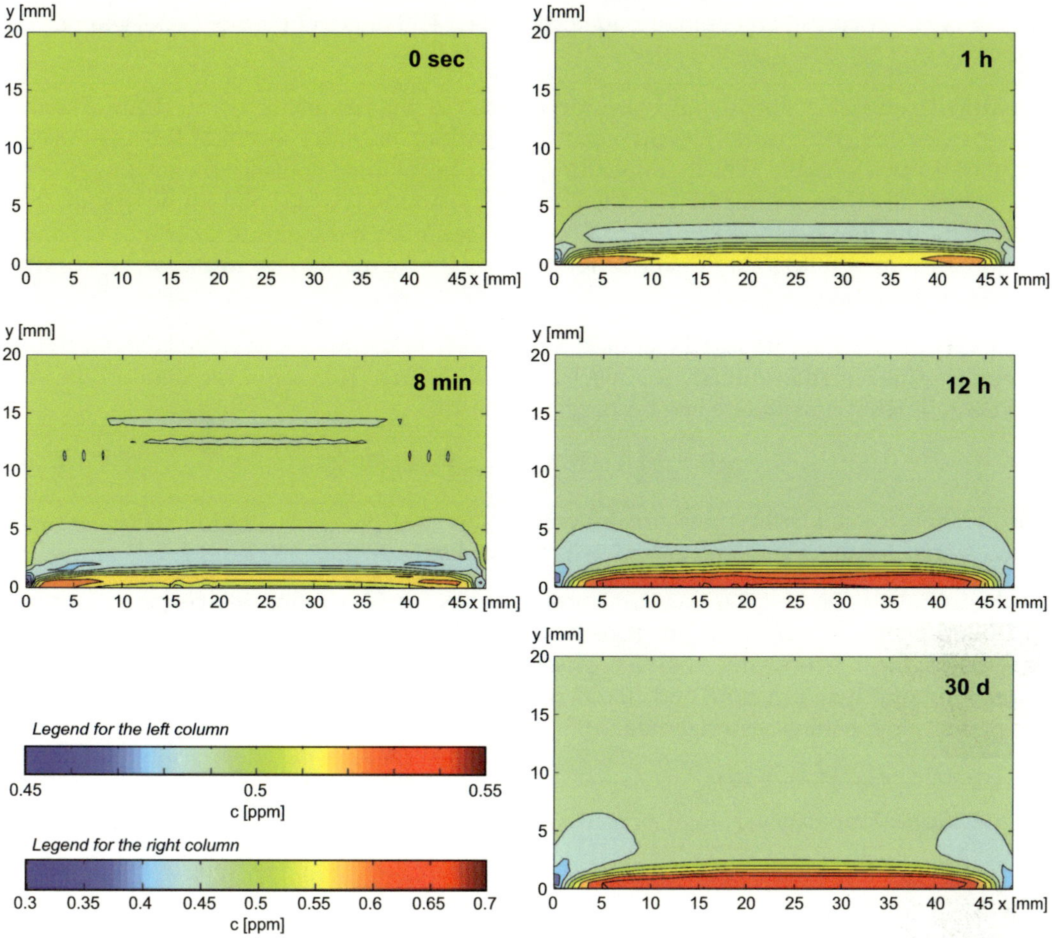

Fig. 33. Predicted evolution of the local hydrogen concentration in the vicinity of the welding line. Initial, homogeneous hydgrogen concentration: 0.5 ppm.

by successive H-diffusion to regions of high residual stresses. This phenomenon may lead to significant local H-concentrations and to fatal embrittlement. Moreover, boundary conditions, such as air moisture and hydrogen within (Zn-)coatings can emphasize this effect, which, nevertheless, causes failure like e.g. cold cracking. Such defects often occur within the bulk of the material and, therefore, are not visible to the naked eye. Here theoretical simulations can help to identify domains of increased failure sensitivity.

6.3 Coupling of weld-pool-dynamics calculations and solid state FE-analysis

A crucial precondition for the above illustrated FE-investigation is the a priori knowledge of the weld pool geometry. However, local temperature distribution and resulting varying surface tension (additionally influenced by steel composition, e.g. S and N content) directly effect the weld pool flow conditions, (Mills et al., 1998). Therefore, varying process parameter as well as different steel grades may lead to considerable changes of the weld pool geometries, primarily w.r.t. thickness and depth.

In particular the weld pool depth is a priori defined by the parameter $z^{(2/3))}$ (LASIM makro provided by ANSYS®) or by f_i (GAUSSian distribution). In order to cancel this assumption simulations are desirable, which - depending on the heat input - consider the melting process (and eventually the vaporization) and weld pool dynamics, i.e. the solid-liquid system (and eventually the liquid-vapour system). Then, the resulting temperature distribution can be used in a separate solid state FE-simulation to calculate the mechanical response, i.e. residual stresses or distortions.

Starting point of most computational fluid dynamics (CFD) simulations is the NAVIER-STOKE equation. This equation directly follows by considering the balance of momentum, Eq. (2), and the following constitutive law for the stress tensor:

$$\underline{\underline{\sigma}} = -p\,\underline{\underline{\mathbf{I}}} + \nu \mathrm{Tr}(\underline{\boldsymbol{\nabla}}\,\underline{\mathbf{v}})_{\mathrm{sym}}\,\underline{\underline{\mathbf{I}}} + 2\mu(\underline{\boldsymbol{\nabla}}\,\underline{\mathbf{v}})\,. \tag{53}$$

The symbols p, ν, μ identify pressure, kinematic and shear viscosity. Furthermore the index $(\cdot)_{\mathrm{sym}}$ stands for the symmetric part of velocity gradient.

At this point we present various conceptual studies for illustrating the coupling between CFD- and FE-simulations. For this reason we use the CFD-software package ANSYS CFX® as well as the FE-software ANSYS® classic. By considering the welding process, the transition point between solid and liquid is defined by the rapid, jump-wise change of the temperature-depending material data for:

- dynamic viscosity $\eta = \nu/\rho$,
- specific heat capacity c_{p},
- heat conductivity κ and
- mass density ρ.

The transition between the liquid and gaseous state must be also specified, especially to realistically model the free surface melt/air (e.g. the melt pool curvature). For this reason we define a volume of Length×Width×Heigth = L×W×H= 50 × 35 × 26.5 mm (half model), consisting of the steel sheet (L×W×H = 50 × 35 × 1.5 mm), the air above (H=20 mm) and the air below (H=5 mm). After meshing the materials properties must be assigned to each element in order to define the different materials within the volume. Here - as a first approximation - constant standard values are used for the elements representing the air. Such element-wise assignment can be realized by the definition of element-specific volume fractions (and partial pressures). It holds:

$$\Xi_{\mathrm{iron}}(h) = \begin{cases} 1 & \text{if} \quad -1.5 \leq h \leq 0 \\ 0 & \text{if} \quad h < -1.5 \quad \text{or} \quad h > 0 \end{cases} . \tag{54}$$

and vice versa for $\Xi_{\mathrm{air}}(h)$. Note, that the origin of the coordinate system in Eq. (54) is defined a the upper surface of the metallic sheet. Within the thin boundary volume between liquid and air, in which a linear interpolation is performed between the jump-wise changing properties, the heat is supplied via the GAUSSian distribution ($\sigma = 1.4$ mm, $P = 0.5\,\eta\,P_0 = 843$ W). Here the centre of distribution changes according to the welding velocity $v_{\mathrm{s}} = 0.625$ m/min. Figure

34 and 35 (1st row) illustrate the calculated melt pool geometry (including main flow lines) as well as the resulting temperature distribution within the steel sheet (for $t = 2.1$ s).

In order to couple CFD- and FE-analysis *in one direction* the following strategy is applied: **(a)** The temperature distribution of the steel sheet, resulting from the CFD-simulation, is mapped to a new mesh (mostly with larger element size), which will be later used in the FE-analysis. **(b)** The mapped temperatures of the steel sheet are exported to data files for each time step; **(c)** The mechanical FE-analysis (calculation of stresses and strains) is performed by reading the temperatures from the corresponding data files for each time step.

Figure 35 (2nd row, left) shows the imported temperature distribution in ANSYS® classic accordingly to the results of ANSYS CFX® in Figure 35 (1st row). The lower right panel of Figure 35 represents the results of the mechanical FE-simulation (distortions are amplified by factor 5). Here the mechanical boundary condition is chosen such, that all nodes of the upper and lower side at $x = W/2$ are fixed ($\underline{\mathbf{u}} = 0$). Furthermore, the discretization of the mesh is obviously chosen too large. Thus inaccurate stresses are obtained in some regions, which are justifiable due to the conceptual character of the study. Moreover, the laser-welding specific phenomenon of keyhole formation is not considered within the above explained CFD-analysis. Therefore the presented strategy is more appropriate for 'classical' welding processes, such as TIG-, MIG-, or MAG welding. Nevertheless, the results clearly show the potential for coupling weld-pool-dynamics simulations with solid state FE-analysis.

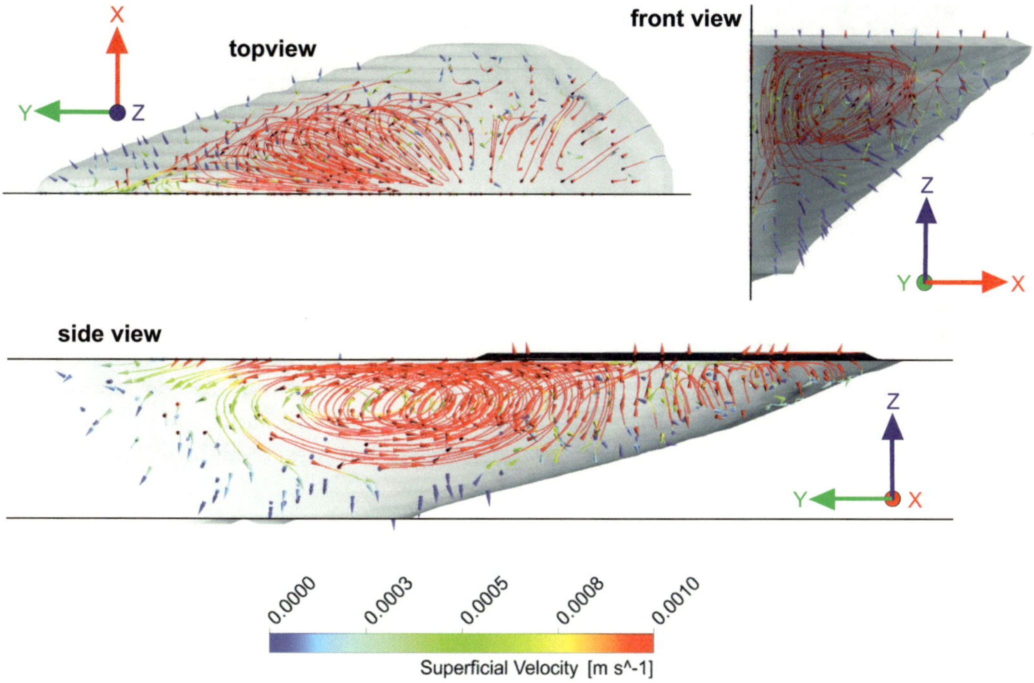

Fig. 34. Weld pool geometry and main flow lines calculated by ANSYS CFX®. Clearly to see is the MARANGONI convection due to the temperature-depending surface tension, (Lee et al., 1998).

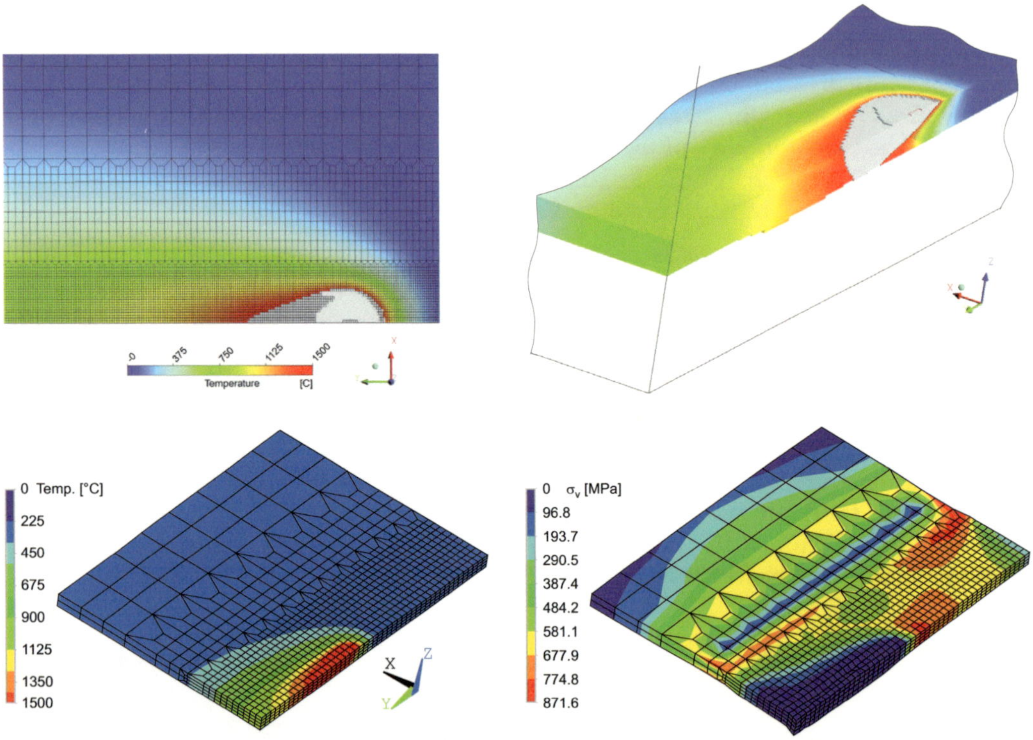

Fig. 35. *1st row*: Temperature distribution calculated with ANSYS CFX®. Here the moving heat source is applied within the boundary layer air/steel. *2nd row (left)*:Temperature distribution imported from CFX® to ANSYS® classic. *2nd row (right)*: Calculated VON MISES stress distribution during welding.

7. Conclusion and summary

In the present work the laser welding process, usually found during continuous manufacturing of steels, was investigated from the experimental and numerical point-of-view. We restricted to the example of multiphase, as-rolled steels and presented the following items in detail:

1. Explanation of the continuum mechanical and thermodynamical framework required for modelling of welding;
2. Introduction into the mathematical, numerical framework typically used for FE-simulations;
3. Derivation of all required materials data, including experimental testing and optimization by minimizing the mean-square deviation;
4. Presentation of thermal and mechanical FE-simulations of the laser welding process for TRIP 700 (as-rolled state);

5. Performing of welding experiments and comparison of the experimental and theoretical results; and
6. Discussion of continuative, welding-related phenomena and tasks for future investigations.

Great importance is attached to the determination of required materials data, since simulations can only be as reliable as the underlying input data. Here all caloric materials data were taken from literature or databases. Only mechanical data, in particular temperature depending stress-strain curves, are rarely available. Therefore one-dimensional tensile tests were performed at different temperatures by using the Gleeble3500 equipment.

All simulations were realized by means of the FE-program package ANSYS®. For the experimental investigations material were removed from production and joined at the laser test bench under manufacturing conditions. The comparison of the experimental and theoretical results shows partially excellent agreements; only the calculated lateral distortions are overestimated but also qualitative reproduce the experimental curve. Phase transformations during welding are not considered within the simulations. Although micro-structural changes do not appear to strongly influent the mechanical response of TRIP 700, this phenomenon is worthy to model in future work. Especially, the finite cooling time during welding lead to a shift of jumps of the (stationary) curves in Figure 10. Additionally micro-morphological changes (e.g. grain coarsening within the HAZ) and partial trans-formations occur. These changes may not affect the mechanical response but yield modified materials properties, such as loss of hardness, which could be relevant for subsequent materials processing.

8. Appendix: Thermal boundary conditions

A. Radiation and natural convection

We start with the established relation for heat transfer via convection over the boundary of the body:

$$\underline{\mathbf{q}}_n = \alpha A(T - T_0) \cdot \underline{\mathbf{n}} \,. \tag{A1}$$

For radiation the heat transfer is quantified by the STEFAN-BOLTZMANN law, viz.:

$$\underline{\mathbf{q}}_S = \sigma_{SB}\epsilon A(T^4 - T_0^4) \cdot \underline{\mathbf{n}} \,. \tag{A2}$$

By combining the equations one relation for both effects follows, which contains an effective heat transfer coefficient α_{eff}:

$$\underline{\mathbf{q}}_{n,\text{tot}} = \underline{\mathbf{q}}_n + \underline{\mathbf{q}}_S = \alpha_{\text{eff}}(T)A(T - T_0) \cdot \underline{\mathbf{n}} \quad , \quad \alpha_{\text{eff}}(T) = (\alpha + \sigma_{SB}\epsilon)\frac{T^4 - T_0^4}{T - T_0} \tag{A3}$$

with $\alpha = 5.0$ W/(m^{2}K): convective heat transfer coefficient[25]; $\sigma_{SB} = 5.67 \cdot 10^{-8}$ W/(m^{2}K^4): STEFAN-BOLTZMANN constant; $\epsilon = 0.5$: emission coefficient[25]; A: area of radiation; $\underline{\mathbf{n}}$: normal vector; T_0: reference temperature.

[25] These values are ad-hoc assumed according to standard literature.

B. Forced convection

Figure 25 (left) illustrates the supply of shielding gas (Helium) via a capillary tube contrary to the joining direction during welding. By assuming a setting angle of $\alpha = 45°$, tube diameter of $D = 1.8$ mm and the typical volume flux of $\dot{V} = 10$ l/min the following Helium velocity parallel to the sheet surface result, see also Figure 36:

$$v_x = v_s \sin(\alpha) = \frac{4\dot{V}}{\pi D^2} \sin(\alpha) = 46.3\,\text{m/s}\,. \tag{A4}$$

By means of the kinematic viscosity $\nu = 104.5$ mm^2/s (Helium), the characteristic length $L = 50$ mm, the conductibility of temperature $a = 153 \cdot 10^{-6}$ m^2/s and the heat conductivity of fluid $\kappa_{He} = 0.14$ W/(mK) the following dimensionless numbers hold:

$$\text{Re} = \frac{v_x L}{\nu} = 22.2 \quad \text{and} \quad \text{Pr} = \frac{\nu}{a} = 0.68\,. \tag{A5}$$

The REYNOLDS number Re characterizes the turbulent behaviour of the fluid flow, whereas the PRANDTL number describes the thickness of the temperature boundary layer. In the present case laminar conditions hold; consequently we use the relation, (Grothe & Feldhusen, 2007):

$$\text{Nu} = \frac{\alpha L}{\lambda} = 0.332\,\sqrt{\text{Re}}\,\sqrt[3]{\text{Pr}}\,. \tag{A6}$$

By using Eq. (A6) the modified heat transfer coefficient $\alpha = 2300$ W/(m^{2}K) directly follows. This value holds for the joining regions at the upper surface during welding. During cooling the supply of yielding gas is cancelled, thus natural convection including radiation takes place, cf. Figure 36.

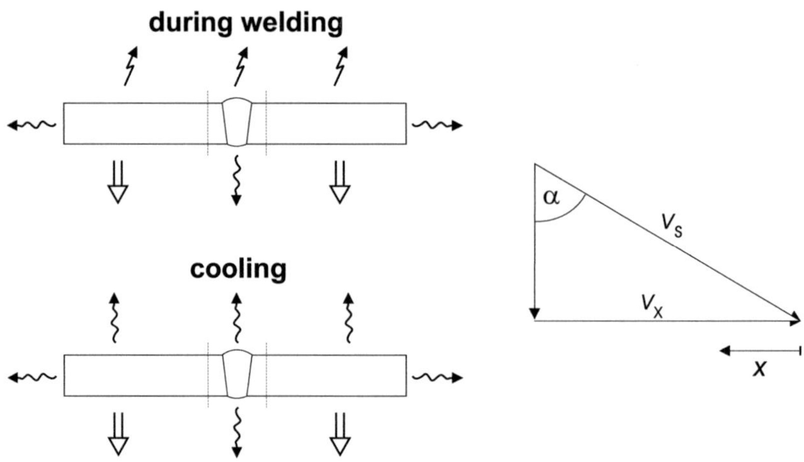

Fig. 36. Different thermal boundary conditions during welding and cooling (the used symbols are explained in Figure 8). *Right*: velocity triangle to determine the parallel contribution v_x.

C. Heat conduction by thermal contact

Beyond the lateral distance from the welding seam of $3-4$ mm the lower side of the sheet lies on a steel surface. Thus the thermal contact steel-to-steel must be considered. Such situation is typically described within the literature by means of the quantity of Thermal Contact Resistance, R_{TC}, or conductance, C_{TC}. The following relation holds:

$$\underline{\mathbf{q}} = \frac{A\,\Delta T}{R_{TC}} \cdot \underline{\mathbf{n}} \quad \text{and} \quad C_{TC} = \frac{1}{R_{TC}} \,. \tag{A7}$$

Values for C_{TC} and R_{TC} can be found in literature for typical materials combinations, but note, there is a strong dependence to the contact pressure. Obviously the heat transfer coefficient α can be directly replaced by C_{TC} in case of thermal contact. Based on (Shojaefard & Goudarzi, 2008) the following value is used in the simulations:

$$R_{TC} = 0.33\,\frac{\mathrm{m^2K}}{\mathrm{kW}} \quad \Rightarrow \quad \alpha = \frac{1}{R_{TC}} \approx 3000\,\frac{\mathrm{W}}{\mathrm{m^2K}} \,. \tag{A8}$$

9. Acknowledgement

The technical expertise of M. Maurer and T. Hagemann guaranteed the successes of the experimental work. Furthermore fruitful discussions with K. Göhler, M. Dubke, C. Trachternach, M. Koch and S. Wischmann (ThyssenKrupp Steel Europe, Duisburg, Germany); A. Junk and C. Groth (CADFEM Deutschland GmbH, Hannover, Germany); O. Voss (ThyssenKrupp Tailored Blanks S.A. de C.V. Puebla, Mexico); W. Schulz (Fraunhofer Institute for Laser Technology, Aachen, Germany), and R. Ossenbrink (Chair of Joining and Welding Technology, Brandenburg University of Technology, Cottbus, Germany) are gratefully acknowledged.

10. References

ANSYS® Release 11.0 SP1 (2007). Copyright SAS IP Inc., Canonsburg.

ANSYS Inc. (2007). *Theory Reference for ANSYS and ANSYS Workbench.* ANSYS Release 11.0, SAS IP Inc., Canonsburg.

Avrami, M. (1939). Kinetics of Phase Change. I General Theory. *J. Chem. Phys.*, Vol. 7, 1103–1112.

Avrami, M. (1940). Kinetics of Phase Change. II Transformation-Time Relations for Random Distribution of Nuclei, *J. Chem. Phys.*, Vol. 8, 212–224.

Barth, G. & Groß, U. (2004). *Untersuchung der thermophysikalischen Stoffeigenschaften von Stahl bei Temperaturen bis zu 1600 °C*, report of the TU Bergakademie Freiberg, Freiberg, Germany.

Cahn, J.W. & Hilliard, J.E. (1958). Free energy of a Nonuniform System. I. Interfacial Free Energy. *J. Chem. Phys.*, Vol. 28, No. 2, 258–267.

Cahn, J.W. (1961). On Spinodal Decomposition. *Acta Metall.*, Vol. 9, 795–801.

Darken, L.S. & Smith, R.P. (1949). Behavior of hydrogen in steel during and after immersion in acid. *Corrosion*, Vol. 5, 1–16.

De, A.; Maiti, S.K.; Walsh, C.A. & Bhadeshia, H.K.D.H. (2003). Finite element simulation of laser spot welding. *Sci. Technol. Weld. Joi.*, Vol. 8, No. 5, 377–384.

Dean, D.; Yu, L.; Hisashi, S.; Masakazu, S. & Hidekazu, M. (2003). Numerical simulation of residual stress and deformation considering phase transformation effect. *Trans. JWRI*, Vol. 32, No. 2, 325–333.

Dean, D. & Hidekazu, M. (2006). Prediction of welding residual stress in multi-pass butt-welded modified 9Cr-1Mo steel pipe considering phase transformation effects. *Comp. Mater. Sci.*, Vol. 37, 209–219.

Dean, D.; Hidekazu, M. & Wei, L. (2008). Numerical and experimental investigations on welding residual stress in multi-pass butt-welded austenitic stainless steel pipe. *Comp. Mater. Sci.*, Vol. 42, 234–244.

Dean, D. & Hidekazu, M. (2008). Finite element analysis of temperature field, microstructure and residual stress in multi-pass butt-welded 2.25Cr-1Mo steel pipes. *Comp. Mater. Sci.*, Vol. 43, 681–695.

DIN EN 10002-1 (2001). Zugversuch Teil 1: Prüfverfahren bei Raumtemperatur. Deutsches Institut für Normung e.V., Beuth Verlag GmbH, Berlin.

Dong, Z.B. & Wei, Y.H. (2008). Three dimensional modeling weld solidification cracks in multipass welding. *Theor. Appl. Fract. Mech.*, Vol. 46, 156–165.

Elmer, J.W.; Wong, J. & Ressler, T. (2001). Spatially Resolved X-Ray Diffraction Mapping of Phase Transformations in the Heat-Affected Zone of Carbon-Manganese Steel Arc Welds. *Metall. Mater. Trans. A*, Vol. 32A, 1175–1187.

Fan, H.G. & Kovacevic, R. (1999). Droplet formation, detachment, and impingement on the molten pool in gas metal arc welding. *Metall. Mater. Trans. B*, Vol. B30, 791–801.

Feldmann, C. (2008). *Thermoelemente*, talk manuscript, University of Applied Scienes and Arts Dortmund, url: http://ln.iuk.fh-dortmund.de/~gebhard/STA/vortraege/ (last access: Mai 2011).

Francis, J.A.; Bhadeshia, H.K.D.H. & Withers, P.J. (2007). Welding residual stresses in ferritic power plant steels. *Mater. Sci. Tech. Ser.*, Vol. 23, No. 9, 1009–1020.

Gross, D. & Seelig, T. (2006). *Fracture Mechanics*, Springer, ISBN: 978-3-540-37113-7, Berlin Heidelberg New York.

Grothe, K.-H. & Feldhusen, J. (2007). *Dubbel Taschenbuch für den Maschinenbau*, 22nd edition, Springer, ISBN 978-3-540-49714-1, Berlin Heidelberg New York, page D32 ff..

Hemmer, H. & Grong, Ø. (1999). A process model for the heat-affected zone microstructure evolution in duplex stainless steel weldments: Part I. the model. *Metall. Mater. Trans. A*, Vol. 30A, 2915–2929.

Hemmer, H.; Grong, Ø & Klokkehaug, S. (2000). A process model for the heat-affected zone microstructure evolution in duplex stainless steel weldments: Part II. Application to electron beam welding *Metall. Mater. Trans. A*, Vol. 31A, 1035–1048.

JMatPro 6.0. (2011), Sente Software Ltd., Surrey Technology Centre, 40 Occam Road, GU2 7YG, United Kingdom.

Junk, A. & Groth, C. (2004). Schweißprozesssimulation mit dem SST, Integration neuer Ansätze und bewährter Teilmodelle in die FE-Simulation des Laserstrahlschweißens. *Proceedings of the 22nd CAD-FEM Users' Meeting*, 13 pages, ISBN: 3-937523-01-4, Dresden, November 2004, CADFEM, Grafing.

Kang, S.-H. & Im, Y.-T. (1987). Three dimensional thermo-elastic-plastic finite element modeling of quenching process of plain-carbon steel in couple with phase transformation. *Int. J. Mech. Sci.*, Vol. 49, 423–439.

Khiabani, A.C. & Sadrnejad, S.A. (2009). Finite element evaluation of residual stresses in thick plates *Int. J. Mech. Mater. Des.*, Vol. 5, No. 3, 253–261.

Ki, H.; Mohanty, P.S. & Mazumder, J. (2002). Modeling of laser keyhole welding: Part I. mathematical modeling, numerical methodology, role of recoil pressure, multiple reflections, and free surface evolution. *Metall. Mater. Trans. A*, Vol. 33A, No. 6, 1817–1830.

Ki, H.; Mohanty, P.S. & Mazumder, J. (2002a). Modeling of laser keyhole welding: Part II. simulation of keyhole evolution, velocity, temperature profile, and experimental verification. *Metall. Mater. Trans. A*, Vol. 33A, No. 6, 1831–1842.

Koistinen, D.P. & Marburger, R.E. (1959). A general equation prescribing the extent of the austenite-martensite transformation in pure iron-carbon alloys and plain carbon steels. *Acta Metall.*, Vol. 7, 59–60.

Komanduri, R. & Hou, Z.B. (2000). Thermal Analysis of the Arc Welding Process: Part I. General Solutions. *Metall. Mater. Trans. B*, Vol. 31B, 1353–1370.

Kotowski, J. (1998). *Experimentelle und numerische Ermittlung der Energieeinkopplung während des Laserstrahlschweiï£¡ens von Tailored Blanks*, diploma thesis, TU Braunschweig.

Krom, A.H.M. & Bakker, A.D. (2000). Hydrogen Trapping Models in Steel. *Metall. Mater. Trans. B*, Vol. 31B, 1475–1482.

Leblond, J.B. & Devaux, J. (1984). A new kinetic model for anisothermal metallurgical transformations in steels including effect of austenite grain size. *Acta Metall.*, Vol. 32, No. 1, 137–146.

Lee, P.D.; Quested, P.N. & McLean, M. (1998). Modelling of Marangoni Effects in Electron Beam Melting. *Phil. Trans. R. Soc. Lond. A*, Vol. 356, 1027–1043.

Maroef, I.; Olson, D.L.; Eberhart, M. & Edwards, G.R. (2002). Hydrogen trapping in ferritic steel weld metal. *Int. Mater. Rev.*, Vol. 47, No. 4, 191–223.

McNabb, A. & Foster, P.K. (1963). A New Analysis of the Diffusion of Hydrogen in Iron and Ferritic Steels. *Trans. Metall. Soc. AIME*, Vol. 227, 618–627.

Mills, K.C.; Keene, B.J.; Brooks, R.F. & Shirali, A. (1998). Marangoni Effects in Welding. *Phil. Trans. R. Soc. Lond. A*, Vol. 356, 911–925.

Müller, I. (1984). *Thermodynamics*, Pitman Publishing Inc., ISBN: 0-273-08577-8, London.

Müller, I. (1999). *Grundzüge der Thermodynamik*, Springer (2nd edition), ISBN: 3-540-64703-1, Berlin Heidelberg New York.

Müller, I. & Müller, W.H. (2009). *Fundamentals of Thermodynamics and Applications*, Springer, ISBN: 978-3-540-74645-4, Berlin Heidelberg.

Oriani, R.A. (1970). The diffusion and trapping of hydrogen in steel. *Acta Metall.*, Vol. 18, 147–157.

Ossenbrink, R. & Michailov, V. G. (2007). *Thermomechanical Numerical Simulation with the Maximum Temperature Austenisation Cooling Time Model (STAAZ)*, in: Mathematical Modelling of Weld Phenomena 8 (ed. Cerjak, H.H.), Verlag der Technischen Universität Graz, Graz (Austria), ISBN: 978-3-902465-69-6, 357–372.

Ossenbrink, R. (2008). *Thermomechanische Schweißsimulation unter Berücksichtigung von Gefügeumwandlungen*, PhD thesis, Brandenburg University of Technology, Shaker Verlag, ISBN: 978-3-8322-8131-1, Aachen.

Phanikumar, G.; Dutta, P. & Chattopadhyay, K. (2004). Computational modeling of laser welding of Cu-Ni dissimilar couple. *Metall. Mater. Trans. B*, Vol. 35B, No. 2, 339–350.

Pfeiffer, S. & Schulz, W. (2009). Structural validation of different approaches for the modelling of welding distortion of steel-based parts. Final Report, project no. P708, FOSTA - Forschungsvereinigung Stahlanwendung e.V., Düsseldorf.

Richter, F. (1973). *Die wichtigsten physikalischen Eigenschaften von 52 Eisenwerkstoffen.* Mitteilungen aus dem Forschungsinstitut der Mannesmann AG, Stahleisen - Sonderberichte, Heft 8, Stahleisen Dï£¡sseldorf m.b.H., Düsseldorf.

Serebrinsky, A.; Carter, E.A. & Ortiz, M. (2004). A quantum-mechanically informed continuum model of hydrogen embrittlement. *J. Mech. Phys. Solids*, Vol. 52, 2403–2430.

Shojaefard, M.H. & Goudarzi, K. (2008). The Numerical Estimation of Thermal Contact Resistance in Contacting Surfaces. *American Journal of Applied Sciences*, Vol. 5, No. 11, 1566–1571.

Sloan, S.W. (1987). Substepping schemes for the numerical integration of elastoplastic stress-strain relations. *Int. J. Numer. Meth. Eng.*, Vol. 24, 893–911.

Spina, R.; Tricarico, L.; Basile, G. & Sibillano, T. (2007). Thermo-mechanical modeling of laser welding of AA5083 sheets. *J. Mater. Process. Technol.*, Vol. 191, 215–219.

Stouffer, D.C. & Dame, L.T. (1996). *Inelastic Deformations of Metals*, John Wiley & Sons Inc., ISBN: 0-471-02143-1, New York.

Thiessen, R.G.; Sietsman, J.; Palmer, T.A.; Elmer, J.W. & Richardson, I.M. (2007). Phase-field modelling and synchrotron validation of phase transformations in martensitic dual-phase steel. *Acta Mater.*, Vol. 55, 601–614.

Tian, Y.; Wang, C.; Zhu, D. & Zhou, Y. (2008). Finite element modeling of electron beam welding of a large complex Al alloy structure by parallel computations. *J. Mater. Process. Technol.*, Vol. 199, 41–48.

Timmins, P.F. (1997). *Solutions to Hydrogen Attack in Steels.* ASM International, Materials Park , OH (USA), ISBN: 0-87170-597-4.

Umemoto, M.; Horiuchi, K. & Tamura, I. (1982). Transformation Kinetics of Bainit during Isothermal Holding and Continuous Cooling. *Trans. Iron Steel Inst. Jpn.*, Vol. 22, 854–861.

Umemoto, M.; Horiuchi, K. & Tamura, I. (1983). Pearlite Transformation during Continuous Cooling and Its Relation to Isothermal Transformation. *Trans. Iron Steel Inst. Jpn.*, Vol. 23, 690–695.

Voss, O. (2001). *Untersuchung relevanter Einflussgrößen auf die numerische Schweißsimulation*, PhD thesis, TU Braunschweig, Shaker Verlag, ISBN: 3-8265-9119-4, Aachen.

Wang, Y.; Shi, Q. & Tsai, H.L. (2001). Modeling of the effects of surface-active elements on flow patterns and weld penetration. *Metall. Mater. Trans. B*, Vol. 32B, 145–161.

Wang, Y. & Tsai, H.L. (2001). Effects of surface active elements on weld pool fluid flow and weld penetration in gas metal arc welding. *Metall. Mater. Trans. B*, Vol. 32B, 501–515.

Wriggers, P. (2000). *Nichtlineare Finite-Elemente-Methoden*, Springer, ISBN: 3-540-67747-X, Berlin Heidelberg New York.

Zener, C. (1949). Theory of Growth of Spherical Precipitates from Solid Solution. *J. App. Phys.*, Vol. 20, 950–953.

Zhou J. & Tsai, H.L. (2008). Modelling of Transport Phenomena in Hybrid Laser-MIG Keyhole Welding. *Int. J. Heat Mass Transf.*, Vol. 51, 4353–4366.

Part 2

Biomaterials

6

Collagen - Modified Layered Silicate Biomaterials for Regenerative Medicine of Bone Tissue

Zina Vuluga[1], Catalina-Gabriela Potarniche[2], Madalina Georgiana Albu[3], Viorica Trandafir[4], Dana Iordachescu[5] and Eugeniu Vasile[5]

[1]National Research and Development Institute for Chemistry and Petrochemistry-ICECHIM, Bucharest

[2]Department of Mechanical and Manufacturing Engineering Aalborg University, Aalborg,

[3]National Research and Development Institute for Textile and Leather, Bucharest,

[4]University of Bucharest, Research Center for Biochemistry and Molecular Biology, Bucharest,

[5]METAV CD, Bucharest,

[1,3,4,5]Romania

[2]Denmark

1. Introduction

The interdisciplinary field of polymer nanocomposite biomaterials brings together researchers from polymer science, biology, materials and biomedical engineering, chemistry and physics. In order to understand the need of biomaterials in our life, first we need to define and understand the specific terms that link these fields together. It is obvious that, due to the fact that the field of biomaterials is quite a new interdisciplinary one, the same term can be found with several explanations.

"The application of the principles of natural sciences, especially of biology, biochemistry and physiology to clinical medicine" was defined as **biomedicine** (Miller & O'Toole, 2005, as cited in Medical Dictionary, ndc; Picket, 2000, as cited in Medical Dictionary, ndc).

Collaboration between fields leads to the generation and use of new terminology and definitions that are used across each discipline in part. For example, the traditional definition of biomaterials has changed in time (Wu et al., 2010) as these materials find their use in a variety of medical and nonmedical technologies that are inspired by biology (Huebsch & Mooney, 2009; Williams, 2009; Wu et al., 2010). In 1987, The European Society of Biomaterials (Williams, 1987) defines the biomaterial as "a non viable material used in a medical device, intended to interact with biological systems." Later on, in 1999 (Williams, 1999), in the dictionary of biomaterial science, the definition of biomaterial was changed, saying that it represents "a material intended to interface with biological systems to evaluate, treat, augment or replace any tissue, organ or function of the body" (Williams,

2009). In a 2003 medical encyclopedia (Miller & O'Toole, 2005 as cited in Medical Dictionary, ndb) the authors combine the 1999 and 2002 definitions saying that the biomaterial can be defined as "any substance (other than a drug), synthetic or natural, that can be used as a system or part of a system that treats, augments, or replaces any tissue, organ, or function of the body; especially, material suitable for use in prostheses that will be in contact with living tissue". In 2005, modern medicine (Segen, 2005 as cited in Medical Dictionary, ndb) gives two definitions for biomaterial as being - "1. any synthetic material or device - e.g. implant or prosthesis - intended to treat, enhance or replace an aging or malfunctioning–or cosmetically unacceptable — native tissue, organ or function in the body, bioengineering, breast implants, hybrid artificial pancreas, Shiley valve, teflon, total hip replacement and 2. a biomaterial used for its structural, not biological properties - e.g., collagen in cosmetics, carbohydrates modified by biotechnology to be used as lubricants for biomedical applications or as bulking agents in food industry." The latest definition in 2009 (Williams, 2009), which includes all the previous ideas, states that "A biomaterial is a substance that has been engineered to take a form which, alone or as part of a complex system, is used to direct, by control of interactions with components of living systems, the course of any therapeutic or diagnostic procedure, in human or veterinary medicine." Facing this changes in time, makes it difficult to classify polymer nanocomposite biomaterials into materials which are intended for medical devices and which implies their use for bio-technological purposes (e.g., renewable resources) (Wu et al., 2010). Polymeric biomaterials are developed in different shapes and for different purposes in order to substitute and repair biological tissues (Hench & Polak, 2002; Ma & Elisseeff, 2005, as cited in Wu et al., 2010). New attempts to design self-assembled and smart nanocomposite biomaterials are conducted to fulfil some external requirements such as optic, temperature, mechanic, electric and magnetic fields (Hirst et al., 2008; Peppas et al., 2006; Stuart et al. 2010). These requirements are very important when designing smart implants and drug delivery systems as well as new bio-technologies including biosensors, *in vitro* diagnostics, cell culture matrices, contrast agents and bioassays (Hirst et al., 2008; Stuart et al. 2010; Wu et al., 2010).

Initially the implants were intended mainly to replace affected (i.e., broken bones, heart, vein, liver, pancrease) or malfunctioning body parts which people cannot live without. Subsequently, the industry of implants developed fast and now there are implants which are used for reconstructing and aesthetic purposes such as breast implants, bone facial implants and skin implants. According to the type of replaced tissue, they are divided into hard and soft implants. Hard tissue implants are usually made of metals or ceramics (bone implants, dental sealant, and joints) while soft tissue implants are made of polymers (blood vessels, skin, plastic surgery).

Essential components of modern medicine have become orthopaedic implants (represented by metallic implants as well as bone cements) (Widmer, 2001).

In recent years, orthopaedic implants and prostheses have imposed an increasing demand on the materials used. The more complex the function of the implant becomes, the more complex the requirements of the constructional material are. The success or failure of any device is often dependent on the choice of material as it is on the configurational and functional design.

The requirements that modern bone implants are expected to meet can be divided into three groups:

1. The biocompatibility between the material and the surrounding environment.
2. The mechanical and physical properties necessary to achieve the desired function.
3. The relative ease of fabrication and supply of the required components (Hoeppner & Chandrasekaran, 1994).

These requirements derive from the bone complex structure and properties. Bone is a complex and dynamic living tissue, which from morphological point of view can be divided into two types: cortical or compact bone and trabecular or cancellous or spongy bone. The mass of the skeleton is composed of 80% compact bones and 20% spongy bones. Cortical and trabecular bones contain the same cells and extracellular matrix (ECM) components, except that they are organized in a different way. The trabecular bone consists of a porous matrix with interconnected columns filled with bone marrow and is being responsible for metabolic functions of the bone, while the cortical bone contains fewer spaces, forms the external layer of all bones and provides them protection and load-bearing capabilities (Baron, 2003, as cited in Wilson, 2011).

Extracellular matrix is the most abundant component in bone tissue and is composed from 30% of an organic phase of collagen and of other proteins and from 70% of an inorganic phase of calcium phosphate. The organic phase provides elasticity and strength, while the inorganic phase gives hardness and load-bearing capabilities (Mundy et al. 2003, as cited in Wilson, 2011; Robey & Boskey, 2008, as cited in Wilson, 2011).

Remodelling is a physiological process in which the bones are being built, resorbed and then rebuilt again and that process is carried out and carefully controlled by a variety of cell types. Osteoblasts synthesize and mineralize the bone matrix which is maintained by osteocytes, and as required, is resorbed by osteoclasts (Wilson, 2011).

A biomaterial to be used as scaffold in tissue engineering must satisfy a number of requirements. These requirements include biocompatibility, biodegradation to non-toxic products within the period required for the application, processability to complicated shapes with appropriate porosity, ability to support cell growth and proliferation, ability to incorporate cells, growth factors, appropriate mechanical properties, as well as maintaining mechanical strength during the tissue regeneration process. All these requirements need to be fulfilled by implanted materials in order to achieve "biointergration". However, the term of biointegration is not completely accepted by researchers in the medical field (Szmukler-Moncler & Dubruille, 1990). Dubruille et al. did not agree to define biointegration as "describing the direct biochemical bonding of bone to an implant surface". For this situation they propose a new term "osseointegration" since the "biointegration" does not refer only to the integration with bone. Even though, we will find that biointegration term is still used and universally accepted as referring to the bioactivity of implanted material with bone tissue (Blokhuis et al., 2000).

Biocompatibility was defined as being "the property of being biologically compatible by not producing a toxic, injurious, or immunological response in living tissue" (Picket, 2000, as cited in Medical Dictionary, nda).

In biology, the term "cell growth" is used in two different ways. In the context of reproduction of living cells the phrase "cell growth" is used for the idea of cell reproduction meaning growth in cell numbers. During cell reproduction the "parental" cell divides into two "daughter" cells. In other contexts, "cell growth" refers to increases in cell size (Conlon &

Raff, 1999). The first definition of the cell growth is also attributed to the word "proliferation" which is defined as "to grow or multiply by rapidly producing new tissue, parts, cells, or offspring" (Picket, 2000, as cited in Medical Dictionary, ndd) when we refer only to cells. So, in material science in order to differentiate between the two concepts, we refer to cell growth when the size of the cell increases and use the term proliferation when the number of the cells increases.

Another important issue that needs to be considered, when obtaining an implant, is body response. Upon implantation, cell adhesion and fibrous tissue encapsulation takes place as a defensive response of the body (Recum & Strauss, 1998, as cited in Shin & Schoenfisch, 2006; Wilson & Gifford, 2005). The wound healing response initiates as an immune response. There were differentiated four distinct stages of the wound healing which are met starting after few seconds until complete healing: hemostasis, inflammation, repair, and encapsulation. Bacterial infection that can occur due to bacterial adhesion is a serious problem that affects implanted materials during healing or in the case of extended periods of time (Dankert et al., 1986, as cited in Shin & Schoenfisch, 2006; Sawan & Manivannan, 2000, as cited in Shin & Schoenfisch, 2006). In order to overcome this problem, strategies are being developed based on control release of an anticoagulant or antimicrobial agent from implanted material (Christiansen et al., 2011; Hendricks et al., 2000; Shin & Schoenfisch, 2006; Williams, 2009).

Drug delivery systems (DDS) are new types of drug carriers, designed to improve the pharmacological and therapeutic properties of administrated drugs (Allen & Cullis, 2004). Such systems are designed according to the type of administration. There are transdermal drug delivery systems (TDDS) used for external administration of drugs and liposomal and targeted drug delivery systems (LTDDS) for internal administration. TDDS present the advantages of avoiding hepatic first pass metabolism, of decreasing gastrointestinal effect and improving compliance, while LTDDS are used to reduce the dose and increase the activity of the drug at the precise site (Kshirsagar, 2000). DDS can be classified after their release mechanism into (Lager & Peppas, 1981):

1. Diffusion Controlled Systems
   a. Reservoirs (membrane devices)
   b. Matrices (monolithic devices)
2. Chemically Controlled Systems
   c. Bioerodible systems
   d. Pendant Chain systems
3. Swelling Controlled Systems
4. Magnetically Controlled Systems

The dose released can differ according to the type of the system used and can be: extended release, slow release and sustained release (Kshirsagar, 2000).

Generally, DDS present zero-order kinetics and they should be inert and biocompatible with the environment. As compared to the conventional way of drug administration, DDS have the advantages of being more comfortable for the user, easily administrated, possess a high drug concentration, have good mechanical strength, safe and free from leaks. They also have biocompatible, non-toxic, nonmutagenic, non-carcinogenic, non-teratogenic and nonimmunogenic properties (Lager & Peppas, 1981).

One of the mostly researched biomaterials used in various fields of medicine is natural polymer-collagen. Collagen is the most abundant protein in the body (skin, bones, teeth, tendons, cartilage, basement membrane, cornea, etc.). It can be used by different processing techniques in large various molecular structures (micro and nanostructures) as powder, injectable solutions, films, membrane and matrices (sponges).

The use of collagen as biomaterial, biocompatible and bioresorbable for the connective tissues prosthesis in which collagen is the basic protein is very well known.

To use collagen as scaffold in the recovery of bone tissue modifications in the structure and composition of matrix we need to obtain osteoinductive and osteoconductive effect. This can be achieved by the formation of biocomposites with SiO_2, TiO_2, clay, hydroxyapatite, etc (Christiansen et al., 2011; Vuluga et al., 2008a). Osteoinduction is the effect that appears in bone healing and involves the recruitment and stimulation of immature cells to develop into preosteoblast cells. Osteoconduction refers to bone growth on an implanted material, including bone conduction on its surface or down into pores, channels or pipes (Albrektsson & Johansson, 2001). Osteoconductive effect depends not only on biological factors, but also on the response to a foreign material. Bone conduction is not possible on certain materials, like some metals (Cu, Ag) (Albrektsson, 1995, as cited in Albrektsson & Johansson, 2001), while osteoinductive effect was observed on a large number of biomaterials (Barradas et al., 2011).

The layered silicate improves the thermal and mechanical properties of collagen and has a biostimulative effect on cellular metabolism. The structure of natural silicate would be like the structure of the living tissue because of the ionic exchange properties in aqueous medium just like the physiological phenomena in the human body where permanent information exchanges are present. Silicates surface can be modified with different bioactive substances and obtained nanohybrids can be dispersed in a collagen matrix (Potarniche et al., 2010; Potarniche et al. 2011). Depending on the concentration and the method of the silicate incorporation in the collagen protein, nanostructured systems with different morphologies, capable to control release of active principles (antibiotics, anti-inflammatory, enzymes, etc.) can be obtained. These types of nanocomposites can be used in regenerative medicine of bone tissue (Christiansen et al., 2011).

Polymer/inorganic nanocomposites are materials with unique properties which have found applications in large fields of activity: aircraft industry, automobile, packaging, construction, electronics, electrical engineering and last but not least in medicine and pharmaceutical industry. To obtain nanocomposites hybrids, the clay minerals (hydrated layered silicates) are preferred due to their high mechanical and chemical strength and low cost. The biomedical properties of clays are known since antiquity, these being in principal antiseptic, bactericides, scar action, antitoxic, without microbial germs. The special surface physical and chemical properties, such as high specific surface, layer charge, swelling capacity, make clay and clay minerals to be useful for human health in both external and "per os" (internal) applications. Clays can be used as active principles or excipients in pharmaceutical formulas, cosmetic medicine, in treating intestinal disorders and as drugs, being good adsorbents and mucostabilizers (Choy et al., 2007).

In medicine, synthetic polymers are also frequently used, not because of their biological activity, but because they are relatively easy to obtain, they are inert to body fluids, they are cheap and available. From the synthetic polymers used in biomedicine, the copolymers of maleic anhydride (MA) were selected, because of their "drug-delivery" properties. In the field of medical and pharmaceutical applications, MA copolymers are used as components of some polymer - drug systems with controlled release (Gupta et al., 2002). Using MA copolymers for pharmaceutical and biomedical applications has advantages related to the structure, activity and their properties. MA copolymers have, in general, a relatively well characterized and reproducible alternating structure, hydrophilic or hydrophobic character and charge density can be varied by appropriate choice of comonomer (comonomers) (Edlund & Albertsson, 2002). Although polyanhydrides are best to the drug delivery application, they were not used in orthopedic implantation due to low load bearing and mechanical properties. However, MA copolymers with considerable enhanced mechanical properties were specially designed for the orthopedic applications such as poly (anhydride-e-imides) which can be used as scaffold for bone tissue engineering application (Sabir et al., 2009).

In this chapter, we intend to present the obtaining and properties of new types of nanocomposites based on natural polymer (collagen) and bioactive compounds (natural layered silicate modified with maleic copolymers) which can be used as biomaterial, scaffold for bone tissue regeneration.

Binary nanocomposites of layered silicate/maleic copolymers can contain different bioactive substances which can be released in time and at low and controlled concentrations (bone growth factors, titania or carbon aerogels, hydrolyzed collagen, anticancer compounds, antimicrobial antioxidants, etc.).

Obtaining of collagen/modified layered silicate nanocomposites implies:

- layered silicate purification for usage in the biomedical field;
- modification of the layered silicate surface (organophilization) with maleic copolymers thus obtaining binary nanocomposites;
- uniform dispersion of binary nanocomposites in collagen gel thus obtaining ternary nanocomposites from which, by lyophilization, spongy matrices can be obtained.

The binary nanocomposites have high thermal stability which can allow the usage of some bioactive substances activated by local heating. These nanocomposites are compatible with collagen gel in such away biocompatible ternary nanocomposites matrices useful for regenerative medicine of bone tissue can be obtained.

2. Materials and characterization methods

2.1 Materials

2.1.1 Collagen gel

Collagen gel (CG) is a polydisperse colloidal system which consists in molecules with native triple helix conformation and it is extensively used to obtain collagen-based biomaterials. Collagen gel is defined as a system with intermediate properties which vary between a liquid and a solid. It is characterized by structural viscosity, pseudoplastic behaviour and its average molecular weight is about 300000 Da, a value corresponding to the native collagen molecule weight (Trandafir et al., 2007).

Mild chemical and/or enzymatic treatments allow the extraction of collagen from calf hide into aqueous medium with preservation of triple helical structure of molecules, of microfibrils and fibrils. The collagen obtained is a gel, type I fibrillar collagen.

Type I collagen gel at a concentration of 2.3% and pH = 3 is obtained from calf hide by the current technology used in the Collagen Department in Leather and Footwear Research Institute, Bucharest, Romania (Albu, 2011; Budrugeac et al., 2003; Siapi et al., 2005). The obtained gel is physically-chemically characterized by 2.17% protein substance and 0.13% ash contents. The circular dichroism results show the technology used allows obtaining type I collagen with a high degree of purity and with triple helical structure (Albu, 2011). Absence of fat from this gel shows the high degree of purity. We use 1.2% gels to obtain collagen biomaterials.

2.1.2 Purified bentonite (natural layered silicate)

Purified bentonite (natural layered silicate) characterized by 82% montmorillonite (MMT), by basal spacing, d_{001} = 14.7 Å, by 8% weight loss in the range of temperature 20 ÷ 200 °C (TGA) and by $R_{1,0}$=0%; $R_{0,5}$=0,2%; $R_{0,063}$=65,8%; $R_{0,04}$=97,1%, $R_{0,025}$=100%, residue on sieve is obtained by purification of Romanian sodium bentonite.

2.1.3 Maleic copolymers

Maleic copolymers having variable hydrophobicity (maleic anhydride copolymers with a hydrophilic comonomer - vinyl acetate and a hydrophobic comonomer - methyl methacrylate), are obtained in anhydrous form by "Petru Poni" Institute of Macromolecular Chemistry, Romanian Academy, Iasi.

The monomers, catalysts, and solvents are carefully purified before usage, according to already known methods (Caze & Loucheux, 1975 as cited in Chitanu et al., 2006; Riddick & Burger, 1970 as cited in Chitanu et al., 2006).

The MA-VA copolymers of maleic anhydride (MA) with vinyl acetate (VA), (Chitanu et al., 2006) and MA-MMA copolymers with methyl methacrylate (MMA), (Chitanu et al., 2007) are synthesized by free radical polymerization, purified by extraction and washed with diethyl ether. The composition is estimated by conductometric titration with aqueous 0.1N NaOH in 1:1 acetone-water mixture and the molecular weight (M_w) by viscosimetric measurements in acetone at 30 °C.

The chemical structures of maleic copolymers used are presented in Fig. 1 and 2.

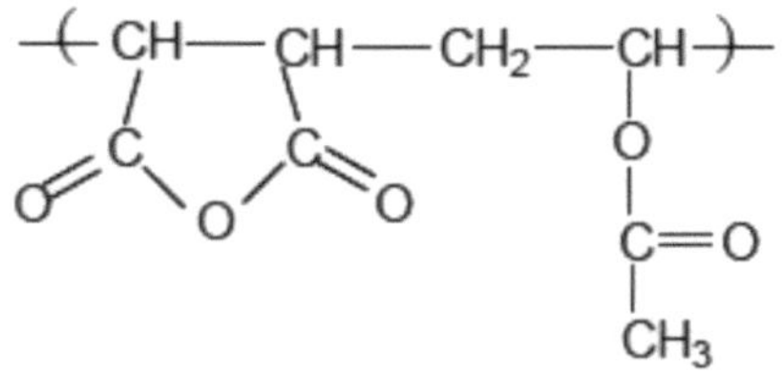

Fig. 1. Chemical structure of MA-VA copolymer.

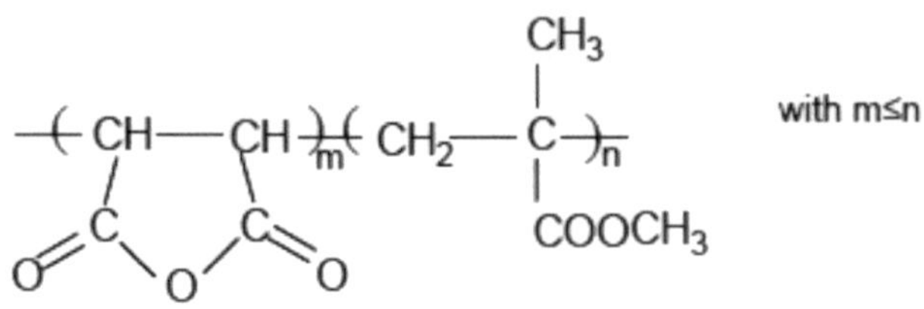

Fig. 2. Chemical structure of MA-MMA copolymer.

The characteristics of MA-MMA maleic copolymers with hydrophobic comonomer and of MA-VA maleic copolymers with hydrophilic comonomer are presented in Table 1.

Characteristics	MA-VA	MA-MMA
Average viscosimetric molecular weight, M_w	96000 g/mole	49000 g/mole
Acid index	0.481 g NaOH/ g copolymer	0.2535 g NaOH/ g copolymer
Composition	maleic anhydride : vinyl acetate 1 : 0.77	maleic anhydride : methyl methacrilate 1 : 2.17

Table 1. The characteristics of maleic copolymers.

2.2 Methods of physical-chemical and structural characterization

Physical-chemical and structural properties of purified layered silicate, of binary nanocomposites based on layered silicate/maleic copolymers and of ternary nanocomposites of collagen/layered silicate/maleic copolymers are determined by using different techniques described below.

2.2.1 X-ray diffraction

X-ray diffraction technique is used to determine the basal spacing. Difractograms are recorded on automatic system using a DRON-20 difractometer, with a horizontal goniometer and a scintillation counter. We used as radiation source, $CoK_{\bar{\alpha}}$ (λ = 1.7902 Å) filtered with Fe in order to remove k_{β} component from the Bragg-Brentano system (in reflexion mode). *Basal spacing* is determined using Bragg equation[1]:

$$n\lambda = 2d \sin\theta,\ d = \frac{n\lambda}{2\sin\theta} \quad (1)$$

where n represents the reflexion order, λ is the wavelength of X-rays, θ is the diffraction angle, d is the distance between the planes of crystalline network which produces the diffraction.

2.2.2 Scanning electron microscopy (SEM)

Nanocomposite particles shape and dimension are observed using a scanning electron microscope and microanalyzed with an energy dispersive X-ray spectrometer (EDAX). The samples are studied at different magnitude orders using a HITACHI scanning electron microscope model S-2600N equipped with an energy dispersive X-ray spectrometer (EDAX).

2.2.3 Dynamic diffusion of light with laser source (DLS)

Using a Zeta Seizer MALVERN UK, the average diameter of submicron particles is determined in aqueous suspension ranging 3 nm ÷1 µm.

2.2.4 Thermal analyses

Thermal behavior of the obtained materials is analyzed by determining the sample weight loss when heated at a constant rate (TGA), depending on temperature. By differential thermal analyses (DTA) we determine the temperature at which we get the highest decomposition rate. Thermal tests are performed on THERMAL ANALYST DUPONT 2100 using:

- heating rate: 20 °C/min;
- in air with the flow rate of 50 cm^3/min;
- temperature range: 0 ÷ 700 °C.

2.3 Biocompatibility

Biocompatibility properties are evaluated on already obtained binary/ternary systems using *in vitro* tests. Biocomposite matrices are cut into small samples (1 cm^2) and are sterilized in ultraviolete light on the both sides. The samples are sown with osteoblast cells from G 292 cellular line at an initially cellular density of 3.5×10^5 cells/plate. The obtained cultures are hatched at 37 °C in moist atmosphere with 5% CO_2, then monitored from cytomorphologically point of view at 24 and 72 hours and from cellular viability point of view after 72 hours since hatching. After 24 hours, the matrices are observed under a Nikon TS 100 microscope in phase contrast and pictures are taken using a Nikon Cooplix 4500 digital camera. After 72 hours, the cultures are stained with hypericin and pictures are taken.

3. Bentonite purification

Knowing that the bentonite may contain different percentages (5÷40%) of quartz and other impurities, it is necessary to purify the natural layered silicate (bentonite). These impurities must be removed because in the surface modification process, in order to ensure the compatibility between the silicate and polymer matrix, they act as a sterile, hindering the process (Carrado, 2006).

Purification is made by dispersing the bentonite in distilled water at 60 ÷90 °C. Quartz and other hydrophobic impurities are separated by decantation.

Small amounts of bentonite (approximately 5 g) are added into 100 ml of hot distilled water (60÷90 °C) under mixing conditions. The mixing is followed in a ball mill for 2 hours until the suspension becomes homogenous. The obtained mixture is allowed to settle for 72 hours, then the upper layer is separated (the purified suspension) and dried for 24 hours at

45 °C. Purified bentonite results (PB), which is ground to obtain a powder with < 60 μm size particles.

Bentonite morphology, before and after purification, is highlighted through X-ray diffraction (XRD), thermo-gravimetrical analyses (TGA) and scanning electron microscopy (SEM).

From X-ray diffractions we notice that small angle diffraction peaks present a higher surface area in purified bentonite as compared with the initial one, proving an increase of MMT concentration. We accepte that MMT and cristobalite are the two main components in the bentonite analysed (without taking into account the impurity components). We calculate the concentration of MMT in initial bentonite based on the maximum diffraction intensities: 77% montmorillonite in initial bentonite and 82% montmorillonite in purified bentonite (Vuluga et al., 2008b).

The TGA results (Table 2) are in accordance with X-ray diffraction results. We notice that, after purification the mass loss on the second step of decomposition decreases, while the mass loss on the third step of decomposition increases proving the decrease of the impurities concentration and the increase of the amount of montmorillonite.

Sample	Weight loss on each decomposition step (%)			
	I 20÷200°C	II 200÷400°C	III 400÷550°C	IV 550÷700°C
Initial bentonite (B)	6.8	0.6	0.4	1
Purified bentonite (PB)	8.3	-	-	1.4

Table 2. TGA results for bentonite before and after purification.

From scanning electron microscopy images (Fig. 3 and 4) we notice the morphology of purified powder with submicron dimensions and with uniform arrangement of chemical elements as compared to the morphology of initial powder where chemical elements are relatively uniformly dispersed.

Fig. 3. Secondary electron images (SEI) and distribution of X-rays for initial bentonite.

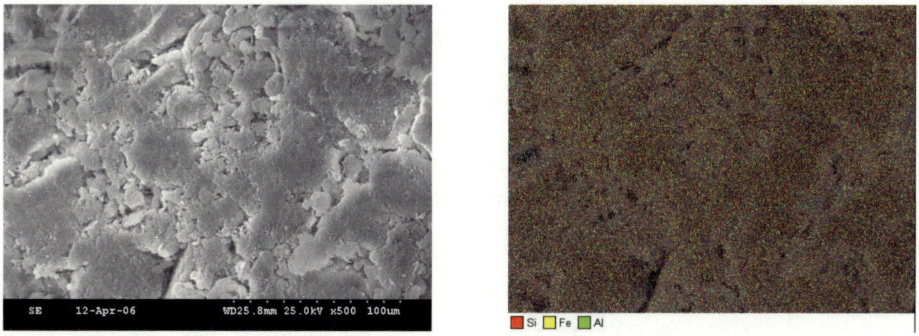

Fig. 4. Secondary electron images (SEI) and distribution of X-rays for purified bentonite.

4. Collagen/ modified layered silicate nanocomposites

Polymer/inorganic nanocomposites have attracted great recent attention and possessed a special position in constructing nano-assemblies due to their unique microstructure, outstanding properties and particular versatility. Nanocomposite materials can achieve much better properties than just the sum of their components as a result of interfacial interaction between the matrix and filler particles. It is the nature and degree of such interactions that play a pivotal role on the characteristics of resulted nanocomposites such as solubility, optical properties, electrical and mechanical aspects, biocompatibility, biodegradability. The organic-inorganic composites/hybrids represent a recent topic in which bibliographic references are still few. The year 2000 could be considered as the "starting point" in this field. Research for obtaining new types of bio-nanocomposites based on collagen in solid physical forms of matrices or membranes in which inorganic components can be hydroxyapatite, biovitroceramica, TiO_2, SiO_2, natural layered silicate as it is or modified with hydrolysed collagen, antioxidants has developed (Ficai et al. 2010a; Albu et al., 2009; Baia et al., 2008; Piticescu et al., 2008; Potarniche et al., 2010; Vuluga et al., 2007; Vuluga et al., 2008a). Organic compounds that can interact with collagen are represented by a series of biological active substances or other biocompatible polymers, their compatibility being previously tested with collagen gel or colloidal solutions. In such phases, the nanocomposites are obtained: membranes by free drying at 25 °C or matrices by freeze-drying process (lyophilization). According to the type of obtained nanocomposite, the amount of non collagen components should be: in case of matrices between 5÷20% insoluble powders and 30÷50% water soluble substances; in case of membranes between 1÷2% insoluble powders and 10-30% water soluble substances. Exceeding these amounts negatively influences the collagen matrix and membrane specific stability structure as well as the mechanical and hydrophilic properties. (Olteanu et al., 2008; Trandafir at al., 2007, Ficai et al., 2010b, Lungu et al., 2007, Titorencu et al., 2010).

Using hydrolyzed collagen modified layered silicates we obtain collagen based nanocomposites which can be used in healing varicose ulcer, the silicate improving the regenerative action of type I collagen on conjunctive tissue, reducing the healing time of ulcerous wound (Trandafir et al., 2007).

By introducing maleic copolymers as structural modifiers of layered silicates leads to binary stable hybrids compatible with collagen gel, a process which forms a liquid ternary nanocomposite resistant to further processing, for example to lyophilization.

4.1 Layered silicate/maleic copolymers binary nanocomposites

Until now, there has not been shown any interest in obtaining systems/composites/hybrids based on copolymers of anhydride or maleic acid with clays/bentonites. Literature is mostly made of patents which refer to synthesis or uses of systems in which the bentonite and the polymer are mixed without being associated with the composites or hybrids formation. One of the first areas, in which they are mentioned as systems of bentonite and maleic copolymers or polymers in general, is conducting drilling fluids for oil secundary recovery (Al-Marhoun et al., 1988; Gleason & Brase, 1985; Hale & Lawson, 1988; Hayes, 2005; Jarm et al., 1988; Libor et al., 1987; Martinko, 1987; Shaarpour, 2004). In medical field, different systems of bentonite/maleic polymers are obtained for dentistry uses (Heathman & Ravi, 2005), for parasites control (mites, moths) and water or dirt rejection (De Sloovere et al., 2004) and as dirt retention agents recommended for hand cleaning pastes (Florescu et al., 1994). In conclusion, we can say that the composite/organo-inorganic hybrids based on bentonite and maleic copolymers are a promising open field.

The natural layered silicate/maleic copolymer binary nanocomposites are obtained by solution intercalation method (Ray & Okamoto, 2003). According to the chemical structure of maleic partner, the interaction between inorganic and organic components is performed at different ratios (1: 1; 1: 2 and 2: 1), in a mixture of 1:1 ethanol- water. Thus, we obtain maleic copolymer modified layered silicate binary nanocomposites: PB/ MA-MMA (1: 1; 1: 2 and 2: 1) and PB/ MA-VA (1: 1; 1: 2 and 2: 1).

From particle size analysis of resulted binary nanocomposites (Fig. 5 and 6) we notice:

- the PB/ MA-MMA nanocomposite presents particle size dimensions at a range of 3000÷6000 nm;
- the PB/ MA-VA nanocomposite presents a similar behavior as that obtained with PB/ MA-MMA the difference being the particle size is bigger (at a range of 4000÷13000 nm);
- the higher the maleic copolymer concentration is the bigger the particle size; this is valid up to an approximately 50% concentration as compared to silicate particle size; an over 50% concentration is followed by a decrease in particle size up to a close value of unmodified silicate.

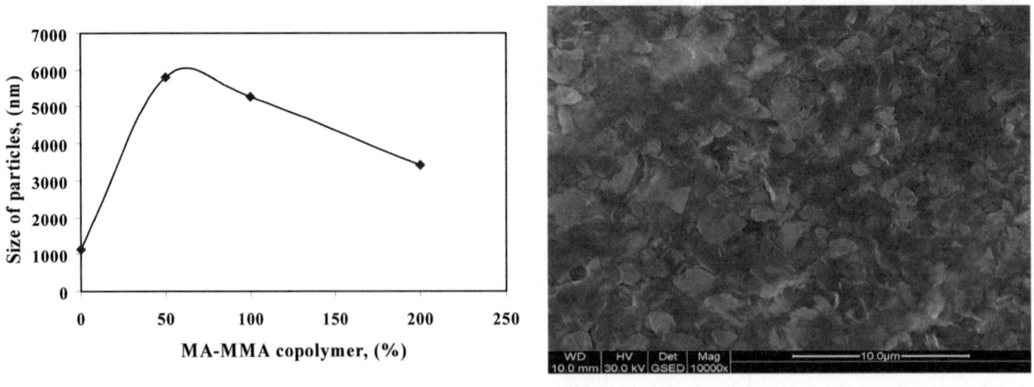

Fig. 5. Particle size and morphology for PB/ MA-MMA binary nanocomposites.

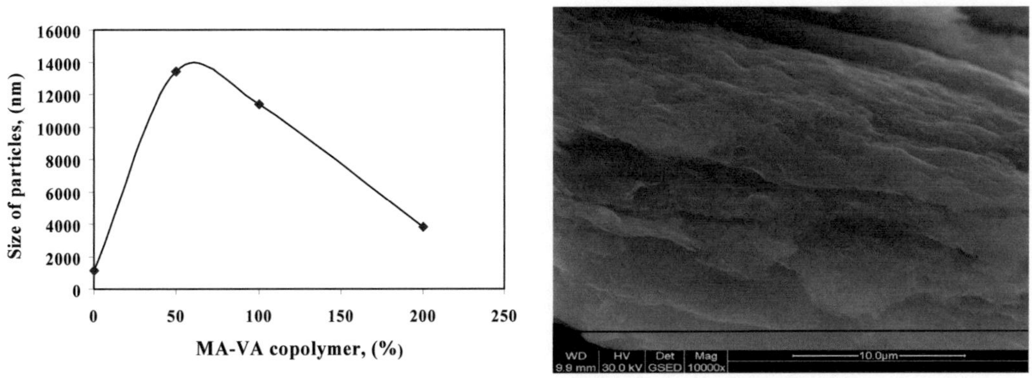

Fig. 6. Particle size and morphology for PB/ MA-VA binary nanocomposites.

From X-ray diffractions (Fig. 7 and 8) we notice that silicate structure changes in a specific way: a peak disappears at 2 θ = 23°, which is silicate characteristic peak and shapes very well the peak at approximately 2 θ = 3°. This proves the interaction between silicate and maleic copolymers as well as an organized structure. Intercalated nanocomposites are obtained with a lamellar ordered structure. Maleic copolymer with hydrophobic comonomer (MA-MMA) interacts better with silicate, the best structure is obtained at 1:1 silicate/copolymer ratio. In the case of maleic copolymer with hydrophilic comonomer (MA-VA), the best structure is obtained at a low amount of copolymer, meaning at 2:1 ratio. These two types of layered silicate/ maleic copolymer binary nanocomposites are selected to obtain ternary nanocomposites.

From thermal stability point of view, binary nanocomposites thermogrames present differences according to hydrophilic or hydrophobic comonomer used to modify layered silicate.

As compared to purified bentonite, at which the maximum rate of decomposition is on the first step, at a temperature of 74 °C, the organophilized bentonite presents a shift of maximum decomposition rate towards higher temperatures (220÷225°C for PB/ MA-VA nanocomposites and 220 ÷365 °C for PB/ MA-MMA nanocomposites), on the second step of decomposition, similar to maleic copolymers (Table 3). An increase of maleic copolymer concentration (from 50% to approximately 200% reported to bentonite) reflects into an increase of weight loss at the temperature range of 200 ÷500 °C. The obtained results show the intercalation of maleic copolymers between the silicate layers. Thermal stability of obtained binary nanocomposites correlates with the degree of interaction between silicate and copolymer. We notice that in the case of PB/ MA-MMA nanocomposites the best thermal stability is obtained at 1:1 ratio while the best thermal stability for the PB/ MA-VA nanocomposites is obtained at 2:1 ratio. These results are in accordance with X-ray diffraction results.

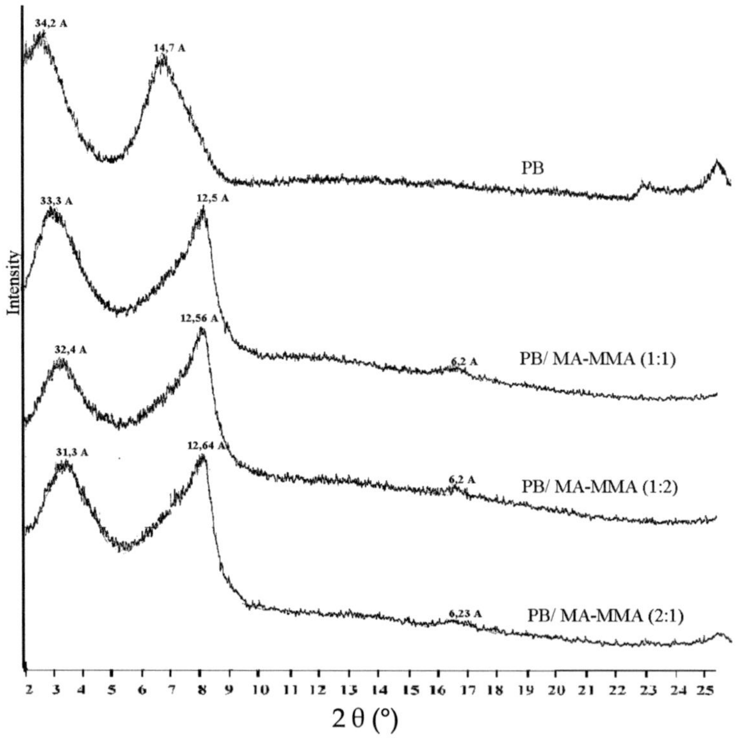

Fig. 7. X-ray diffractions for PB/ MA-MMA nanocomposites at different ratios.

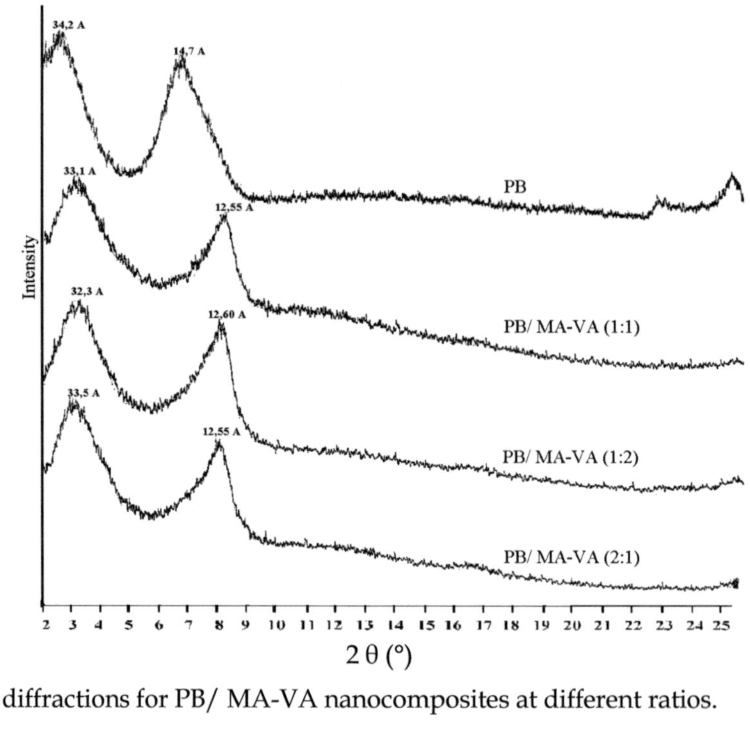

Fig. 8. X-ray diffractions for PB/ MA-VA nanocomposites at different ratios.

Sample	Weight loss on the each decomposition step (%)/ T_{rdmax} (°C)			
	I 20 ÷200 °C	II 200÷ 400 °C	III 400÷ 550 °C	IV 550 ÷700 °C
PB	8.3/ 74	-	-	1.4
MA-MMA	4.9	84.3/ 321	8.1	2.7
PB/ MA-MMA (1:1)	10.7	38.5/ 367	6.7	1.6
PB/ MA-MMA (1:2)	10.4	52.1/ 365	8.7	1.3
PB/ MA-MMA (2:1)	8.4	22.1/ 218	7.5	5.4
MA-VA	7.6	61.3/ 258	27.9	0.9
PB/ MA-VA (1:1)	7.4	35.6/ 223	14.4	3.3
PB/ MA-VA (1:2)	7.1	27.2/ 220	8.2	6
PB/ MA-VA (2:1)	7.4	25.2/ 226	4.5	7.9

T_{rdmax} Temperature at the maximum rate on the first decomposition step

Table 3. TGA results for PB/ Maleic copolymer binary nanocomposites as compared to TGA results for maleic copolymers (MA-MMA and MA-VA).

4.2 Collagen/ layered silicate/ maleic copolymer nanocomposites

The layered silicate can improve thermal stability of collagen and depending on the collagen morphology (gel or colloidal solutions) and on the method used for silicate dispersion, a variety of lamellar structures and morphologies – intercalated or exfoliated- can be obtained (Vuluga et al., 2007).

It is well known that the interaction of collagen with maleic copolymers takes place prevalently through weak forces and, to a little extent, through electrostatic interaction between the carboxylic groups of copolymer and the aminic groups of collagen (Anghelescu-Dogaru et al., 2004; Chitanu et al., 2002).

Maleic anhydride functionalized polymers ensure organosilicate dispersion and exfoliation into the polymer matrix (i.e. polypropylene) due to the interaction that may occur between maleic anhydride and hydroxyl groups on the silicate surface (Faisant et al., 1998; Vuluga et al., 2008b; Utracki, 2002).

Considering the results of these studies, it is reasonable to believe that a high adhesion to the collagen/modified layered silicate interface due to the interaction between the components leads to a uniform dispersion of modified layered silicate into collagen matrices and to improved properties.

A dispersion of 0.5% of layered silicate modified with maleic copolymer in water: alcohol (1:1) mixture is added, when stirring, at room temperature, into a solution of collagen with 1.2% dried substance to obtain ternary nanocomposites. The different types of nanocomposites obtained in liquid form (which contain 5-10% of MMT reported to collagen dried substance) are crosslinked with 0.1% formaldehide, reported to collagen dried substance and are dried by lyophilization to obtain spongious matrices. The freeze-drying process is an advantageous conditioning technique which consists of drying frozen samples by ice sublimation into vacuum. To maintain the triple helix conformation of collagen molecules we use the following process: the liquid ternary nanocomposites are cast in stainless steell plates, cooling to -40 ^{0}C (4 hours), kept up for 4 hours, then freeze-dried at -

40^{0}C and 0.12 mbar for 10 hours, subsequently heated to +20 ^{0}C at a rate of 3 ^{0}C/hour (20 hours) at 0.12 mbar, heated again (6 hours) to 30 ^{0}C at a rate of 2 ^{0}C/hour and finally freeze-dried at the same temperature at 0.01 mbar for 4 hours, using the Christ Model Delta 2–24 LSC freeze-dryer (Germany). During the 48 hours of freeze-drying we obtain: CG/ PB/MA-VA and CG/ PB/MA-MMA spongious matrices.

We study both the influence of binary nanocomposite type and of the collagen solution pH upon ternary nanocomposite properties. Although the normal human body pH is 7.2 we still study the nanocomposite properties at an acidic pH of 3 knowing that in acidic environment collagen molecules are positively charged by the amine groups thus being able to interact more easily with the negatively charged layered silicates.

Both ternary nanocomposites (CG/PB/MA-MMA and CG/PB/MA-VA) present a partially exfoliated lamellar structure (Fig. 9). In the nanocomposites with hydrophobic comonomer (MMA) the degree of interaction is stronger at basic pH, thus we obtain partially exfoliated nanostructured biomaterials. When we use hydrophilic comonomer (VA), the degree of interaction is stronger at acidic pH. For biocompatibility tests, we select the nanocomposite with hydrophobic comonomer. The strong degree of interaction for the hydrophobic comonomer reflects itself in an improved thermal stability (Table 4).

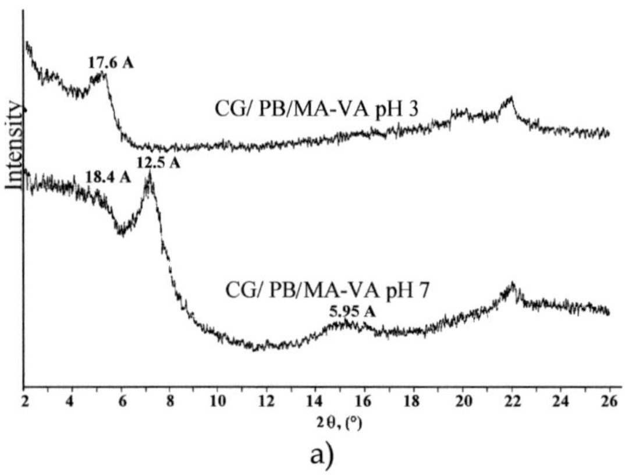

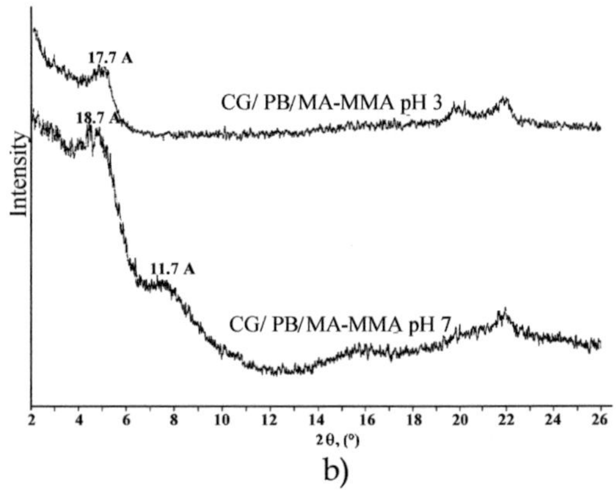

Fig. 9. X-ray diffraction patterns for ternary nanocomposites with hydrophilic (a) or hydrophobic (b) comonomer, at acidic and basic pH.

As compared to collagen, the CG/ PB/MA-MMA nanocomposites are more thermally stable. Thermal stability for the ternary nanocomposites is higher at basic pH and with hydrophobic maleic comonomer (Table 4).

The nanocomposites CG/PB/MA-MMA (pH 7) present a spongious structure with interconnected macro, micro and nano sized pores, with nanostructured regions (Fig. 10 a). Collagen morphologic structure presented in secondary electron images (SEI) shows regions with large communicating pores as well as interconnected ones. We also notice the morphologic homogeneity of the collagen fibers (Fig. 10 b). Even though the amount of maleic copolymer modified layered silicate in ternary nanocomposites is less than the amount of collagen, a compact fibrous structure can be observed, which can be the reason of an increase of thermal stability.

Sample	Residue at 700 °C, (%)	Weight loss on each decomposition step (%)/T_{vmax} (°C)		
		I 0 ÷250 °C	II 250 ÷500 °C	III 500 ÷700 °C
Collagen Matrix pH 7.2	1.6	14.3/ 147	50.2/ 337	33.9/ 600
CG/ PB/ MA-VA nanocomposite matrix pH 3	16.3	13.1/ 171	40.8/ 344	29.8/ 588
CG/ PB/ MA-VA nanocomposite matrix pH 7	18.9	16.3/ 171; 207	39.3/ 345	25.2/ 605
CG/ PB/ MA-MMA nanocomposite matrix pH 3	15.4	11.6/ 173	48.9/ 349	24.1/ 585
CG/ PB/ MA-MMA nanocomposite matrix pH 7	23.4	9.1/ 87; 172	41.6/ 351	25.2/ 586

$T_{vmax,}$ temperature at maximum rate of decomposition, (°C)

Table 4. Thermal stability for the ternary nanocomposite matrices.

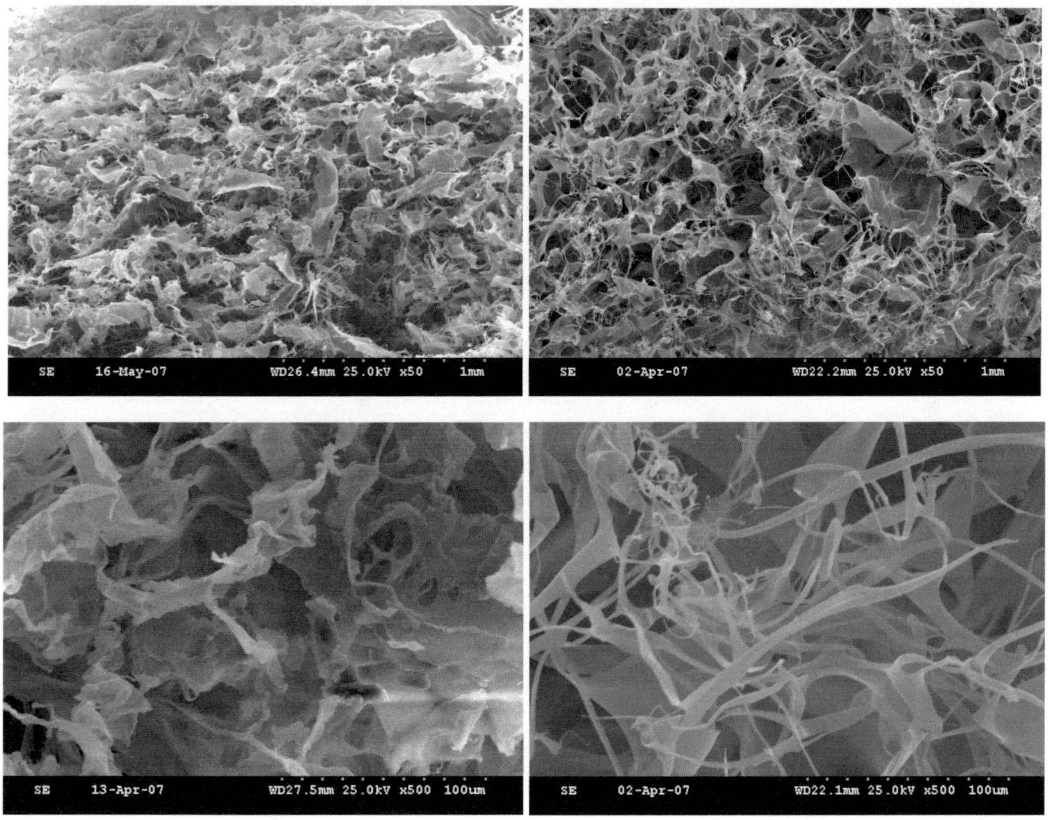

a) b)

Fig. 10. Morphologic structure of (a) CG/ PB/ MA-MMA nanocomposites pH 7, compared to (b) collagen morphology.

The spongious matrix of CG/ PB/ MA-MMA nanocomposite was *in vitro* tested on osteoblast cell cultures. *In vitro* test results are presented in Table 5. After 24 and 72 hours since hatching, the cells proliferate on matrix surface, the viability being of 97%, very close to the reference value (control sample). These results are also supported by microscopy observation of cellular density (Fig. 11). The collagen/layered silicate/MA-MMA ternary nanocomposite does not present any citotoxic response and the cells present normal phenotype.

Sample	Optical density DO 550nm	Viability from control sample, (%)
CONTROL	1.0263	100%
CG/ PB- MA-MMA nanocomposite matrix pH 7	0.9954	97%

Table 5. Proliferation and osteoblast cells viability hatched on matrix of CG/ PB/ MA-MMA nanocomposites.

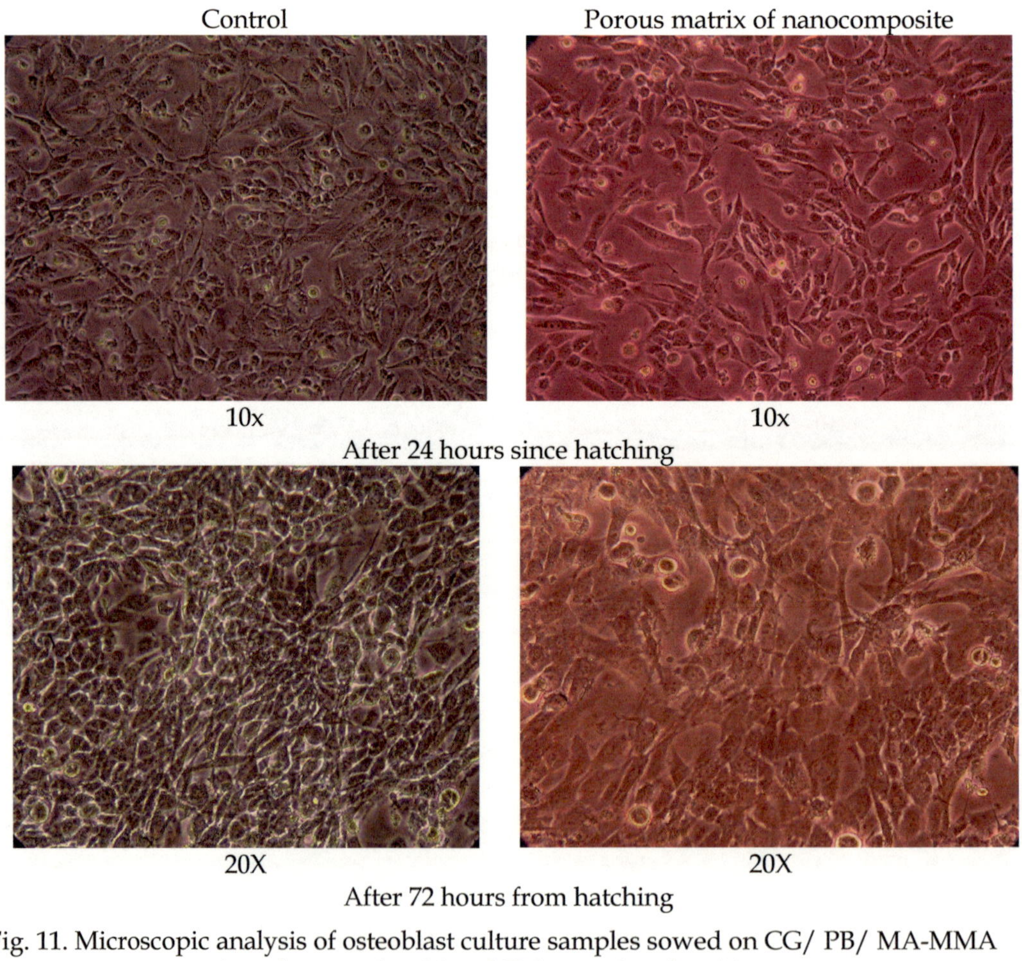

Fig. 11. Microscopic analysis of osteoblast culture samples sowed on CG/ PB/ MA-MMA ternary nanocomposite substrate after 24 and 72 hours since hatching.

5. Conclusions

The key results consist in obtaining new types of collagen nanostructured biomaterials in the form of spongious, microporous matrices, nontoxic and biocompatible with osteoblast cells. The preparation of the new nanocomposites is based on the use of two natural components: collagen, which contributes to bone regeneration and natural layered silicate (montmorillonite) which improves the thermal resistance of collagen, able to release in time an active substance at low and controlled concentrations. Maleic copolymers favour the silicate dispersion in the matrix of collagen and the obtaining of intercalated partially exfoliated nanocomposites with disordered lamellar structure. The nanocomposites in form of microporous matrix (scaffold) obtained by dispersion of binary nanocomposites in collagen gel have a spongy structure which contains macro and micro interconnected nanopores, similar with extra cellular matrix, which allows the penetration of the physiological fluids, necessary for the growth of cells. In comparison with collagen matrix, CG/ PB-MA-MMA nanocomposites have improved thermal stability.

6. Acknowledgment

The financial support of MATNANTECH Program of the Romanian Ministry of Education and Research, by means of CEEX project no. 16/2005-2007, to achieve this contribution is gratefully acknowledged.

The authors are gratefully acknowledged for their collaboration with C. Radovici for XRD, C. Nistor for DLS and S. Serban for TGA from National Research and Development Institute for Chemistry and Petrochemistry-ICECHIM, Bucharest and to "Petru Poni" Institute of Macromolecular Chemistry, Romanian Academy, Iasi team which provided maleic copolymers.

7. References

Al-Marhoun, M.A. & Rahman, S.S. (1988). Optimizing the Properties of Water-Based Polymer Drilling Fluids for Penetrating Formations with Electrolyte Influx. *Erdoel, Erdgas, Kohle*, Vol. 104, No. 7-8, (September 1988), pp. (318-323), ISSN 0179-3187

Albrektsson, T. (1995). Principles of Osseointegration, In: *Color Atlas and text of Dental and maxillofacial implantology*, J.A. Hobkirk, K. Watson (Ed.), pp. (9–19), Mosby-Wolfe, ISBN 9780723417866, London, United Kingdom

Albrektsson, T. & Johansson, C. (2001). Osteoinduction, Osteoconduction and Osseointegration. *European Spine Journal*, Vol. 10, No. 2, (October 2001), pp. (96–101), ISSN 1432-0932

Albu, M.G.; Ghica, M.V.; Giurginca, M.; Trandafir, V.; Popa, L. & Cotrut, C. (2009). Spectral characteristics and antioxidant properties of tannic acid immobilized on collagen drug-delivery systems. *Revista de Chimie – Bucharest*, Vol.60, No.7, (July 2009), pp. 666-672, ISSN 0034-7752

Albu, M.G. (2011). *Collagen gels and matrices for biomedical applications,* LAP LAMBERT Academic Publishing GmbH & Co. KG, ISBN 978-3-8443-3057-1, Saarbrucken, Germany

Allen, T.M. & Cullis, P.R. (2004). Drug Delivery Systems: Entering the Mainstream. *Science,* Vol. 303, No. 5665, (March 2004), pp. (1818-1822), ISSN 1095-9203

Anghelescu-Dogaru, A.G., Popescu, I., Chitanu, G.C. (2004). Maleic Polyelectrolytes as Ecologically Favourable Additives in Chrome Tanning Process. *Journal of Environmental Protection and Ecology*, Vol. 5, No. 2, (April 2004), pp. (265-270), ISSN 1311-5065

Baia, L., Baia, M., Danciu, V., Albu, M.G., Cosoveanu, V., Iordachescu, D., Trandafir, V. (2008). *Type I collagen-TiO2 aerogel based biocomposites*. Journal of Optoelectronics and Advanced Materials, Vol. 10, No. 4, (April 2008), pp. (933-936), ISSN 1454-4164.

Baron, R. (2003). General Principles of Bone Biology, In: *Primer on the Metabolic Bone Disease and Disorders of Mineral Metabolism*, M.J. Favus (Ed.), pp. (1-8), American Society for Bone and Mineral Research, ISBN 978-0-974-47820-3, Washington DC, USA

Barradas, A.M.C., Yuan, H., Blitterswijk, C.A. & Habibovic, P. (2011). Osteoinductive Biomaterials: Current Knowledge of Properties, Experimental Models and Biological Mechanisms. *European Cells and Materials*, Vol. 21, (May 2011), pp. (407 - 429), ISSN 1473-2262

Blokhuis, T.J., Termaat, M.F., den Boer, F.C., Patka, Peter., Bakker, F.C. & Haarman H.J.Th.M. (2000). Properties of Calcium Phosphate Ceramics in Relation to Their In Vivo Behavior. *The Journal of Trauma: Injury, Infection, and Critical Care*, Vol. 48, No. 1, (January 2000), pp. (179-186), ISSN 1079-6061

Budrugeac, P., Trandafir, V. & Albu, M.G. (2003). The Effect of the Hydration Degree on the Hydrotermal and Thermo-oxidative Stability of some Collageneous Matrices. *Journal of Thermal Analysis and Calorimetry*, Vol. 72, No. 2, (July 2006), pp. (581-585), ISSN 1388-6150

Carrado, K.A. (2006). Synthetic Clay Minerals and Purification of Natural Clays, In: *Handbook of Clay Science*, Bergaya, F., Theng, B.K.G. & Lagaly, G., 115-139, Elsevier, ISBN 978-0-08-044183-2, Oxford, UK

Caze, C. & Loucheux, C. (1975). Mechanism of Alternating Copolymerization of Vinyl Acetate and Maleic Anhydride. *Journal of Macromolecular Science*, Vol. 9, No. 1, (January 1975), pp. (29-43), ISSN 0022-233x

Chitanu, G.C., Creanga, D., Hirano, T., Badea, N. (2003). Supramolecular structures in complex systems from natural and synthetic polymers. 1 - Interaction between collagen, maleic polyelectrolytes and chromium ions. *Revue Roumaine de Chimie*, Vol. 47, No. 3-4, (March-April 2002), pp. (343-352), ISSN 0035-3930

Chitanu, G.C., Popescu, I. & Carpov, A. (2006). Synthesis and Characterization of Maleic Anhydride Copolymers and Their Derivatives. 2 New Data on the Copolymerization of Maleic Anhydride with Vinyl Acetate. *Revue Roumaine de Chimie*, Vol. 51, No. 9, (September 2006), pp. (923–929), ISSN 0035-3930

Chitanu, G.C., Popescu, I. & Carpov, A. (2007). Synthesis and Characterization of Maleic Anhydride Copolymers and Their Derivatives. 3 Synthesis and Characterization of Maleic Anhydride-Methyl Methacrylate Copolymers. *Revue Roumaine de Chimie*, Vol. 52, No. 1–2, (January - February 2007), pp. (135–141), ISSN 0035-3930

Choy, J.H., Choi, S.J., Oh, J.M. & Park, T. (2007). Clay Minerals and Layered Double Hydroxides for Novel Biological Applications. *Applied Clay Science*, Vol. 36, No. 1-3, (April 2007), pp. (122–132), ISSN 0169-1317

Christiansen, J.D., Potarniche, C.G., Vuluga, Z. & Drozdov, A.D. (2011). Nanomaterials in Biomedical Bpplications, *Proceeding of 2nd International Conference on Wireless*

Communications Vehicular Technology, Information Theory and Aerospace & Electronic Systems Technology IEEE, ISBN 978-8-7923-2962-2, Chennai, India, 28 Feb – 3 May, 2011

Conlon, I. & Raff, M. (1999). Size Control in Animal Development. *Cell*, Vol. 96, No. 2, (January 1999), pp. (235-244), ISSN 0092-8674

Dankert, J., Hogt, A. H. & Feijen, J. (1986). Biomedical Polymers: Bacterial Adhesion, Colonization, and Infection. *Critical Reviews in Biocompatibility*, Vol. 2, No. 3, (June 1986), pp. (219-301), ISSN 0748-5204

De Sloovere, X., Desschans, D., & Sirejacob, G. (April 2004). Composition for Combating/Repelling Insects, Birds, Dirts and Parasites, Patent No. US20040067247, US

Edlund, U. & Albertsson, A.C. (2002). Degradable Polymer Microspheres for Controlled Drug Delivery. In: *Degradable Aliphatic Polyesters*, A.C. Albertsson, pp. (67-112), Springer Berlin, Retrieved from http://www.springerlink.com/content/ce7tj18xutcf7xkq/fulltext.pdf

Faisant, J., Ait-Kadi, A., Bousmina, M. & Deschenes, L. (1998). Morphology, thermomechanical and barrier properties of polypropylene-ethylene vinyl alcohol blends. *Polymer*, Vol. 39, No. 3, (January 1998), pp. (533-545), ISSN 0032-3861

Ficai, A., Andronescu, E., Voicu, G., Ghitulica, C., Vasile, B.S., Ficai, D., Trandafir, V. (2010a). Self-assembled collagen/hydroxyapatite composite materials. Chemical Engineering Journal. Vol. 160, No. 2, (June 2010), pp. (794-800), 1385-8947

Ficai, A., Andronescu, E., Trandafir, V, Ghitulica, C., Voicu, G. (2010b). Collagen/hydroxyapatite composite obtained by electric field orientation. *Materials Letters*, Vol. 64, No. 4, (February 2010), pp. (541-544), 0167-577X.

Florescu, S., Rob, M., Balintescu, G., Diaconescu, E., Golgojan, A. & Enculecsu, I. (September 1994). Produs pentru Curatarea Mainilor, Patent No. RO 108978, Romania

Gleason, P.A. & Brase, I.E. (May 1985). Drilling Mud Dispersants, Patent No. 4518510, US

Gupta, P., Vermani, K. & Garg, S. (2002). Hydrogels: from controlled release to pH-responsive drug delivery. *Drug Discovery Today*, Vol. 7, No. 10, (May 2002), pp. (569-579), ISSN 1359-6446

Hale, A.H. & Lawson, H.F. (April 1988). Well Drilling Fluids and process for Drilling Wells, Patent No. 4740318, US

Hayes, J.R. (August 2005). High Performance Water-Based Mud Systems, Patent No. US20050187113, US

Heathman, J.F. & Ravi, K.M. (September 2005). Methods and Compositions for Compensating for Cement Hydration Volume Reduction, Patent No. US20050204960, US

Hench, L.L. & Polak, J.M. (2002). Third-Generation Biomedical Materials. *Science*, Vol. 295, No. 5557, (February 2002), pp. (1014-1017), ISSN 1095-9203

Hendricks, S.K., Kwok, C., Shen, M., Horbett, T.A., Ratner, B.D. & Bryers, J.D. (2000). Plasma-Deposited Membranes for Controlled Release of Antibiotic to Prevent Bacterial Adhesion and Biofilm Formation. *Journal of Biomedical Materials Research*, Vol. 50, No. 2, (February 2000), pp. (160-170), ISSN 1552-4965

Hirst, A.R., Escuder, B., Miravet, J.F. & Smith, D.K. (2008). High-Tech Applications of Self-Assembling Supramolecular Nanostructured Gel-Phase Materials: From

Regenerative Medicine to Electronic Devices. *Angewandte Chemie International Edition*, Vol. 47, No. 42, (October 2008), pp. (8002-8018), ISSN 1521-3773

Hoeppner, D.W. & Chandrasekaran, V. (1994). Fretting in Orthopaedic Implants: A Review. *Wear*, Vol. 173, No. 1-2, (April 1994), pp. (189-197), ISSN 0043-1648

Huebsch, N. & Mooney, D.J. (2009). Inspiration and Application in the Evolution of Biomaterials. *Nature*, Vol. 462, No. 7272, pp. (426-432), (November 2009), ISSN 0028-0836

Jarm, V., Martinko, B. & Kovac-Filipovic, M. (1988). Synthetic Water-Soluble Polymers as Aqueous Drilling Fluid Additives. *Polimeri*, Vol. 9, No. 11, (November 1988), pp. (276-278), ISSN 0351-1871

Kshirsagar, N.A. (2000). Drug Delivery Systems. *Indian Journal of Pharmacology*, Vol. 32, No. 4, (July-August 2000), pp. (54-61), ISSN 1998-3751

Lager, R.S. & Peppas, N.A. (1981). Present and Future Applications of Biomaterials in Controlled Drug Delivery Systems. *Biomaterials*, Vol. 2, No. 4, (October 1981), pp. (201-214), ISSN 0142-9612

Libor, O., Nagy, G. & Szekely, T. (July 1986). Process for the Preparation of Clay Mineral-Containing Gels with Stabilized Structure and Reversible Water Absorption Ability, Patent No. 4600744, US

Lungu, A.; Albu, M.G. & Trandafir, V. (2007). New biocomposite matrices structures based on collagen and synthetic polymers designed for medical applications. *Materiale Plastice*, Vol.44, No.4, pp. 273-277, ISSN 0025-5289

Ma, P.X. & Elisseeff, J. (Eds.). (2005). *Scaffolding in Tissue Engineering*, Taylor and Francis, ISBN 9781574445213, Florida, United States of America

Martinko, B. (1987). Determination of Thermal Stability of Polymers for Water Base Muds. *Nafta*, Vol. 38, No. 11-12, (November 1987), pp. (679-685), ISSN 0027-755X

Medical Dictionary. (nda). Biocompatibility, In: *The free dictionary*, (2011) Avaliable from: http://medical-dictionary.thefreedictionary.com/biocompatibility

Medical Dictionary. (ndb). Biomaterial, In: *The free dictionary*, (2011) Avaliable from: http://medical-dictionary.thefreedictionary.com/biomaterial

Medical Dictionary. (ndc). Biomedicine, In: *The free dictionary*, (2011) Avaliable from: http://medical-dictionary.thefreedictionary.com/biomedicine

Medical Dictionary. (ndd). Proliferate, In: *The free dictionary*, (2011) Avaliable from: http://medical-dictionary.thefreedictionary.com/proliferate

Miller, K. & O'Toole, M.T. (2005). *Encyclopedia and Dictionary of Medicine, Nursing, and Allied Health*, Elsevier, ISBN 978-1-4160-2604-4, Camden, New Jersey

Mundy, G.R., Chen, D. & Oyajobi, B.O. (2003). Bone Remodeling, In: *Primer on the Metabolic Bone Disease and Disorders of Mineral Metabolism*, M.J. Favus, pp. (46-58), American Society for Bone and Mineral Research, ISBN 978-0-974-47820-3, Washington DC, USA

Olteanu, M., Achimescu, D., Vuluga, Z. & Trandafir, V. (2008). Measurement and interpretation of wetting properties of new collagen-silicate biomaterial. *Revue Roumaine de Chimie*, Vol. 53, No. 2, (February 2008), pp. (157–163), ISSN 0035-3930

Peppas, N.A., Hilt, J.Z., Khademhosseini, A. & Langer, R. (2006). Hydrogels in Biology and Medicine: From Molecular Principles to Bionanotechnology. *Advanced Materials*, Vol. 18, No. 11, (May 2006), pp. (1345-1360), ISSN 0935-9648

Picket, J.P. (Ed). (2000). *The American Heritage Dictionary of the English Language*, Houghton Mifflin Company, ISBN 978-0-618-70172-8, Boston, Massachusetts

Piticescu, R.M., Trandafir, V., Danciu, V., Vuluga, Z., Vasile, E. & Iordachescu, D. (2008). Ternary Bio-nanostructured Systems Prepared under High Pressure Conditions, Proceeding of *20th International Symposium on Ceramics in Medicine*, ISBN 978-0-87849-457-6, Nantes, France, 24-26 October, 2008

Potarniche, C.G., Vuluga, Z., Radovici, C., Serban, S., Vuluga, D. M., Ghiurea, M., Purcar, V., Trandafir, V., Iordachescu, D. & Albu, M.G. (2010). Nanocomposites based on Collagen and Na-Montmorillonite Modified with Bioactive Substances. *Materiale Plastice*, Vol. 47, No. 3, (September 2010), pp. (267-273), ISSN 0025-5289

Potarniche, C.G., Vuluga, Z., Donescu, D., Christiansen, J.deC., Eugeniu, V., Radovici, C., Serban, S., Ghiurea, M., Somoghi R. & Beckmann, S. (2011). Morphology Study of Layered Silicate/Chitosan Nanohybrids. *Surface and Interface Analysis*, (May 2011), Early View, Online Version, ISSN 1096-9918

Ray, S.S. & Okamoto, M. (2003). Polymer/layered silicate nanocomposites: a review from preparation to processing, *Progress in Polymer Science*, Vol. 28, No. 11, (August 2003), pp. (1539–1641), ISSN 0079-6700

Recum, A.F. & Strauss, S. (Eds.). (1998). *Handbook of Biomaterials Evaluation: Scientific, Technical, and Clinical Testing of Implant Materials*, Taylor & Francis, ISBN 978-1560324799, Philadelphia, USA

Riddick, J.A. & Burger, W.A. (Eds.). (1970). *Organic Solvents: Physical Properties and Methods of Purification*, Wiley-Interscience, ISBN 9780471927266, New York, USA

Robey, P.G. & Boskey, A.L. (2008). The Composition of Bone, In: *Primer on the Metabolic Bone Disease and Disorders of Mineral Metabolism*, C.J. Rosen (Ed.), pp. (32-36), American Society for Bone and Mineral Research, ISBN 978-0-977-88821-4, Washington DC, USA

Sabir, M.I., Xu, X. & Li, L. (2009). A Review on Biodegradable Polymeric Materials for Bone Tissue Engineering Applications. *Journal of Materials Science*, Vol. 44, No. 21, (November 2009), pp. (5713–5724), ISSN 1573-4803

Sawan, S.P. & Manivannan, G. (Eds.). (2000). *Antimicrobial/Anti-Infective Materials: Principles, Applications and Device*, Technomic Publishing Co., ISBN 9781566767941, Lancaster, USA

Segen, J.C. (2005) *Concise Dictionary of Modern Medicine*, The McGraw-Hill Companies, Inc., ISBN 0838515355, New York, New York

Shaarpour, M. (December 2004). Lost Circulation Material Blend Offering High Fluid Loss with Minimum Solids, Patent No. US20040244978, US

Shin, J.H. & Schoenfisch, M.H. (2006). Improving the Biocompatibility of in Vivo Sensors via Nitric Oxide Release. *Analyst*, Vol. 131, No. 5, (March 2006), pp. (609–615), ISSN 0003-2654

Siapi, E., Mavromoustakos, T., Trandafir, V., Albu, B. & Budrugeac, P. (2005). The use of differential scanning calorimetry to study the effects of gentamycin on fibrous collageneous membranes, *Termochimica Acta*, Vol. 425, No. 1-2, (August 2005), pp. (165-171), ISSN 0040-6031

Stuart, M.A.C., Huck, W.T.S., Genzer, J., Muller, M., Ober, C., Stamm, M., Sukhorukov, G.B., Szleifer, I., Tsukruk, V.V., Urban, M., Winnik, F., Zauscher, S., Luzinov, I. & Minko,

S. (2010). Emerging Applications of Stimuli-Responsive Polymer Materials. *Nature Materials*, Vol. 9, No. 2, (January 2010), pp. (101-113), ISSN 1476-4660

Szmukler-Moncler, S. & Dubruille J.H. (1990). Is Osseointegration a Requirement for Success in Implant Dentistry?, *Clinical Materials*, Vol. 5, No. 2-4, (1990), pp. (201-208), ISSN 0267-6605

Titorencu, I.; Albu, M.G.; Giurginca, M.; Jinga, V.; Antoniac, I.; Trandafir, V.; Cotrut, C.; Miculescu, F. & Simionescu, M. (2010). In *Vitro* Biocompatibility of Human Endothelial Cells with Collagen-Doxycycline Matrices. *Molecular Crystals and Liquid Crystals*, Vol.523, (May 2010), pp. 82/[654] - 96/[668], ISSN 1542-1406

Trandafir, V., Popescu, G., Albu, M.G., Iovu, H. & Georgescu, M. (2007). *Collagen based bioproducts*, Ars Docendi, ISBN 978-973-558-291-3, Bucharest, Romania

Utracki, L. (2002). Thermodynamics of Polymer Blends. In: *Polymer Blends Handbook*, Utracki. L., pp. (123-202), Kluwer Academic Publishers, ISBN 1-4020-1110-5, Dordrecht, Netherlands

Vuluga, Z., Donescu, D., Trandafir, V., Radovici, C. & Serban, S. (2007). L'influence de la Morphologie des Extraits de Collagene sur les Proprietes des Nanocomposites a Silicates Naturels. *Revue Roumaine de Chimie*, Vol. 52, No. 4, (April 2007), pp. (395-404), ISSN 0035-3930

Vuluga, Z., Radovici, C., Serban, S., Potarniche, C. G., Danciu, V., Trandafir, V., Vuluga, D. M. & Vasile, E. (2008a). Titania Modified Layered Silicate for Polymer/Inorganic Nanocomposites. *Molecular Crystals and Liquid Crystals*, Vol. 483, No. 1, (June 2008), pp. (258—265), ISSN 1563-5287

Vuluga, Z., Donescu, D. & Vuluga, D.M. (2008b). *Materiale Compozite Polimerice Termoplastice*, Ars Docendi, ISBN 978-973-558-370-5, Bucuresti, Romania

Widmer, A.F. (2001). New Developments in Diagnosis and Treatment of Infection in Orthopedic Implants. *Clinical Infectious Diseases*, Vol. 33, No. 2, (September 2001), pp. (94–106), ISSN 1058-4838

Williams, D.F. (1987). Definitions in Biomaterials. *Proceedings of a Consensus Conference of the European Society for Biomaterials*, ISBN 978-0-444-42858-5, Chester, England, March 3-5, 1986

Williams, D.F. (1999). *The Williams Dictionary of Biomaterials*, Liverpool University Press, ISBN 978-0-853-23734-4, Liverpool, United Kingdom

Williams, D.F. (2009). On the Nature of Biomaterials. *Biomaterials*, Vol. 30, No. 30, pp. (5897-5909), (October 2009), ISSN 0142-9612

Wilson, G.S. & Gifford, R. (2005). Biosensors for Real-Time in Vivo Measurements. *Biosensors and Bioelectronics*, Vol. 20, No. 12, (June 2005), pp. (2388-2403), ISSN 0956-5663

Wilson T. (2011). *Effects of Silica Based Biomaterials on Bone Marrow Derived Cells. Material aspects of bone regeneration*, Annales Universitatis Turkuensis, Retrieved from http://www.doria.fi/bitstream/handle/10024/67471/AnnalesD956Wilson.pdf?sequence=1

Wu, C.J., Gaharwar, A.K., Schexnailder, P.J. & Schmidt, G. (2010). Development of Biomedical Polymer-Silicate Nanocomposites: A Materials Science Perspective. *Materials*, Vol. 3, No. 5, (April 2010), pp. (2986-3005), ISSN 1996-1944

7

Tailored and Functionalized Magnetite Particles for Biomedical and Industrial Applications

Anamaria Durdureanu-Angheluta[1,2],
Mariana Pinteala[1] and Bogdan C. Simionescu[1,2]
[1]*Centre of Advanced Research in Bionanoconjugates and Biopolymers, "Petru Poni" Institute of Macromolecular Chemistry of Romanian Academy, Iasi,*
[2]*Department of Natural and Synthetic Polymers, "Gh. Asachi" Technical University of Iasi, Iasi,*
Romania

1. Introduction

Magnetic particles have a significant role in nanotechnology due to their surface properties and their applicability in physical and chemical processes like ionic exchange, specific complexation, biocompatibility and bioactivity, capacity of selection and transport for cells and chemical compounds (Safarik & Safarikova, 2002).

Magnetite is an interesting superparamagnetic nanomaterial, considered as a challenge by the modern research related magnetic applications, due to its high susceptibility at oxidation as compared to other magnetic compounds (Cornell & Schwertmann, 1996).

The name "superparamagnetic" refers to those particles that in presence of magnetic field are attracted and in absence of the magnetic field the particles don't have residual magnetism. The importance of magnetite particles is related to their most important properties - magnetic and catalytic. These properties are strongly dependent on the selected method of preparation.

These magnetic particles are used in many applications that involve their immobilization and transport in the presence of magnetic field, or magnetically tagged biological entities due to the intrinsic penetrability of magnetic fields into human tissue.

The chapter describes the most important approaches in the preparation of magnetite particles, as presented in the literature. An introspective view of biomedical, industrial and catalytic applications of magnetite micro- and nanoparticles is also reported.

The magnetite particles are usually attracted to each other. Electrostatic or steric stabilization represent the main step in obtaining "core-shell" magnetic particles with magnetite core and a shell formed by different surfactants. We summarize here some examples and also some data previously reported by our group (figure 1) on the preparation of magnetite particles coated by different Si-containing compounds (monomers and polymers) (Durdureanu-Angheluta et al., 2008; Durdureanu-Angheluta et al., 2009; Durdureanu-Angheluta et al., 2010; Pricop et al., 2010).

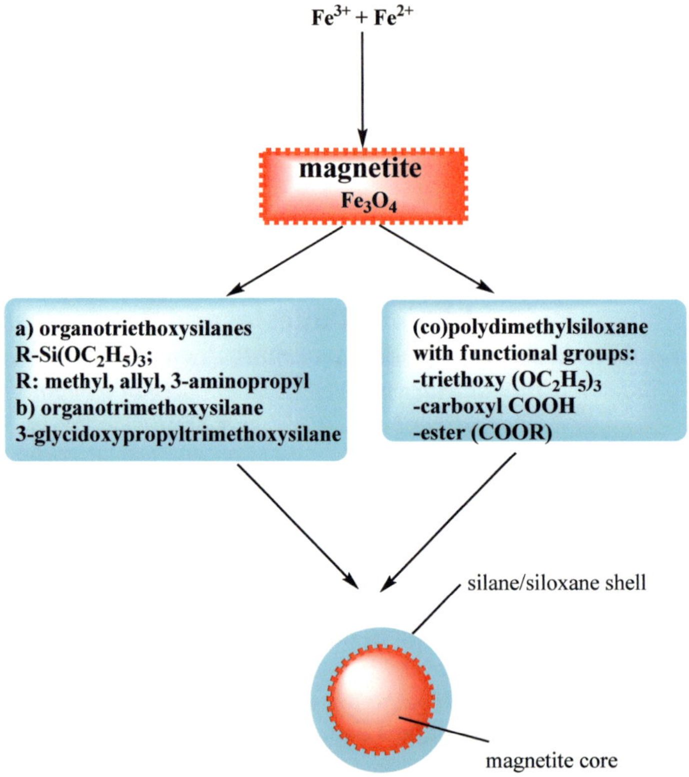

Fig. 1. Schematic representation of the preparation of core-shell magnetite particles with different Si-containing compounds.

2. Synthesis methods

Magnetite particles are quite intriguing, due to their catalytic and magnetic properties strongly dependent on the chosen synthesis method. Several methods are described in the literature, but in the present manuscript we are going to focus on the most common ones: iron salts co-precipitation (Massart, 1981; Martinez-Mera et al., 2007), sol-gel reaction, micro-emulsion (Woo et al., 2003), reaction in mass without solvent (Ye et al., 2006), polyols process (Feldmann & Jungk, 2001), decomposition of iron pentacarbonyl (Shafi et al., 2001), etc.

2.1 Co-precipitation

The synthesis by co-precipitation is the easiest way to obtain iron oxides like magnetite (Fe_3O_4) (Laurent et al., 2008) or like maghemite (γ-Fe_2O_3) and has shown to be a highly economic and versatile method. We used the method described by Massart (Massart, 1981) involving the co-precipitation of two iron salts $FeCl_3 \cdot 6H_2O$ and $FeCl_2 \cdot 4H_2O$, both prepared in HCl, with the consequent addition of a NH_4OH solution, vigorous stirring and under anoxic conditions, at room temperature. The shift of the initially orange color of the solution to black indicates the formation of magnetite particles. The surface of magnetite particles, in contact with a neutral aqueous solution is able to adsorb OH^- and H^-, respectively, giving an OH- rich surface (Durdureanu-Angheluta et al., 2008).

The reaction for obtaining magnetite is given below:

$$2FeCl_3 + FeCl_2 + 8NH_3 + 4H_2O \rightarrow Fe_3O_4 + 8NH_4Cl \quad (1)$$

The disadvantage of the co-precipitation process is that the control of particle size distribution is limited, due to kinetic factors that are controlling the growth of the crystal. Two stages are involved in the co-precipitation process (Cornell & Schwertmann, 1996; Boistelle et al., 1988; Sugimoto, 2003; Cornell et al., 1991; Gribanow et al., 1990): nucleation that occurs when the concentration of the species reaches critical supersaturation, and a slow growth of the nuclei by diffusion of the solutes to the surface of the crystal. To produce monodisperse iron oxide nanoparticles, these two stages should be separated. Nucleation should be avoided during the period of growth (Tartaj et al., 2006).

An alternative route to obtain magnetite particles by co-precipitation is the method described by Kim et al. (Kim et al., 2001) and Koneracka et al. (Koneracka et al., 2002) involving the preparation of the two solutions of ferric chloride ($FeCl_3 \cdot 6H_2O$) and ferrous sulphate ($FeSO_4$) by dissolution in a HCl solution at room temperature under vigorous stirring using a mechanical stirrer. The solutions are mixed and an aqueous dispersion of particles is obtained after adding a solution of NaOH under sonication or under vigorous mechanical stirring.

Magnetic particles are prepared, also, using $FeCl_3 \cdot 6H_2O$ and $FeSO_4 \cdot 7H_2O$ salts in aqueous solutions, both added in a polymeric starch matrix at 80^0C under nitrogen atmosphere and vigorous stirring (Jiang et al., 2009). NaOH should be added to the mixture until reaching a basic pH (9-11) and the black precipitate is neutralized with HCl.

2.2 Sol-gel reaction

The reaction takes place at the water/oil interface. An aqueous solution containing Fe^{2+}/Fe^{3+} salts is added to an organic solvent containing the stabilizer. The sol-gel reaction is performed at room temperature and is based on the hydroxylation and condensation of molecular precursors in solution and leads to a three-dimensional metal oxide network, the wet gel. Heat treatments are further needed to acquire the final crystalline state (Liu et al., 1997; Kojima et al., 1997). The properties of the gel are dependent on the structure created during the sol stage of the sol-gel process.

Sol-gel nanocomposite materials (Fe_xO_y-SiO_2) were obtained using alkoxide and aqueous routes (Raileanu et al., 2003). The structure and properties of the prepared particles were compared for different precursors of silica (tetramethoxysilane, methyltriethoxysilane, colloidal silica solution, etc.).

Magnetite-based ferrofluids were synthesized through a modified sol–gel method (Gamarra et al., 2005). They precipitated iron oxyhydroxy in water in the presence of a surfactant (Renex), and next they partiallly reduced Fe^{3+} to Fe^{2+} by a mild drying process in N_2 atmosphere, leading to the formation of magnetite.

2.3 Free-solvent synthesis

The free-solvent synthesis of magnetite is an economical and non-toxic method. The solid iron salts ($FeCl_3 \cdot 6H_2O$ and $FeCl_2 \cdot 4H_2O$) and a solid base (NaOH) are used directly with the surfactant and mixed together using a mortar and a pestle (Ye et al., 2006). The synthesis procedure takes place at room temperature in a glove box filled with argon gas. This method is easy, without any advanced techniques and toxic solvents. The reaction duration is about a few minutes and the high yield magnetite obtained is monodisperse.

2.4 Other methods

Polyol process

Polyols are known to reduce metal salts to metal particles and to be good solvents for various inorganic compounds. This is why the synthesis of nano- and microparticles with well-defined shapes and controlled sizes can be performed by the polyol process, also (Fievet et al., 1989). The metal precursor is heated to a given temperature which cannot be higher than the boiling point of the polyols, with generation of intermediates that are reduced to metal nuclei and nucleate to form metal particles. Polyols act like solvents due to their high dielectric constants and like surfactants, preventing the particles aggregation. Their high boiling points offer the advantages of operating on a large temperature interval (from 25^{0}C up to the boiling point). The "core-shell" PVP (poly(vinylpyrrolidone))-nanoparticles can be obtained in one-pot polyol process by reduction of $Fe(acac)_3$ with 1,2-hexadecanediol in the presence of the PVP polymer surfactant in octyl ether (Liu et al., 2007). These particles are promising for biomedical applications like MRI agents and biosensors. The magnetite nanoparticles prepared by this method have a long-term stability and are easily coated by various materials to generate multifunctional nanoparticles. Non-aggregated magnetite nanoparticles have been already synthesized in liquid polyols at elevated temperature (Cai et al., 2007). Highly performance magnetite nanoparticles have been also obtained by a modified polyol process. For that, only one iron rich precursor has been used and no further reducing agent and surfactants were required. The magnetite nanoparticles can be coated in situ by hydrophilic polyol ligands, and the hydrophilic triethylene glycol magnetic nanoparticles are easily dispersed in water and other polar solvents. This method is a good alternative to produce superparamagnetic water-soluble magnetic nanoparticles and can replace the co-precipitation method.

Micro-emulsion

Emulsions are thermodynamically stable and consist of two different immiscible liquids (like oil - organic solvent and water) (Bagwe et al., 2001). The emulsion is stabilized by adding a surfactant. The size of magnetite particles synthesized by micro-emulsion method is several nanometers in diameter (López Pérez et al., 1997), except in some specific cases (20-30 nm) (Hirai et al., 1997). This method is difficult to control, the obtained yield is low and the particles are polydisperse. The commonly used method is the one that involves water-in-oil micro-emulsion and also includes a reverse micelle system (Ganguli et al., 2003). For the synthesis of Fe_3O_4, the precipitation technique consisting of alkalization of a solution of metal salt and hydrolysis in micro-emulsions (Boal et al., 2004; Hirai et al., 1997) is commonly used, as well as biosynthetic routes (Matsunaga, 1998).

Magnetite/silica particles were prepared by reverse-micelle micro-emulsions in an organic phase such as cyclohexane and heptane (Yang et al., 2004; Santra et al., 2001).

Fe_3O_4 nanoparticles in poly(organosilsesquioxane) with sizes of 4-15 nm were prepared by the one-pot synthesis using reverse micelle method (Ervithayasuporn et al., 2009). The ferrofluid droplets were in situ encapsulated, via polycondensation of molecularly self-assembled octenyltrimethoxysilane.

Sonochemical decomposition of iron pentacarbonyl

Magnetite nanoparticles were obtained by a sonochemical micro-emulsion polymerization process of n-butyl methacrylate (BMA) monomer (Teo et al., 2009). Iron tri(acetylacetonate), 1,2-tetradecanediol, oleic acid, and oleyl amine were mixed in benzyl ether, under vigorous stirring. The mixture was heated gradually under nitrogen atmosphere and the black-colored mixture was then cooled to room temperature by removing the heat source. The pre-formed particles were then encapsulated within the host poly(BMA) latex particles with 120 nm diameter and low size dispersion. The distribution of magnetite particles over the polymer particle population and within each polymer particle was nevertheless rather heterogeneous and the ratio of magnetite particles per polymer particle was determined as being equal to 50.

3. Influence of Fe^{2+} onto magnetite properties

The stoechiometric factor of Fe^{2+} and Fe^{3+} is important for the magnetic properties of the obtained particles (coercivity, crystallinity, sorbtion capacity, etc.) (Gorski et al., 2010). The Fe^{2+} chemical species direct the magnetite particles synthesis kinetics and their composition (Tronc et al., 1992). The Fe^{2+}/Fe^{3+} molar ratio is highly important if one focusses on obtaining magnetite particles with specific properties. Thus, a $Fe^{2+}/Fe^{3+}<0.1$ molar ratio is too small to achieve a stable solution. In this context, if the content of chemical species of Fe^{2+} is low, goethite (α-FeO(OH)) is obtained as the only stable product. The use of the $Fe^{2+}/Fe^{3+}>0.1$ molar ratio is favorable for obtaining magnetite instead of goethite.

The influence of the Fe^{2+}/Fe^{3+} molar ratio on the characteristics of magnetite particles obtained by co-precipitation of Fe^{2+} and Fe^{3+} (composition, size, morphology and magnetic properties) was studied by Jolivet (Jolivet et al., 1992). Chemical species in different proportions were precipitated with ammonia solution to pH~11. The analysis of the products obtained with different Fe^{2+}/Fe^{3+} molar ratios, in the range 0.10-0.50, concluded that for values lower than 0.30 two different phases coexist:

(a) the first phase contains particles of 4 nm in diameter, with oxyhydroxy (FeO(OH)) surface functional groups and low Fe^{2+} content, reflected by the ratio Fe^{2+}/Fe^{3+}~0.07;

(b) the second phase is characterized by increased content of Fe^{2+} (Fe^{2+}/Fe^{3+}~0.33 molar ratio), the final product is magnetite with increased particle sizes. It was noticed that the share of this phase increases with increasing the Fe^{2+}/Fe^{3+} molar ratio. Thus, for values greater than 0.35, the product is found only in the second phase.

In conclusion, for a 0.5 Fe^{2+}/Fe^{3+} molar ratio homogenous magnetite particles of uniform size and composition are more likely to be obtained. The order of addition of ionic species (Fe^{2+} and Fe^{3+}) in co-precipitation reaction does not influence the final characteristics (size, composition) of the obtained particles.

The influence of ionic species concentrations on the properties of magnetite particles was also followed. It was noted that the Fe^{2+}/Fe^{3+} molar ratio was a determining factor in obtaining sub-micron sizes, while by increasing the ratio, the mean diameter of the magnetic particles increased, but unfortunately the yield decreased (Babes et al., 1999).

4. Sterically functionalized magnetite core-shell particles

Magnetite surface chemistry

Magnetite particles act as Lewis acids in aqueous systems and coordinate water or hydroxyl groups. This is why the particles surface chemistry is highly dependent of pH value; at low pH the surface of the particles is protonated and at high pH the surface is negatively charged (Figure 2). The hydroxyl groups from magnetite surface are the reactive parts, which can react with acid or base. The HO- groups are reacting with other organic or inorganic anions and may adsorb protons or cations, also.

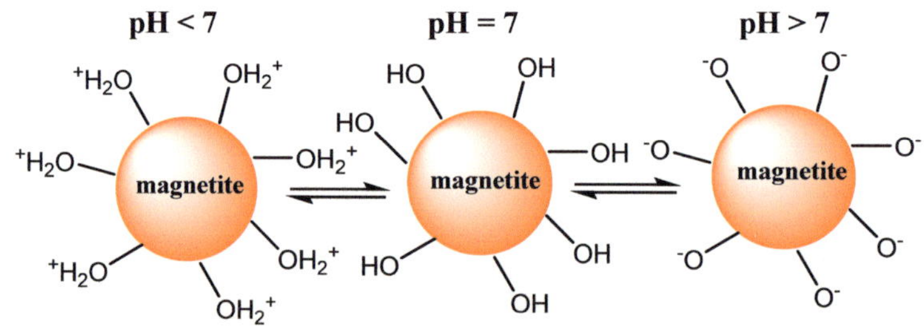

Fig. 2. Magnetite particles behavior at low/high pH.

Magnetic ferrofluids based on magnetite particles have a high capacity of agglomeration due to the fact that two types of forces co-exist: attraction and repulsive forces. The stabilization of the magnetic particles requires the control of these forces. To obtain stable magnetic dispersions, the attraction forces between magnetic particles must be overcomed by electrostatic or steric stabilization (Figure 3).

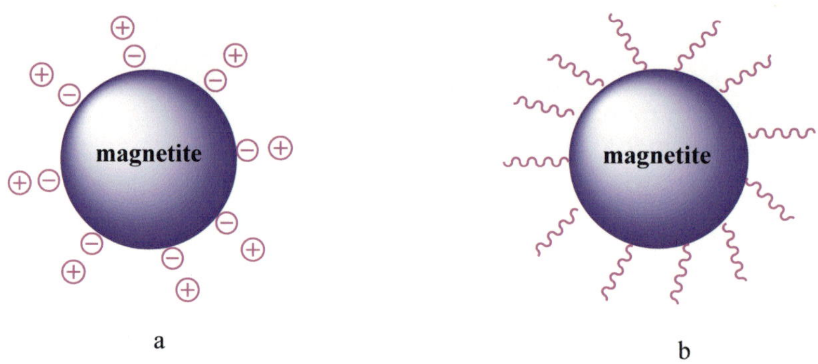

Fig. 3. Electrostatic (a) and steric (b) stabilization of the particles.

In order to stabilize magnetite particles, monomers/polymers with functional groups that can bind to the surface of the particles and act as surfactants can be used. The rest of the chain is solvated in dispersion medium or in a fluid. The process, known to be entropic or steric, refers to the inhibition of particles aggregation by an entropic force, which appears when the particles are closed to each other. As surfactants monomers, homopolymers, block copolymers and polymers with terminal functional groups can be successfully used. After the coating process, these particles are re-suspended in proper solvents and form homogenous suspensions named ferrofluids (Shen et al., 2004).

Magnetite particles precursors	**Coating agent**	**Diameter (nm)**	**Type of linkage**	**Reference**
$Fe(CO)_5$	Polyethylene, Polystyrene, Polyisobutylene - all three functionalized with tetraethylenepentamine	8-50	Hydrogen bond	Burke et al., 2002
$FeCl_2 \bullet 4H_2O$ $FeCl_3 \bullet 6H_2O$	Oleic acid, Stearic acid, Linoleic acid	5.7-9.3		Korolev et al., 2002
$FeSO_4$ $FeCl_3$	N-oleoylsarcosine	8.1-20.7		Xu et al., 2004
$FeCl_2 \bullet 4H_2O$ $FeCl_3 \bullet 6H_2O$	Double shell with oleic acid - Pluronic F-127, and then entrapped the doxorubicin	193 (DLS) 11 (TEM) 9.2 (XRD)		Jain et al., 2005
$FeCl_3 \bullet 6H_2O$ $FeSO_4 \bullet 6H_2O$	Xylan	25.26±0.42 μm		Andriola Silva et al., 2007
$FeCl_2 \bullet 4H_2O$ $FeCl_3 \bullet 6H_2O$	Carboxyethylsilanetriol sodium salt solution	10 (magnetite) 100 (core-shell particles)		Kobayashi et al., 2008
$FeCl_2 \bullet 4H_2O$ $FeCl_3 \bullet 6H_2O$	Ricinoleic acid and Poly(lactic-co-glycolic) acid (PLGA)	140-230		Furlan et al., 2010
$FeCl_2 \bullet 4H_2O$ $FeCl_3 \bullet 6H_2O$	Double shell: silica (SiO_2/Fe_3O_4) and aluminosilicate	100		Chang et al., 2009
$FeCl_3$ Na_2SO_3	Ethoxy-PEGsilane N-hydroxysuccinimide PEG-fluorescein Folic acid	10 (magnetite)	Covalent bond Covalent bond Hydrogen bond	Zhang et al., 2002
$FeCl_2 \bullet 4H_2O$ $FeCl_3 \bullet 6H_2O$	3-aminopropyltrimethoxy silane (APTMS) and monomethoxy-poly(ethylene glycol) (MPEG) $(CH_3CH_2O)_3$-Si-MPEG	200	Covalent bond	Kim et al., 2002
$FeCl_3 \bullet 6H_2O$ $FeSO_4 \bullet 7H_2O$	3-aminopropyltrimethoxy silane (APTES) and Polyethylene glycol diacid (HOOC–PEG–COOH)	40-60		Feng et al., 2008
$FeSO_4 \bullet 7H_2O$	Tetraethoxyorthosilicate (TEOS) $(OEt)_4Si$ 3-aminopropyl) triethoxysilane (APTES) CS_2, NaOH and 2-propanol	50-80		Girginova et al., 2010

Table 1. Core-shell magnetite particles.

The synthesis methods for magnetic particles with magnetite core and monomer/polymer shell, named core-shell magnetic particles, are presented in the literature (Table 1). The surfactant used must have functional groups to interact by hydrogen or covalent bonds (Figure 4) with the hydroxilic groups from pre-formed magnetite surface (trialkoxysilanes or compounds with carboxylic groups) (Kazufumi et al., 2008; Wormuth, 2001; Mondini et al., 2008; Shukla et al., 2007; Wilson et al., 2005) and other functional groups to permit their dispersing in a transport fluid (aminic, epoxy, vinyl groups, etc.).

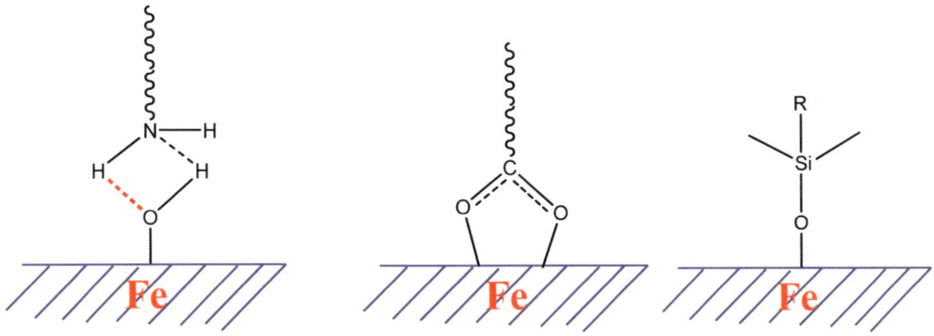

(a) (b)

Fig. 4. Schematic representation of hydrogen bond (a) and covalent bond (b) between core and shell.

5. Applications of micro- and nanomagnetite particles

The high interest for magnetic particles (ferrofluids, nanospheres and microspheres) can be explained due to their various applications, e.g., in magnetic storage devices or magnetic cell separations (Goya et al., 2003). The magnetic particles were also tested for cancer treatment and as drug and nucleic acids carriers (Lubbe et al., 1996).

Top domains for multiple applications of micro- and nanoparticles with magnetic properties are biomedical applications, catalysis and industrial applications.

5.1 Biomedical applications

Magnetite nanoparticles represent an intriguing material for various researchers, due to their biomedical applications in the treatment of solid tumours (Gomez-Lopera, 2006) or contrast agents (Frija et al., 2004). In order to be efficient in biological applications, colloidal suspensions of magnetic particles must have long-time stability and the magnetic core must respond to an external magnetic field that directs the particle to a desired location. Colloidal stability is influenced by the size, structure and composition of the particle with a narrow dimensional controlled polydispersity. All these requirements lead to magnetic fluids that flow through the capillaries that provide a maximum diameter of 1.4 μm. The colloidal stability of magnetite particles is insured by coating the magnetic core with a monomer or (co)polymer, as surfactant, to have the particles dispersed in a carrier fluid.

Magnetite (Fe_3O_4) nanoparticles that exhibit excellent magnetic saturation (~78 emu/g) are desirable for these applications due to the strong ferromagnetic behaviour, lower sensitivity to oxidation and relatively low toxicity as compared to many other materials (e.g., iron, nickel and cobalt) (O'Brien, 2003). Magnetite has been intensively studied in the past decades due to its application in biotechnology and biomedicine: cancer treatment (Jordan et al, 2001, Jordan et al., 1999, Herrera et al., 2008), sensors, catalysis, storage media, clinical diagnosis and treatment of some diseases (Torchilin et al., 2001, Jordan et al., 2001).

Especially for biomedical application, the particles must present some specific properties like:

1. Magnetic properties

In order to respond to an external magnetic field magnetic particles must have a large saturation magnetization, an important characteristic for targeted drug delivery systems and contrast agents for MRI (magnetic resonance imaging) (Bulte, 2006, Modo et al., 2005, Burtea et al., 2005, Boutry et al., 2006, Babes et al., 1999, Sonvico et al., 2005, Corot et al., 2006, Corot et al., 2007)). In the presence of an external magnetic field the magnetic moments are oriented with the field (Figure 5). Magnetic nanoparticles modified with organic molecules and used for biomedical applications can be magnetically controlled by applying an external magnetic field.

2. Stability in colloidal dispersions

The stability of the particles can be enhanced by coating with biocompatible surfactants (Ma et al., 2003) capable of interactions with hydroxyl groups on the magnetite surface and ensuring the ferrofluid stability. The coating agent should have specific functional groups for further functionalization for specific applications.

3. Nano- sized dimensions

The magnetic particles with biomedical applications must have controllable sizes ranging from a few nanometres up to tens of nanometres, smaller than those of a cell (10–100 μm) (Pankhurst et al., 2003), a virus (20–450 nm), a protein (5–50 nm) or a gene (2 nm wide and 10–100 nm long) because they can 'get close' to a biological entity of interest. Magnetic particles encapsulated into a polymer matrix with dimensions more than 1 μm or sub-micron is more effective in the bloodstream through the veins or arteries (Lazaro et al., 2005).

4. Oxidation stability

Metals such as Fe, Co and Ni, which oxidize easily, are not recommended in biomedical uses due to the higher probability of oxidizing inside human bodies. This problem can be solved using magnetite particles that not only present a greater stability to oxidation, but can be coated with different specific surfactants, increasing their resistance to oxidation.

The polymers used as surfactants can be synthetic or natural molecules able to yield stable colloidal suspensions designed for drug delivery. Nevertheless, synthetic polymers have the advantage of high purity and reproducibility over natural polymers.

The colloidal magnetic particles obtained for biomedical application (Asmatulu et al., 2005) as targeted drug delivery will have to pass through a few mechanisms of releasing drug molecules: diffusion, degradation of polymeric shell, swelling of the shell followed by diffusion of active principle outside the magnetic particle. The diffusion involves drug molecules dissolving in body fluids around or within the particles and migration of the drug away from the particles. The degradation mechanism occurs when the polymer chains are hydrolysed into lower molecular weight species and release the drug trapped by the polymer chains. The coated particles, when inside the body, swell to increase pressure, and the drug diffuses from the polymer network into the body.

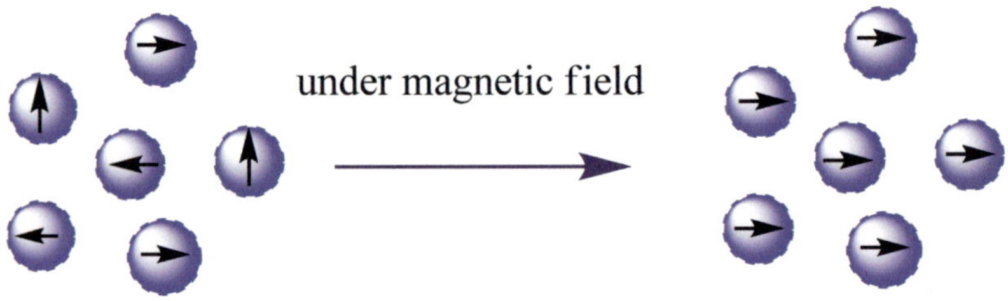

Fig. 5. Magnetite particles in presence/in absence of magnetic field.

Magnetic nanoparticles coated with different organic compounds offer a high potential for numerous biomedical applications, such as cell separation (Kuhara et al., 2004), automated DNA extraction (Yoza et al., 2003), gene targeting (Plank et al., 2003), drug delivery (Lazaro et al., 2005), magnetic resonance imaging (Glech et al., 2005), and hyperthermia (Pardoe et al., 2003). After coating with an antibody, they can be used for immunoassays (Chemla et al., 2000) or small substance recovery (Tanaka et al., 2004). DNA or oligonucleotides immobilized on magnetic particles have been used for DNA hybridization analyses to identify organisms (Matsunaga et al., 2001) and single-nucleotide polymorphism analyses for human blood (Maruyama et al., 2004).

Magnetic particles coated with immunospecific agents have been successfully bound to red blood cells (Molday et al., 1982, Tibbe et al., 1999), lung cancer cells (Kularatne et al., 2002), bacteria (Morisada et al., 2002), urological cancer cells (Zigeuner et al., 2003). Coupling of homing peptide labeled with 5-carboxyl-fluorescein (FAM-A54) and magnetite coated with starch chains functionalized with homing peptide was performed by Schiffs' reaction (Jiang et al., 2009). This magnetic fluid has great potential applications in diagnostics and therapeutics of human tumours.

Nano-sized iron oxide particles (Fe_3O_4) were coated with a lipid layer and then functionalized with biotin for further protein ligands. These particles were then used for binding streptavidin-fluorescein isothiocyanate generating a conjugate used to investigate their magnetic and fluorescent activity into cells (Beckera et al., 2007).

The magnetic particle application for hyperthermia goes back to 1957 (Gilchrist et al., 1957). The authors heated up various damaged tissue samples with 20–100 nm size particles of γ -Fe_2O_3 exposed to a 1.2 MHz magnetic field. Since then the scientists have been developing numerous methods and schemes using different types of magnetic materials, different field strengths and frequencies of encapsulation and delivery of the particles (Mosso et al., 1973, Rand et al., 1976, Gordon et al., 1979, Rand et al., 1981, Borrelli et al., 1984, Hase et al., 1990, Suzuki et al., 1990, Chan et al., 1993, Matsuki et al., 1994, Mitsumori et al., 1994, Suzuki et al., 1995, Moroz et al., 2001, Jones et al., 2002). The hyperthermia procedure involves the dispersion of the magnetic particles throughout the target tissue, followed by applying an AC magnetic field with specific parameters (frequency, strength); the magnetization of the particles is continuously reversed, which translates into a conversion from magnetic to thermal energy and causes the heating of particles. The tumour is destroyed if the temperature is maintained above the therapeutic threshold of 42^{0}C for 30 min or more.

The bisphosphonate modified magnetite nanoparticles were used as agents to remove uranyl ions (e.g., UO_2^{2+}) in vivo from water or blood (Wang et al., 2006). The authors synthesized a bisphosphonate derivative, DA-BP, which modifies the magnetite nanoparticles for the successful decorporation of the radioactive metal from blood due to the chelating effect (Fukuda, 2005). Starting from this study, a protocol that allows the detection and recovery of other heavy metal toxins from the biological systems has been developed and perfected.

Coated magnetite particles with magnetic resonance, fluorescent imaging and drug delivery applications were obtained (Kim et al., 2008). The particles obtained are monodisperse, with dimensions smaller than 100 nm, coated with silica shell using cetyltrimethylammonium bromide as surfactant for the transfer of hydrophobic nanocrystals to aqueous media and also as organic template in the sol-gel reaction. The silica-Fe_3O_4 particles were functionalized with PEG for further biomedical applications in cancer diagnosis and therapy.

Magnetite particles (nanoparticles with 5–300 nm or microparticles with 300–5,000 nm) have been surface functionalized and used as biosensors to recognize specific molecular targets. Three types of biosensors, (i) magnetic relaxation witches, (ii) magnetic particle relaxation sensors, and (iii) magnetoresistive sensors, were coated with different biosensing principles and magnetic materials for instrumentation (Josephson, 2009). Magnetic relaxation switch assay-sensors are based on the effects of the magnetic particles exert on water proton relaxation rates. Magnetic particle relaxation sensors determine the relaxation of the magnetic moment within the magnetic particle. Magnetoresistive sensors detect the presence of magnetic particles on the surface of electronic devices and are sensitive to changes in magnetic fields on their surface.

Magnetic resonance imaging (MRI) is a non-invasive technique without radiations exposure that can provide cross-sectional images from inside the solid materials and living organisms (Kim et al., 2003, Richardson et al., 2005). Magnetite nanoparticles are suitable for MRI contrast (Bean et al., 1959, Koenig et al., 1987). Magnetite particles have been proposed for oral use as magnetic resonance contrast agents and magnetic markers to study the gastrointestinal motility (Briggs et al., 1997, Ferreira et al., 2004). The gastric secretions include pepsin, mucus, and HCl (Paulev, 1999-2000) and magnetite particle dissolution may take place during their passing through the stomach. This could reduce the signal for MRI or for monitoring gastrointestinal motility. The problem that appears in this domain is to protect compounds from gastric environment and the xylan is a good candidate for this issue. Xylans are polysaccharides made from units of xylose and can be found as structural components in the cell walls of some green algae. Xylan-coated magnetic microparticles were obtained (Andriola Silva et al., 2007) to protect magnetite from gastric dissolution.

5.2 Applications in catalysis

The aim of the catalysis is to search new catalysts that can be recovered when their activity is ended (Sankaranarayanapillai et al., 2010). The magnetic dispersions are attractive in separation applications as they offer high surface area and can be functionalized to apply to different molecular or cellular species. The magnetic fluids can be used as magnetically separable nanocatalytic systems that combine the advantages of homogeneous and heterogeneous catalysis (Lu et al., 2004). Novel thermoreversible magnetic fluids based on magnetite (Fe_3O_4) coated with a covalently anchored polymeric shell of poly(2-methoxyethyl methacrylate) (PMEMA) were synthesized by surface-initiated ATRP (Gelbrich et al., 2006). The coated particles form stable dispersions in methanol at temperatures above an upper critical solution temperature.

Magnetically recoverable heterogenized nanoparticle supported chiral Ru complexes were obtained and used in highly enantioselective asymmetric hydrogenation of aromatic ketones (Hu et al., 2005). The catalysts can be recycled by magnetic decantation and used for asymmetric hydrogenation for up to 14 times without loss of activity and enantioselectivity.

5.3 Industrial applications

Ferofluids have numerous applications in rotary and linear seals as magneto transformers, sensors and pressure transducers (flow speed), inertial sensors (acceleration, inclination, and gravity) (Vekas, 2008), passive and active bearings, vibration dampers, linear and rotary drives, and chemical industry (microtechnologies) as chips, etc. Among the most important and commercial applications are developed magnetofluids rotating seals, which are found today in a variety of equipments, in electronics, nuclear industry, biotechnology, aeronautics and space applications.

Magnetofluids sealing advantages compared with mechanical seals are: sealing without leaking, life time (~ 5 years), only viscous friction, zero contamination, field operation (high vacuum (10-8 mbar) at approx. 10 bar), relatively simple construction and low manufacturing cost (www.roseal.topnet.ro).

For seals in dynamic regime, ferrofluids must obey a series of requirements (www.roseal.topnet.ro): high saturation magnetization (up to 50 kA/m (approx. 600 g)), physical properties adapted to operating conditions, especially in gas pressure and temperature sealed, high colloidal stability in intense and powerful magnetic field, Newtonian flow properties even in intense magnetic field.

Ferrofluids should operate at 150^0C in continuous system or at temperatures of 200^0C in intermittent system. They should also work in winter conditions (-20^0C) or in space environment (-55^0C) and to resist at nuclear radiation.

Researchers at NASA (Rosensweig et al., 1985) experienced ferrofluids in a spacecraft control system. While ferrofluids are used often in commercial processes, for the manufacture of CDs to create sophisticated suspension systems for cars, a researcher at

NASA (Kodama) (http://www.interactivearchitecture.org/ferrofluid-sculptures-by-sachiko-kodama.html) choose to use the dynamic qualities of these fluids. Thus, Kodama models their shape using a computer to control the electro-magnetic fields and creating different forms that continually reinvent themselves without the help of video effect. In the printing industry, special inks are used to protect valuable documents, labels, packaging or brand. Magnetic ink contains a pigment that produces magnetic pulses detected by magnetic reading devices, and messages can be read only by exposure to the equipment (http://www.pricon.ro/ro/special.php).

6. XPS results for core-shell magnetic nanoparticles

Magnetite particles coated by different Si-containing compounds (monomers and polymers) were prepared using co-precipitation method or a free-solvent synthesis. The covered magnetite particles were obtained in one step by co-precipitation of iron chloride salts in presence of monomers or (co)polymers adequately functionalized, as surfactants. When the co-precipitation was in the absence of surfactant, the covered particles were obtained in two steps, the first step consisting in the obtaining of magnetite particles and in the second step, the magnetite particles being covered with an adequate monomer or (co)polymer, specific for the desired application. The magnetite particles covered with aminosilane were obtained by replacement of the shell of magnetite-oleic acid particles which were synthesised by a free-solvent method. The average diameters of magnetic nanoparticles are depending on the synthesis method, are less than 1 μm and the smallest dimensions are obtained by free-solvent method (Table 2).

In this chapter we focused on the characterization of core-shell magnetic nanoparticles as determined by X-ray photoelectron spectroscopy (XPS). XPS, also known as Electron Spectroscopy for Chemical Analysis (ESCA), is an elemental analysis technique and used to determine quantitatively the atomic composition and surface chemistry. XPS spectra are obtained by irradiating a material with a beam of X-rays while simultaneously measuring the kinetic energy and number of electrons that escape from the top 1 to 10 nm of the material being analysed (Watts et al., 2003; http://en.wikipedia.org/wiki/X-ray_photoelectron_spectroscopy).

The surface of the magnetite particles modified with aminosilane (Ma-APTES) or with polydimethylsiloxane–carboxy-terminated poly(ethylene oxide) graft copolymer (Ma-PDMSgPEO-COOH) was carefully followed using XPS data.

Hydrophilic magnetite particles (Ma-APTES) were prepared by two methods:

a. Co-precipitation of Fe^{2+} and Fe^{3+} ions in presence of ammoniacal solution according to the Massart method (Massart, 1981). The organic-inorganic "core-shell" magnetic particles (core: magnetite, shell: aminosilane monomer) were obtained by the condensation reaction between the hydroxyl groups on the magnetite surface and the ethoxyl groups from silane structure (Figure 6) (Durdureanu-Angheluta et al., 2008);
b. The replacing, with aminosilane, of oleic acid of magnetite-oleic acid particles which were obtained by free-solvent method (Figure 7, Table 2) (Durdureanu-Angheluta et al., 2011b).

DLS, SEM and AFM analysis indicated the formation of low polydispersity particles, with a spherical morphology and dimensions of nanometers (Table 2).

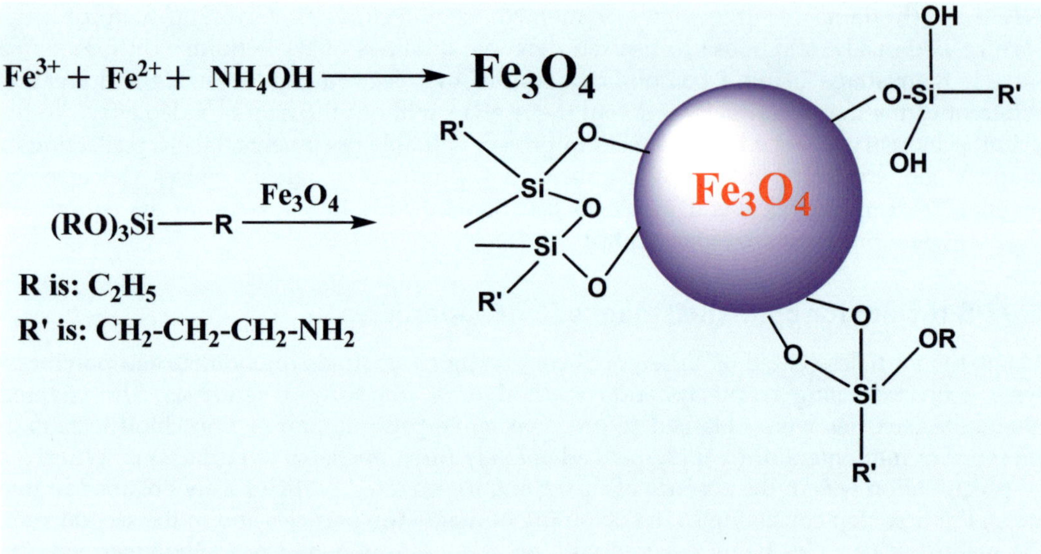

Fig. 6. Schematic principle of covalent bonding between magnetite core and ethoxyl groups of silane monomer (Durdureanu-Angheluta et al., 2008).

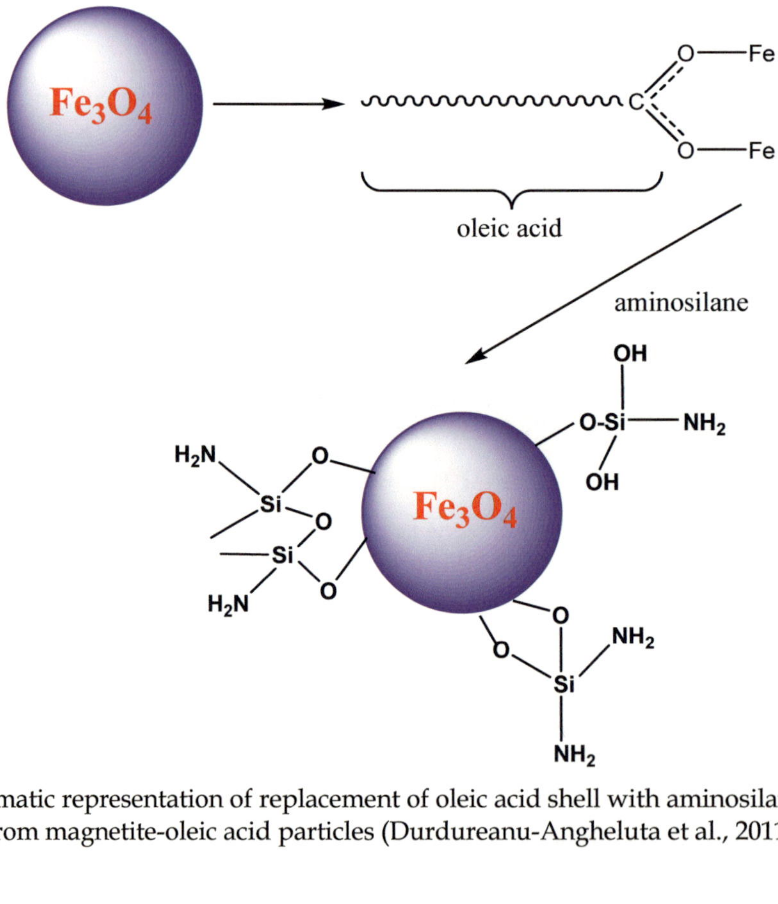

Fig. 7. Schematic representation of replacement of oleic acid shell with aminosilane monomer from magnetite-oleic acid particles (Durdureanu-Angheluta et al., 2011b).

Code sample	Shell type*	Fe^{2+}/Fe^{3+}**	Average diameter (nm) (DLS/AFM/SEM)	Magnetite synthesis method	Reference
Ma-ATES	allyltriethoxysilane	2	496	Co-precipitation	Durdureanu-Angheluta et al., 2008
Ma-MTES	triethoxymethylsilane	2	532	Co-precipitation	Durdureanu-Angheluta et al., 2008
Ma-$APTES_1$	(3-aminopropyl)triethoxysilane	2	500	Co-precipitation	Durdureanu-Angheluta et al., 2008
Ma-GOPS	(3-glycidoxypropyl)trimethoxysilane	2	270	Co-precipitation	Durdureanu-Angheluta et al., 2009
Ma-PDMS-TES	α-triethoxysilil-polydimethylsiloxane	0.5 1	670-800 200-400	One-step co-precipitation Two-step co-precipitation	Durdureanu-Angheluta et al., 2010
Ma-PDMSgE	siloxane functionalized with ester groups	2	173	Co-precipitation	Durdureanu-Angheluta et al., 2011a
Ma-PDMSgPEO-COOH	polydimethylsiloxane–carboxy terminated poly(ethylene oxide) graft copolymer	0.5	500	Co-precipitation	Pricop et al., 2010
Ma-OA	oleic acid	0.15 0.25 0.35	9 15 400	Free solvent synthesis	Durdureanu-Angheluta et al., 2011b
Ma-$APTES_2$	(3-aminopropyl)triethoxysilane	0.15	33	Interchange of shell of Ma-OA	Durdureanu-Angheluta et al., 2011b

* Core of particles: magnetite (Ma)
** Fe^{2+}/Fe^{3+}: molar ratio used in the synthesis process

Table 2. Coated magnetite particles.

The XPS spectra for Ma-APTES$_1$ (Table 2 and Figure 6) were recorded on a XR6 monochromated X-ray source (Thermo Scientific Escalab250Xi), with AlKalpha radiation and a variable 200-900 μm spot size. Charge correction was performed using C1s at 284.8 eV as a reference. The presence of C 1s, O 1s, Si 2p, N 1s, as well Fe 2p peaks in XPS spectra on Ma-APTES$_1$ sample give a proof for the chemical structure of aminopropyltrisiloxy coated magnetite particles. The Si 2p spectrum (Figure 8) has a peak at 102.4 eV, which is typical of silicone atoms, but is also consistent with the oxysilane group expected in the sample.

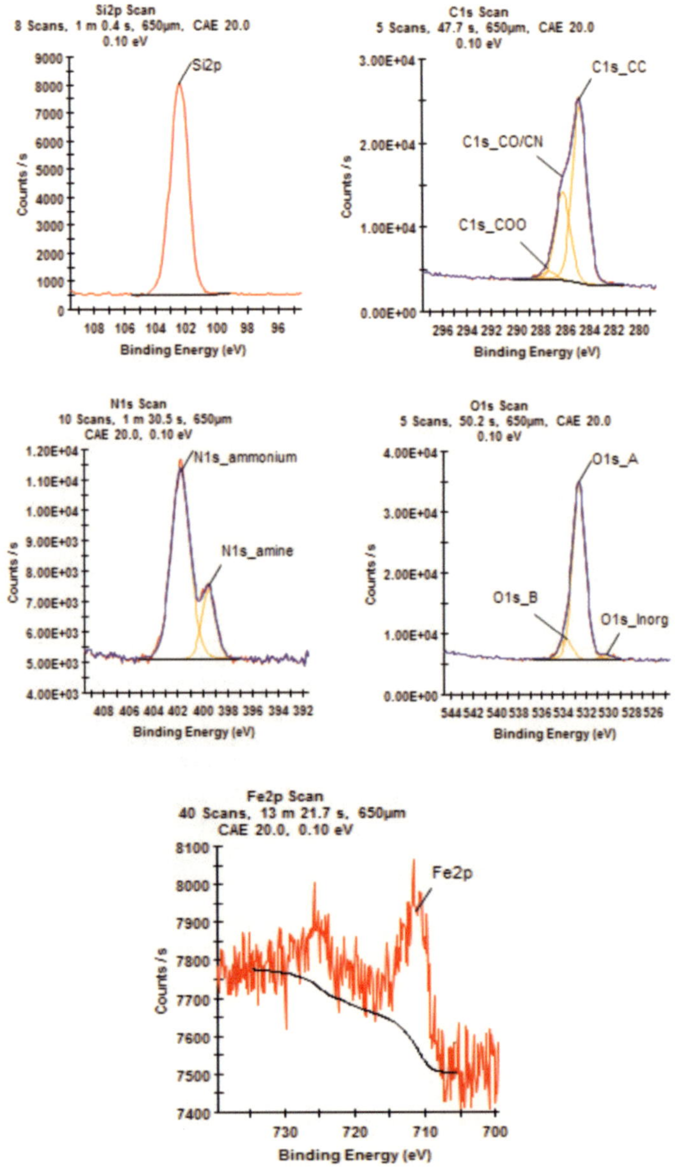

Fig. 8. High-resolution Si 2p, C 1p, N 1s, O 1s and Fe 2p XPS spectra for Ma-APTES$_1$ from Table 2.

C 1s (Figure 8) has two clear peaks, and requires three for a good fit. These have been assigned to C-C, C-O/C-N (should really be at least 2 peaks) and a weak peak tentatively assigned to O-C=O. N 1s (Figure 8) has two peaks, which appear to represent amine and ammonium (or ammonia) in the sample. O 1s (Figure 8) has three peaks, probably from different C-O and Si-O states, and also an inorganic oxide which is presumably the magnetite. Fe 2p (Figure 8) is extremely weak, so it is difficult to confirm the iron chemistry in more detail than "oxide".

The structure of Ma-$APTES_2$ coated magnetite nanoparticles (Table 2) was demonstrated by X-ray photoelectron spectroscopy (XPS), showing the formation of chemical bonds between the surface of magnetite particles and the oxygen atoms of the shells (Durdureanu-Angheluta et al., 2011b). XPS was performed on a KRATOS Axis Nova (Kratos Analytical, Manchester, United Kingdom), using AlKα radiation, with 20 mA current and 15 kV voltage (300 W), and base pressure of 10^{-8} to 10^{-9} Torr in the sample chamber. The incident monochromated X-ray beam was focused on a 0.7 mm x 0.3 mm area of the surface. The XPS survey spectra for the magnetite sample was collected in the range of -5÷1200 eV with a resolution of 1 eV and a pass energy of 160 eV. The high resolution spectra for all the elements identified from the survey spectra were collected using pass energy of 20 eV and a step size of 0.1 eV. The data were analyzed using the Vision Processing software (Vision2 software, Version 2.2.8). A linear background was subtracted before the peak areas were corrected. The binding energy of the C 1s peak was normalized to 285 eV.

In Figure 9 the peak at 283.4 eV belonging to the carbon from C-Si, 285.2 eV, was assigned to the carbon in the aliphatic chain (C-C) and the peak at 286.6 eV was ascribed to the carbon from C-N. The others two small peaks at 287.9 and 289.2 eV belonging to the carboxylate (C=O, COO^-), were in agreement with the data obtained in the previous literature (Zhang et al., 2006), and come from traces of oleic acid. The lack of peak at 290 eV indicates the absence of free oleic acid. The bonding energies at 102.3 and 103.0 eV (Figure 10) were the characteristic peaks from Si-O and Si-C bonds, respectively. Deconvolution of the N 2p peaks (Figure 11) indicated, at 399.3 and 400.8 eV, the presence of amino groups (NH_2) and tertiary amino groups, respectively. From the deconvolution of O 1s peaks (Figure 12) resulted three peaks from magnetite (around 530.8 eV), HO-Fe (around 531.4 eV) (Khurshid et al., 2009) and O-Si oxygen bonds (around 532.4 eV). The peak at around 535 eV could be due to the traces of oleic acid in the layer. The bonding energies at 710.6 and 723.7 eV were the characteristic peaks from Fe $2p_{3/2}$ and Fe $2p_{1/2}$, respectively. Deconvolution of the Fe 2p has shown the presence of Fe^{3+} and Fe^{2+} in both Fe $2p_{2/3}$ and Fe $2p_{1/2}$ regions (Fe^{3+} $2p_{2/3}$/Fe^{3+} $2p_{1/2}$: 710.7 eV/724.4 eV; Fe^{2+} $2p_{2/3}$/Fe^{2+} $2p_{1/2}$: 709.2 eV/722 eV). It should be mention that an energy difference of 13.1 eV between $2p_{3/2}$ and $2p_{1/2}$ peaks indicates that the material has, as dominant phase, Fe_3O_4 and, as secondary phase, siloxy-Fe and carboxylate-Fe bonds (Zhang et al., 2006). Also, XPS results on the Fe^{2+}/Fe^{3+} molar ratio of 0.13 (Figure 13) confirm the Fe^{2+}/Fe^{3+} molar ratio used in the synthesis of the magnetite particles (Ma-OA from Table 2).

Hydrophobic magnetite particles (Ma-PDMSgPEO-COOH) were obtained by co-precipitation of iron salts in water at pH around 11 in the presence of dichloromethane solution of siloxane copolymer (PDMSgPEO-COOH) (Figure 10) (Pricop et al., 2010).

Dimensional analysis of magnetite particles coated with PDMSgPEO-COOH (Table 2) showed core-shell morphology of an approximate spherical shape, with an average

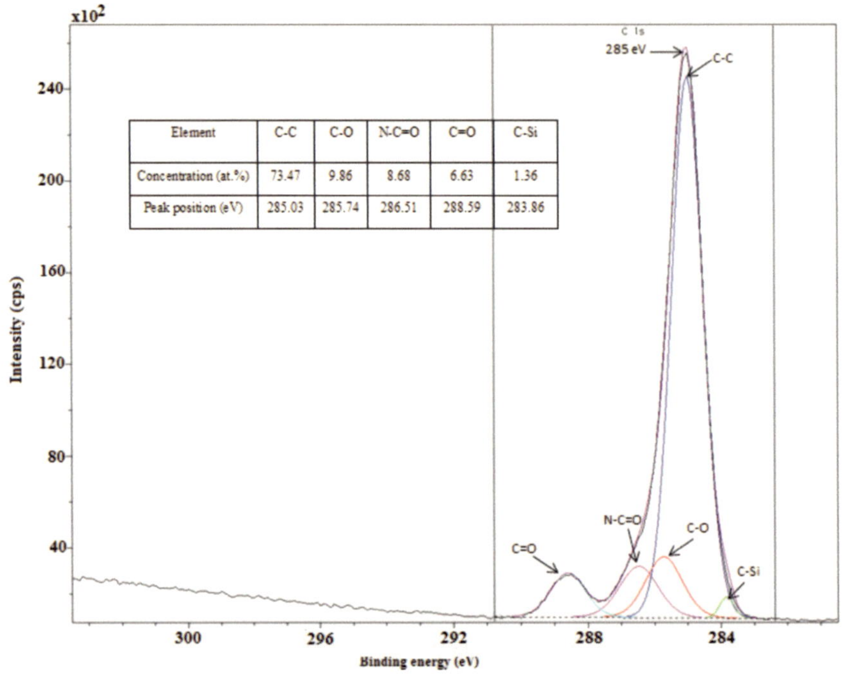

Element	C-C	C-O	N-C=O	C=O	C-Si
Concentration (at.%)	73.47	9.86	8.68	6.63	1.36
Peak position (eV)	285.03	285.74	286.51	288.59	283.86

Fig. 9. High-resolution C 1s XPS spectra for Ma-APTES$_2$ from Table 2 (Durdureanu-Angheluta et al., 2011b).

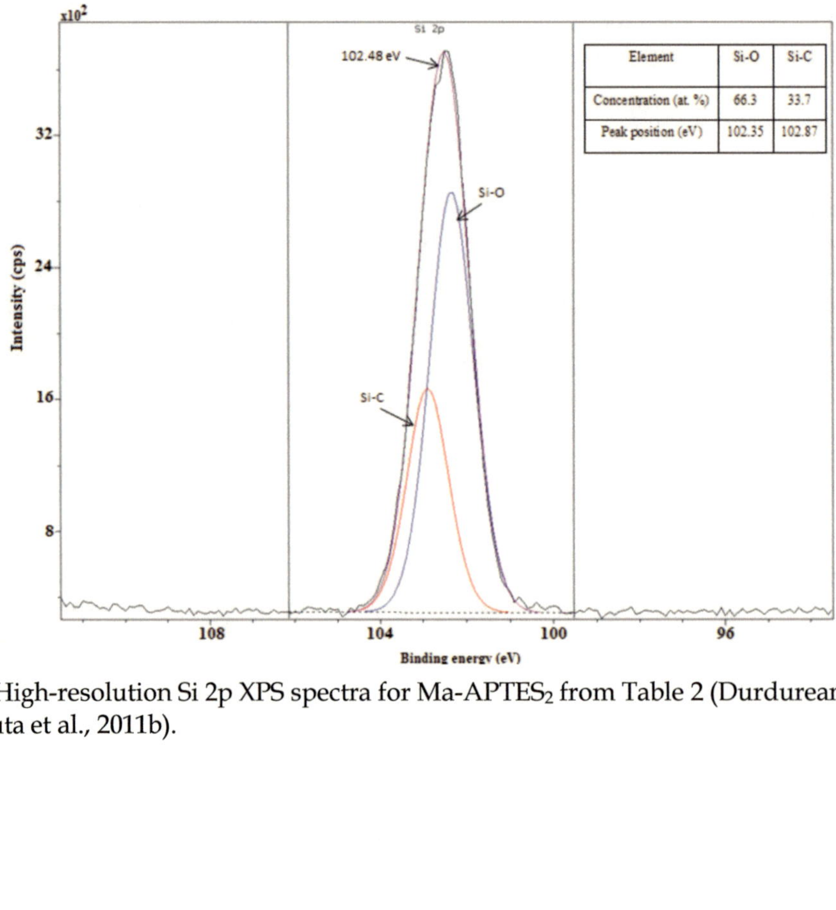

Element	Si-O	Si-C
Concentration (at. %)	66.3	33.7
Peak position (eV)	102.35	102.87

Fig. 10. High-resolution Si 2p XPS spectra for Ma-APTES$_2$ from Table 2 (Durdureanu-Angheluta et al., 2011b).

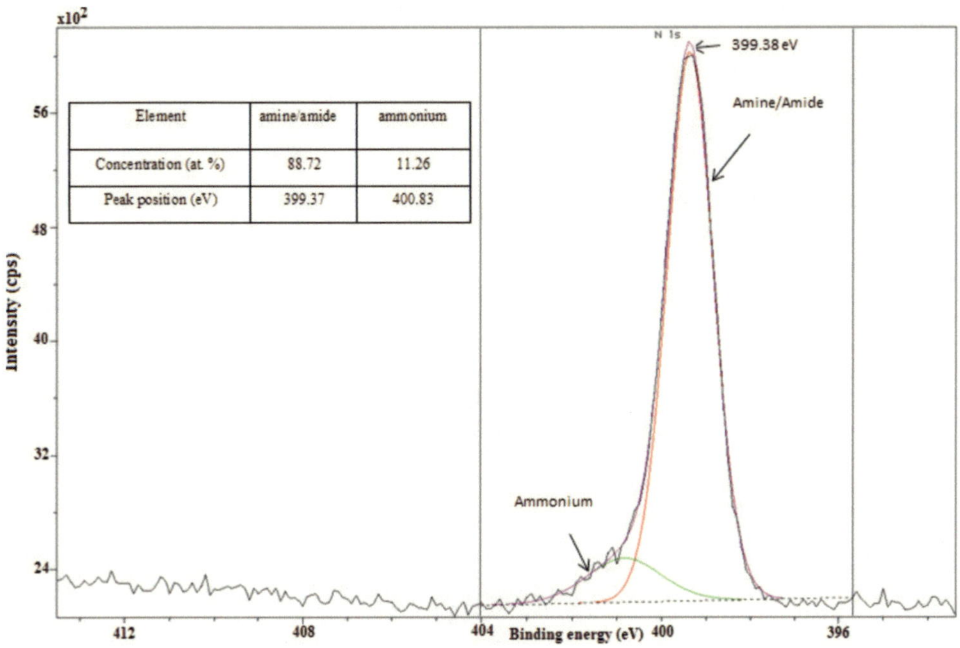

Fig. 11. High-resolution N 1s XPS spectra for Ma-APTES$_2$ from Table 2 (Durdureanu-Angheluta et al., 2011b).

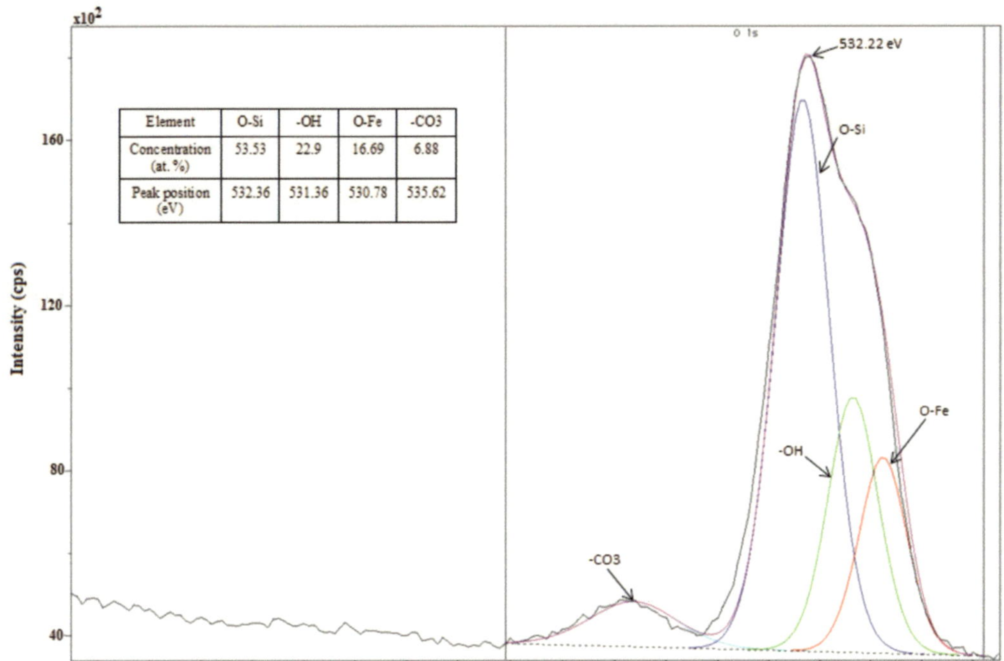

Fig. 12. High-resolution O 1s XPS spectra for Ma-APTES$_2$ from Table 2 (Durdureanu-Angheluta et al., 2011b)

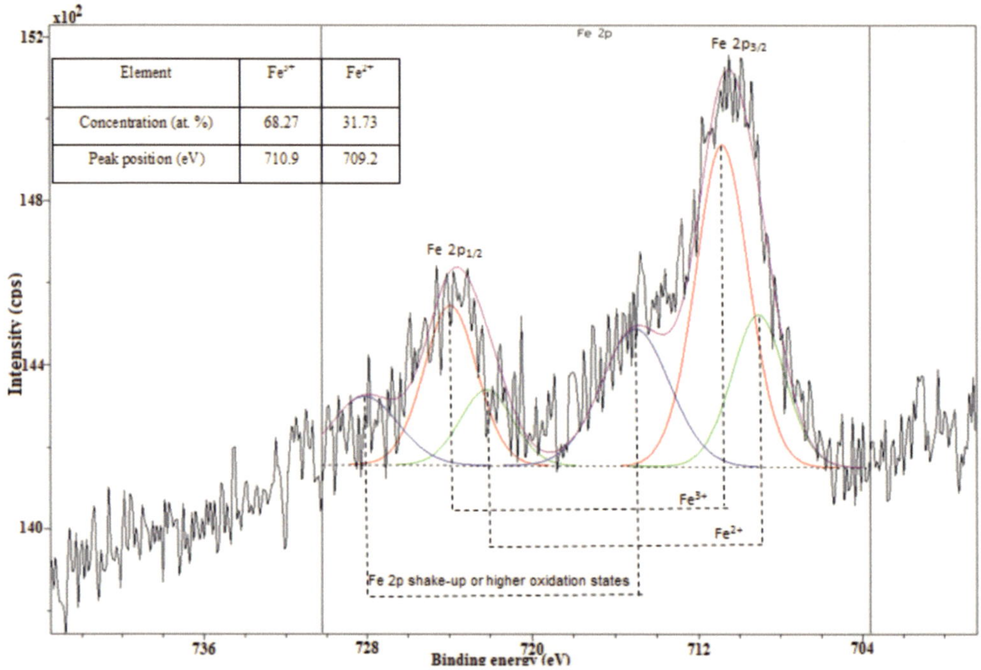

Fig. 13. High-resolution Fe 2p XPS spectra for Ma-APTES$_2$ from Table 2 (Durdureanu-Angheluta et al., 2011b)

diameter of 500 nm. The amphiphilic graft copolymer was proved to be efficient as surfactant in the preparation of magnetite particles with a hydrophobic siloxane shell (Pricop et al., 2010).

Ma-PDMSgPEO-COOH particles were analysed on the AXIS Nova X-ray photoelectron spectrometer built around the AXIS technology. Data reduction and processing was performed using Kratos' Vision 2 Processing software.

XPS scans of Si 2p atoms evidenced the appearance of the peaks at around 100 and 102 eV, characteristic for Si-O and Si-C bonds, respectively (Figure 15). C 1s atoms present specific peaks for CH_3-Si, CH_2-Si, C-O-C, C-C, C=O, O-C=O (Figure 15) in the 282 to 289 eV interval. The deconvolution of O 1s shows the presence of the peak at around 530 eV (Figure 15), corresponding to the metallic oxide and the presence of two peaks, with higher binding energies due to hydroxyl groups or from the coating of the sample. The deconvolution of Fe 2p atoms (Figure 15) presents two peaks at 710.7 and 712.8 eV, attributed to Fe^{3+} $2p_{2/3}$ and Fe^{2+} $2p_{2/3}$, respectively.

The presence of peaks at 724.0 and 726.0 eV of Fe^{3+} $2p_{1/2}$ and Fe^{2+} $2p_{1/2}$, respectively and the difference of 13.2 eV between Fe $2p_{3/2}$ and Fe $2p_{1/2}$ explain the presence of the carboxylate-Fe bonds (Zhang et. al., 2006).

It should be underlined that the XPS results were substantiated by FT-IR data, indicating the formation of the hydrogen or covalent bonds between the iron oxide substrate and the functional group of the surfactant.

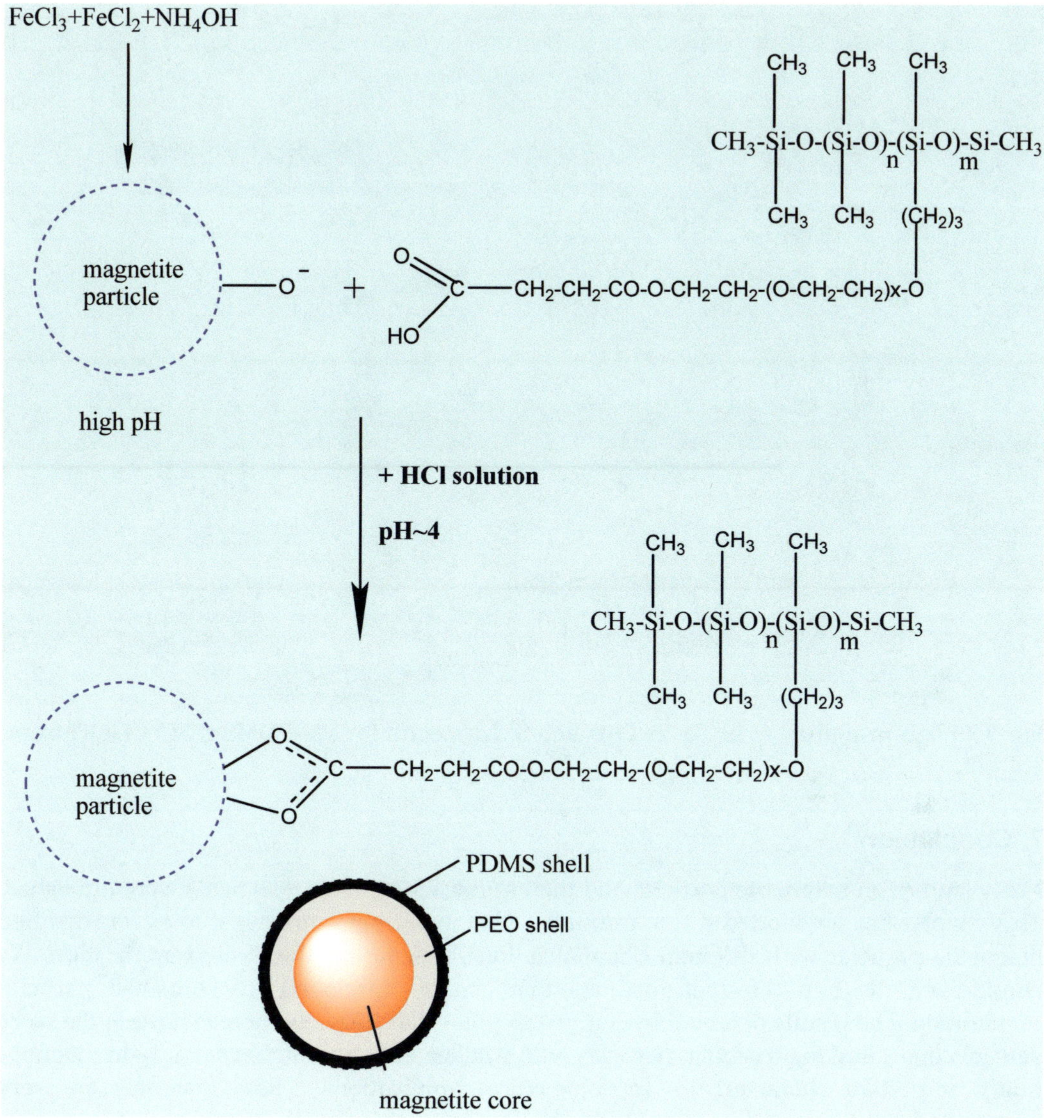

Fig. 14. Schematic principle of hydrogen bonding between magnetic core and the hydrophobic siloxane shell.

The core-shell magnetic particles present superparamagnetic properties and their saturation magnetization values are correlated with those reported by the literature.

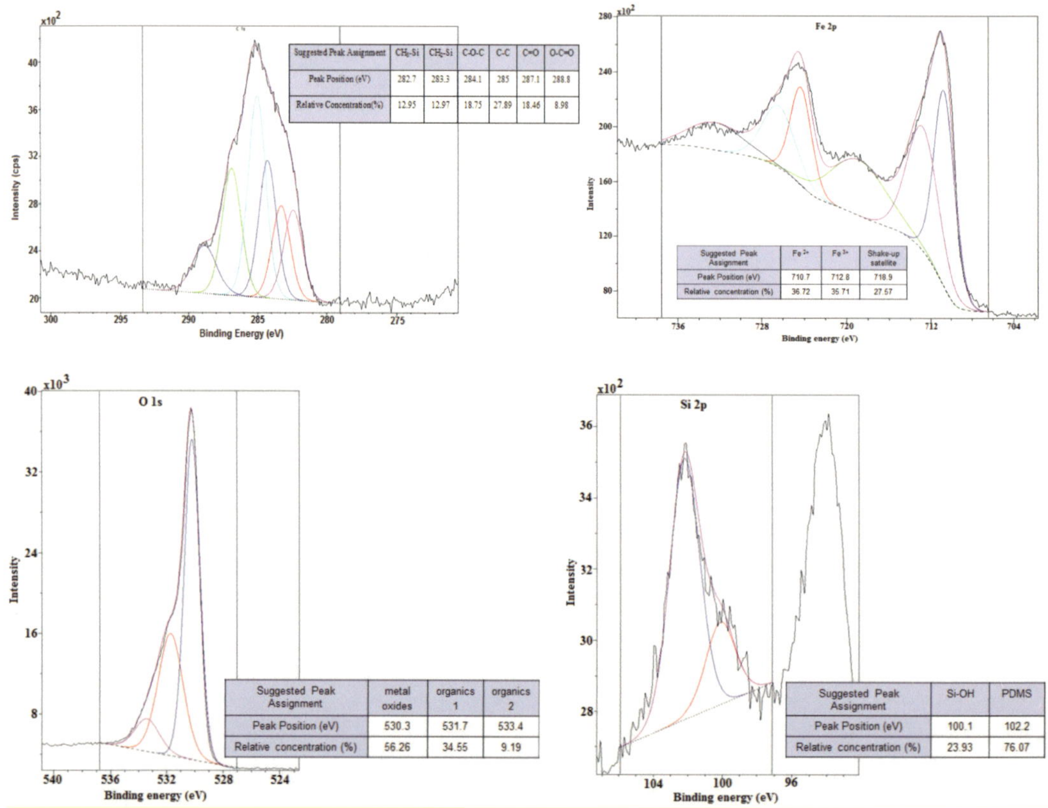

Fig. 15. High-resolution C 1s, Fe 2p, O 1s and Si 2p spectra for Ma-PDMSgPEO-COOH from Table 2.

7. Conclusion

Many studies on magnetite particles and their applications in various fields were published. This chapter has embedded a few examples, also including our own studies on modified magnetite particles with different fictionalizations, specific for the desired application. We emphasized the two different methods that can be employed for core-shell particles preparation. The results obtained by our group point that free-solvent technique is the more versatile one, allowing to obtain particles with smaller sizes and higher stability in magnetic fluids, important characteristics for biomedical applications. These materials are very promising for the immobilization of biologically active compounds.

8. Acknowledgment

This chapter was supported by the Project financed from STRUCTURAL FUNDS within the framework of the Sectorial Operational Programme "Increasing of the Economic Competitiveness", Priority Axis 2 - Operation 2.2.1. - Developing the existing R&D infrastructure and creating a new R&D infrastructure.

9. References

Asmatulu, R.; Zalichb, M.A.; Clausa, R.O. & Riffle, J.S. (2005), Synthesis, characterization and targeting of biodegradable magnetic nanocomposite particles by external magnetic fields. *J. Magn. Magn. Mater.*, 292: 108-119.

Babes, L.; Denizot, B.; Tanguy, G.; Le Jeune, J.J. & Jallet, P. (1999), Synthesis of iron oxide nanoparticles used as MRI contrast agents: A parametric study. *J. Colloid Interface Sci.*, 212(2) : 474-482.

Bagwe, R.P.; Kanicky, J.R; Palla, B.J.; Patanjali, P.K. & Shah, D.O. (2001), Improved drug delivery using microemulsions: rationale, recent progress, and new horizons. *Crit. Rev. Ther. Drug Carrier Syst.*, 18(1): 77-140.

Bean, C.P. & Livingston, J.D. (1959), Superparamagnetism. *J. Appl. Phys.*, 30: 120-129.

Beckera, C.; Hodenius, M.; Blendinger, G.; Sechi, A.; Hieronymus, T.; Muller-Schulte, D.; Schmitz-Rode, T. & Zenke, M. (2007), Uptake of magnetic nanoparticles into cells for cell tracking. *J. Magn. Magn. Mater.*, 311: 234-237.

Bee, A.; Massart, R. & Neveu, S. (1995), Synthesis of very fine magnetite particles. *J. Magn. Magn. Mater.*, 149: 6-9.

Boal, A.K. (2004), Synthesis and application of magnetic nanoparticles, In: *Nanoparticles – Building blocks for nanotechnology*. Rotello V. (ed), pp 1–27, Kluwer Academic/Plenum Publishers, New York.

Boistelle, R. & Astier, J.P. (1988), Crystallization mechanisms in solution. *J. Cryst. Growth*, 90: 14-30.

Borrelli, N.F.; Luderer, A.A.; & Panzarino, J.N. (1984), Hysteresis heating for the treatment of tumours. *Phys. Med. Biol.*, 29: 487–494.

Boutry, S.; Laurent, S.; Elst, L.V. & Muller, R.N. (2006), Specific E-selectin targeting with a superparamagnetic MRI contrast agent. *Contrast Med. Mol. Imaging*, 1(1): 15-22.

Briggs, R.W.; Wu, Z.; Mladinich, C.R.J.; Stoupis, C.; Gauger, J.; Liebig, T.; Ros, P.R.; Ballinger, J.R. & Kubilis, P. (1997), In vivo animal tests of an artifact-free contrast agent for gastrointestinal MRI. *Magn. Reson. Imaging* 15(5): 559-566.

Bulte, J.W. (2006), Intracellular endosomal magnetic labeling of cells, Review. *Methods Mol. Med.*, 124: 419-439.

Burke, N.; Stover, H. & Dawson, F. (2002), Magnetic nanocomposites: preparation and characterization of polymer-coated iron nanoparticles. *Chem. Mater.*, 14: 4752-4761.

Burtea, C.; Laurent, S.; Roch, A.; Elst, L.V. & Muller, R.N. (2005), C-MALISA (cellular magnetic-linked immunosorbent assay), a new application of cellular ELISA for MRI. *J. Inorg. Biochem.*, 99(5): 1135-1144.

Cai, W. & Wan, J. (2007), Facile synthesis of superparamagnetic magnetite nanoparticles in liquid polyols. *J. Colloid Interface Sci.*, 305: 366-370.

Chang, C.F.; Wu, Y.L. & Hou, S.S. (2009), Preparation and characterization of superparamagnetic nanocomposites of aluminosilicate/silica/magnetite. Colloids Surf. A Physicochem. Eng. Aspects, 336: 159-166.

Chemla, Y.R.; Grossman, H.L.; Poon, Y.; McDermott, R.; Stevens, R.; Alper, M.D. & Clarke, J. (2000), Ultrasensitive magnetic biosensor for homogenous assay. *Proc. Natl. Acad. Sci.*, 97: 14268-14272.

Cornell, R.M. & Schertmann, U. (1991). *Iron Oxides in the Laboratory: Preparation and Characterization*. VCH Publishers: Weinheim, Germany.

Cornell, R.M. & Schwertmann, U. (1996). *The Iron Oxides: Structure, Properties, Reactions, Occurrence and Uses,* VCH Verlagsgesell- schaft: Weinheim, New York, ISBN 3527302743.

Corot, C.; Robert, P.; Idee, J.M. & Port, M. (2006), Recent advances in iron oxide nanocrystal technology for medical imaging. *Adv Drug Deliv Rev.* 58(14): 1471-1504.

Corot, C.; Port, M.; Guilbert, I.; Robert, P.; Raynal, I.; Robic, C.; Raynaud, J-S.; Prigent, P.; Dencausse, A. & Idee, J-M. (2007), Superparamagnetic contrast agents. In: *Molecular and cellular MR imaging,* Modo, M.M.J.; Jeff, W.M. & Bulte, J.W.M. (eds.), pp. 59-84, CRC Press; Boca Raton, FL.

Durdureanu-Angheluta, A.; Ardeleanu, R.; Pinteala, M.; Harabagiu, V.; Chiriac, H.; Simionescu, B.C. (2008). Silane covered magnetite particles. preparation and characterisation. *DJNB,* 3: 33–40.

Durdureanu-Angheluta, A.; Stoica, I.; Pinteala, M.; Pricop, L.; Doroftei, F.; Harabagiu, V.; Chiriac, H. & Simionescu, B.C. (2009), Glycidoxypropylsilane-functionalized magnetite as precursor for polymer-covered core-shell magnetic particles. *High Perform. Polym.,* 21: 548-561.

Durdureanu-Angheluta, A.; Pricop, L.; Stoica, I.; Peptu, C-A.; Dascalu, A.; Marangoci, N.; Doroftei, F.; Chiriac, H.; Pinteala, M. & Simionescu, B.C. (2010), Synthesis and characterization of magnetite particles covered with α-triethoxysilil-polydimethylsiloxane. *J. Magn. Magn. Mater.,* 322: 2956-2968.

Durdureanu-Angheluta, A.; Pricop, L.; Pinteala, M. & Simionescu, B.C. (2011a), Covalent coating magnetite with siloxane functionalized with ester groups. *Submitted.*

Durdureanu-Angheluta, A.; Dascalu, A.; Fifere, A.; Coroaba, A.; Chiriac, H.; Pinteala, M.; Simionescu, B.C. (2011b), Progress in the synthesis and characterization of magnetite nanoparticles with amino groups on the surface. *Submitted.*

Ervithayasuporn, V. & Kawakami, Y. (2009), Synthesis and characterization of core–shell type Fe_3O_4 nanoparticles in poly(organosilsesquioxane). *J. Coll. Interf. Sci.,* 332: 389-393.

Feldmann, C. & Jungk, G.O. (2001), Polyol mediated synthesis of nanoscale MS particles (M = Zn, Cd, Hg). *Angew. Chem. Int. Ed.,* 40: 359-362.

Feng, B.; Hong, R.Y.; Wang, L.S.; Guo, L.; Li, H.Z.; Ding, J.; Zheng, Y. & Wei, D.G. (2008), Synthesis of Fe_3O_4/APTES/PEG diacid functionalized magnetic nanoparticles for MR imaging. *Colloids Surf. A: Physicochem. Eng. Aspects,* 328: 52-59.

Ferreira, A.; Carneiro, A.A.O.; Moraes, E.R.; Oliveira, R.B. & Baffa, O. (2004), Study of the magnetic content movement present, in the large intestine. *J. Magn. Magn. Mater.,* 283: 16-21.

Fievet, F.; Lagier, J.P.; Blin, B.; Beaudoin, B. & Figlarz, M. (1989), Homogeneous and heterogeneous nucleations in the polyol process for the preparation of micron and submicron size metal particles. *Solid State Ionics,* 32-33: 198-205.

Frija, G.; Clement, O. & De Kerviler, E. (1994), Overview of contrast enhancement with iron oxides. *Invest. Radiol.,* 29: s75-s77.

Fukuda, S. (2005), Chelating agents used for plutonium and uranium removal in radiation emergency medicine. *Curr. Med. Chem.,* 12: 2765-2770.

Furlan, M.; Kluge, J.; Mazzotti, M. & Lattuada M. (2010), Preparation of biocompatible magnetite–PLGA composite nanoparticles using supercritical fluid extraction of emulsions. *J. of Supercritical Fluids,* 54: 348–356.

Gamarra, L.F.; Britoa, G.E.S.; Pontuschkaa, W.M.; Amarob, E.; Parmac, A.H.C. & Goya G.F. (2005), Biocompatible superparamagnetic iron oxide nanoparticles used for contrast agents: a structural and magnetic study. *J. Magn. Magn. Mater.*, 289: 439–441.

Ganguli, D. & Ganguli, M. (2003), Inorganic particle synthesis via macro- and microemulsions. *Kluwer Academic*, New York.

Gelbrich, T.; Feyen, M. & Schmidt, A.M. (2006), Magnetic thermoresponsive core-shell nanoparticles. *Macromolecules*, 39: 3469-3472.

Gilchrist, R.K.; Medal, R.; Shorey, W.D.; Hanselman, R.C.; Parrott, J.C. & Taylor, C.B. (1957), Selective inductive heating of lymph nodes. *Ann. Surg.*, 146: 596–606.

Girginova, P.I.; Daniel-da-Silva, A.L.; Lopes, C.B.; Figueira, P.; Otero, M.; Amaral, V.S.; Pereira, E. & Trindade, T. (2010), Silica coated magnetite particles for magnetic removal of Hg^{2+} from water. *J. Coll. Inter. Sci.*, 345: 234-240.

Gleich, B. & Weizenecker, J. (2005), Tomographic imaging using the nonlinear response of magnetic particles. *Nature*, 435: 1214–1217.

Gribanov, N.M.; Bibik, E.E.; Buzunov, O.V. & Naumov, V.N. (1990), Physico-chemical regularities of obtaining highly dispersed magnetite by the method of chemical condensation. *J. Magn. Magn. Mater.*, 85: 7-10.

Gomez-Lopera, S.A.; Arias, J.L.; Gallardo, V. & Delgado, A.V. (2006), Colloidal stability of magnetite /poly (lactic acid) core/shell nanoparticles, *Langmuir* 22: 2816-2821.

Gordon, R.T.; Hines, J.R. & Gordon, D. (1979), Intracellular hyperthermia: a biophysical approach to cancer treatment via intracellular temperature and biophysical alterations. *Med. Hypotheses*, 5: 83-102.

Gorski, C.A. & Scherer, M. (2010), Determination of nanoparticulate magnetite stoichiometry by Mössbauer spectroscopy, acidic dissolution, and powder X-ray diffraction: A critical review. *American Mineralogist*, 95: 1017-1026.

Goya, G.F.; Berquo, T.S. & Fonseca, F.C. (2003), Static and dynamic magnetic properties of spherical magnetite nanoparticles. *J. Appl. Phys.*, 94: 3520-3528.

Hase, M.; Sako, M. & Hirota, S. (1990) Experimental study of ferromagnetic induction heating combined with hepatic arterial embolization of liver tumours. *Nippon-Igaku-Hoshasen-Gakkai-Zasshi*, 50: 1402-1414.

Herrera, A.P. ; Rodriguez, M. ; Torres-Lugo, M. & Rinaldi, C. (2008), Multifunctional magnetite nanoparticles coated with fluorescent thermo-responsive polymeric shells. *J. Mater. Chem.*, 18: 855-858.

Hirai, T.; Mizumoto, J.; Shiojiri, S. & Komasawa, I. (1997), Preparation of fe oxide and composite ti-fe oxide ultrafine particles in reverse micellar systems. *J. Chem. Eng. Jpn.*, 30: 938–943.

Hu, A.; Yee, G.T. & Lin, W. (2005), Magnetically recoverable chiral catalysts immobilized on magnetite nanoparticles for asymmetric hydrogenation of aromatic ketones. *J. Am. Chem. Soc.*, 127: 12486-12487.

Jain, T.K.; Morales, M.A.; Sahoo, S.K.; Leslie-Pelecky, D.L. & Labhasetwar, V. (2005), Iron oxide nanoparticles for sustained delivery of anticancer agents. *Molecular Pharmaceutics*, 2(3): 194-205.

Jiang, J-S.; Gan, Z-F.; Yang, Y.; Du, B.; Qian, M. & Zhang P. (2009), A novel magnetic fluid based on starch-coated magnetite nanoparticles functionalized with homing peptide. *J. Nanopart. Res.*, 11: 1321-1330.

Jolivet, J.P.; Belleville, P.; Tronc, E. & Livage, J. (1992), Influence of Fe(II) on the formation of the spinel iron oxide in alkaline medium. *Clays Clay Miner.*, 40(5): 531-539.

Jones, S.K.; Winter, J.W. & Gray, B.N. (2002), Treatment of experimental rabbit liver tumours by selectively targeted hyperthermia. *Int. J. Hyperthermia,* 18: 117-128.

Jordan, A.; Scholz, R.; Wust, P.; Fahling, H. & Felix, R. (1999), Magnetic fluid hyperthermia (MFH): Cancer treatment with AC magnetic field induced excitation of biocompatible superparamagnetic nanoparticles. *J. Magn. Magn. Mater.,* 201: 413-419.

Jordan, A.; Scholz, R.; Maier-Hauff, K.; Johannsen, M.; Wust, P.; Nadobny, J.; Schirra, H.; Schmidt, H.; Deger, S.; Loening, S.; Lanksch, W. & Felix, R. (2001), Presentation of a new magnetic field therapy system for the treatment of human solid tumors with magnetic fluid hyperthermia. *J. Magn. Magn.Mater.,* 225: 118-126.

Koh, I. & Josephson, L. (2009), Magnetic nanoparticle sensors, *Sensors*, 9: 8130-8145.

Kazufumi, O. (2008), (50-3, Aza Samukata Donari, Donari-cho, Awa-sh, Tokushima 06, 7711506, JP) *Pharmaceutical preparation and manufacturing method thereof patent*: WO/2008/139637.

Khurshid, H.; Kim, S.H.; Bonder, M.J.; Colak, L.; Ali, Bakhtyar; Shah, S.I.; Kiick, K.L.; Hadjipanayis, G.C. (2009), Water dispersible Fe/Fe-oxide core-shell structured nanoparticles for potential biomedical applications. *J. Appl. Phys.,* 105: 07B308-07B308-3.

Kim, D.; Toprak, M.; Mikhailova, M.; Zhang, Y.; Bjelke, B.; Kehr J. & Mohammed, M. (2002), Surface modification of superparamagnetic nanoparticles for in vivo bio-medical applications. *Materials Research Society.,* 704: 369-374.

Kim, D.K.; Zhang, Y.; Voit, W.; Rao, K.V. & Muhammed, M. (2001), Synthesis and characterization of surfactant-coated superparamagnetic monodispersed iron oxide nanoparticles. *J. Magn. Magn. Mater.,* 225: 30-36.

Kim, J.; Kim, H.S.; Lee, N.; Kim, T.; Kim, H.; Yu, T.; Song, I.C.; Moon, W.K. & Hyeon, T. (2008), Multifunctional uniform nanoparticles composed of a magnetite nanocrystal core and a mesoporous silica shell for magnetic resonance and fluorescence imaging and for drug delivery. *Angew. Chem. Int. Ed.,* 47: 8438-8441.

Kim, K.W. & Ha, H.K. (2003), MRI for small bowel diseases. *Semin. Ultrasound CT MR,* 24: 387-402.

Kobayashi, Y.; Saeki, S.; Yoshida, M.; Nagao, D. & Konno, M. (2008), Synthesis of spherical submicron-sized magnetite/silica nanocomposite particles. *J. Sol-Gel Sci. Technol.,* 45: 35-41.

Koenig, S.H. & Brown III, R.D. (1987), Relaxometry of magnetic resonance imaging contrast agents. *Magn. Reson. Annu.,* 5: 263-286.

Kojima, K.; Miyazaki, M.; Mizukami, F. & Maeda, K. (1997), Selective formation of spinel iron oxide in thin films by complexing agent-assisted sol-gel processing. *J. Sol-Gel Sci. Technol.,* 8: 77-81.

Koneracka, M.; Kopcansky, P.; Timko, M.; Ramchand, C.N. (2002), Direct binding procedure of proteins and enzymes to fine magnetic particles. *J. Magn. Magn. Mater.,* 252: 409-411.

Korolev, V.; Ramazanova, A. & Blinov, A. (2002), Adsorption of surfactants on the superfine magnetite. *Russian Chemical Bulletin, International Edition,* 51(11): 2044-2049.

Kuhara, M.; Takeyama, H.; Tanaka, T. & Matsunaga, T. (2004), Magnetic cell using antibody binding with protein A expressed on bacterial magnetic particles. *Anal. Chem.,* 76: 6207-6213.

Kularatne, B.Y.; Lorigan, P.; Browne, S.; Suvarna, S.K.; Smith, M.O. & Lawry, J. (2002), Monitoring tumour cells in the peripheral blood of small cell lung cancer patients. *Cytometry*, 50: 160-167.

Laurent, S.; Forge, D.; Port, M.; Roch, A.; Robic, C.; Elst, L.V. & Muller, R.N. (2008), Magnetic iron oxide nanoparticles: synthesis, stabilization, vectorization, physicochemical characterizations, and biological applications. *Chem. Rev.*, 108: 2064-2110.

Lazaro, F.J.; Abadia, A.R.; Romero, M.S.; Gutierrez, L.; Lazaro, J.; Morales, M.P. (2005), Magnetic characterisation of rat muscle tissues after subcutaneous iron dextran injection. *Biochim. Biophys. Acta – Mol. Basis Dis.*, 1740: 434–445.

Liu, H-L.; Ko, S.P.; Wu, J-H.; Jung, M-H.; Min, J.H.; Lee, J.H.; An, B.H. & Kim, Y.K. (2007), One-pot polyol synthesis of monosize PVP-coated sub-5nm Fe_3O_4 nanoparticles for biomedical applications. *J. Magn. Magn. Mater.*, 310: e815-e817.

Liu, X.Q.; Tao, S.W. & Shen, Y.S. (1997), Preparation and characterization of nanocrystalline α-Fe_2O_3 by a sol-gel process. *Sens. Actuators B*, 40: 161-165.

López Pérez, J.A. ; López Quintela M.A. ; Mira, J. ; Rivas, J. & Cherles, S.W. (1997), Advances in the preparation of magnetic nanoparticles by the microemulsion method. *J. Phys. Chem. B*, 101: 8045-8047.

Lu, A.H.; Schmidt, W.; Matoussevitch, N.; Bonnemann, H.; Spliethoff, B.; Tesche, B.; Bill, E.; Kiefer, W. & Schu¨th, F. (2004), Nanoengineering of a magnetically separable hydrogenation catalyst. *Angew. Chem.*, 116: 4403-4406; *Angew. Chem.*, 43: 4303-4306.

Lübbe, A.S.; Bergemann, C.; Riess, H.; Schriever, F.; Reichardt, P.; Possinger, K.; Matthias, M.; Dörken, B.; Herrmann, F.; Gürtler, R.; Hohenberger, P.; Haas, N.; Sohr, R.; Sander, B.; Lemke, A-J.; Ohlendorf, D.; Huhnt, W.; Huhn, D. (1996), Clinical experiences with magnetic drug targeting: a phase I study with 4'-epidoxorubicin in 14 patients with advanced solid tumors. *Cancer Res.*, 56: 4686-4693.

Ma, M.; Zhang, Y.; Yu, W.; Shen, H.; Zhang, H. & Gu, N. (2003), Preparation and characterization of magnetite nanoparticles coated by amino silane. *Colloids and Surfaces A: Physicochem. Eng. Aspects*, 212: 219-226.

Martinez-Mera, I.; Espinosa, M.E.; Perez-Hernandez, R. & Arenas-Alatorre, J. (2007), Synthesis of magnetite nanoparticles without surfactant at room temperature. *Mater. Lett.* , 61: 4447-4451.

Maruyama, K.; Takeyama, H.; Nemoto, E.; Tanaka, T.; Yoda, K. & Matsunaga, T. (2004), Single nucleotide polymorphism detection in aldehyde dehydrogenase 2 (ALDH2) gene using bacterial magnetic particles based on dissociation curve analysis. *Biotechnol. Bioeng.*, 87: 687-694.

Massart, R. (1981), Preparation of aqueous magnetic liquids in alkaline and acidic media. *IEEE Trans Magn.*, 17: 1247-1248.

Matsuki, H.; Yanada, T.; Sato, T.; Murakami, K. & Minakawa, S. (1994), Temperature sensitive amorphous magnetic flakes for intratissue hyperthermia, *Mater. Sci. Eng.*, A181/A182: 1366-1368.

Matsunaga, T. & Takeyama, H. (1998), Biomagnetic nanoparticle formation and application. *Supramolecular Sci.*, 5: 391-394.

Matsunaga, T.; Nakayama, H.; Okochi, M. & Takeyama, H. (2001), Fluorescent detection of cyanobacterial dna using bacterial magnetic particles on a mag-microarray. *Biotechnol. Bioeng.*, 73: 400-405.

Mitsumori, M.; Hiraoka, M.; Shibata, T.; Okuno, Y.; Masunaga, S.; Koishi, M.; Okajima, K.; Nagata, Y.; Nishimura, Y.; Abe, M.; Ohura, K.; Hasegawa, M.; Nagae H. & Ebisawa

Y. (1994), Development of intra-arterial hyperthermia using a dextran–magnetite complex. *Int. J. Hyperthermia*, 10: 785-793.

Modo, M.; Hoehn, M. & Bulte J.W.M. (2005), KCL, Cellular MR imaging. *Mol. Imaging*, 4(3): 143-164.

Molday, R.S. & MacKenzie, D. (1982), Immunospecific ferromagnetic iron–dextran reagents for the labeling and magnetic separation of cells. *J. Immunol. Methods*, 52: 353-367.

Mondini, S.; Cenedese, S.; Marinoni, G.; Molteni, G.; Santo, N.; Bianchi, C.L. & Ponti, A. (2008), One-step synthesis and functionalization of hydroxyl-decorated magnetite nanoparticles. *J. Colloid Interface Sci.*, 322: 173-179.

Morisada, S.; Miyata, N. & Iwahori, K. (2002), Immunomagnetic separation of scum-forming bacteria using polyclonal antibody that recognizes mycolic acids. *J. Microbiol. Methods*, 51: 141-148.

Moroz, P.; Jones, S.K.; Winter, J. & Gray, B.N. (2001), Targeting liver tumors with hyperthermia: ferromagnetic embolization in a rabbit liver tumor model. *J. Surg. Oncol.*, 78: 22-29.

Mosso, J.A. & Rand, R.W. (1973), Ferromagnetic silicone vascular occlusion: A technique for selective infarction of tumors and organs. *Ann. Surg.*, 178: 663–668.

O'Brien, K.W. (2003), *Synthesis of functionalized poly(dimethylsiloxane)s and the preparation of magnetite nanoparticle complexes and dispersions. Ph.D. Dissertation*, Department of chemistry, Virginia Tech.

Pankhurst, Q.A.; Connolly, J.; Jones, S.K. & Dobson J. (2003), Applications of magnetic nanoparticles in biomedicine. *J. Phys. D: Appl. Phys.*, 36: R167-R181.

Pardoe, H.; Clark, P.R.; St Pierre, T.G.; Moroz, P. & Jones, S.K. (2003), A magnetic resonance imaging based method for measurement of tissue iron concentration in liver arterially embolized with ferrimagnetic particles designed for magnetic hyperthermia treatment of tumors. *Magn. Reson. Imaging*, 21: 483-488.

Paulev, P.-E. (1999–2000), Gastrointestinal function and disorders. In: *Textbook in Medical Physiology and Pathophysiology*, Paulev, P.-E. (ed.), Copenhagen Medical Publishers, Copenhagen.

Plank, C.; Schillinger, U.; Scherer, F.; Bergemann, C.; Remy, J.S.; Krotz, F.; Anton, M.; Lausier, J. & Rosenecker, J. (2003), The magnetofection method: Using magnetic force to enhance gene delivery. *Biol. Chem.*, 384: 737-747.

Pricop, L.; Durdureanu-Angheluta, A.; Spulber, M.; Stoica, I.; Fifere, A.; Marangoci, N.L.; Dascalu, A.I.; Tigoianu, R.; Harabagiu, V.; Pinteala, M.; Simionescu, B.C. (2010), Synthesis and micellization of polydimethylsiloxane–carboxy-terminated poly(ethylene oxide) graft copolymer in aqueous and organic media and its application for the synthesis of core-shell magnetite particles. *e-Polymers*, no. 093.

Raileanu, M.; Crisan, M.; Petrache, C.; Crisan, D.; Zaharescu, M. (2003), Fe_2O_3-SiO_2 nanocomposites obtained by different sol-gel routes. *J. Optoelectron. AdV. Mater.*, 5(3): 693-698.

Rand, R.W.; Snyder, M.; Elliott, D.G. & Snow, H.D. (1976), Selective radiofrequency heating of ferrosilicone occluded tissue: a preliminary report. *Bull. Los Angeles Neurol. Soc.*, 41: 154-159.

Rand, R.W.; Snow, H.D.; Elliott, D.G. & Snyder, M. (1981), Thermomagnetic surgery for cancer. *Appl. Biochem. Biotechnol.*, 6: 265-272.

Richardson, J.C.; Bowtell, R.W.; Mader, K.; Melia, C.D. (2005), Pharmaceutical applications of magnetic resonance imaging (MRI). *Adv. Drug Delivery Rev.*, 57: 1191-1209.

Rosensweig, R.E. (1985), *Ferrohidrodynamics*, pp. 344, Cambridge Univ. Press., ISBN 0521 25624 0

Safarik, I. & Safarikova, M. (2002), Magnetic nanoparticles and biosciences. *Monatshefte fur Chemie*, 133: 737-759.

Sankaranarayanapillai, S.; Volker, S.; Werner, R.T. (2010), Magnetically separable nanocatalysts: bridges between homogeneous and heterogeneous catalysis. *Angew. Chem. Int. Ed.*, 49: 3428-3459.

Santra, S.; Tapec, R.; Theodoropoulou, N.; Dobson, J.; Hebard, A. & Tan, W. (2001), Synthesis and characterization of silica-coated iron oxide nanoparticles in microemulsion: the effect of nonionic surfactants. *Langmuir*, 17: 2900-2906.

Shafi, K.V.P.M.; Ulman, A.; Yan, X.; Yang, N.L.; Estournes, C.; White, H. & Rafailovich, M. (2001), Sonochemical synthesis of functionalized amorphous iron oxide nanoparticles. *Langmuir*, 17: 5093-5097.

Shen, X.C.; Fang, X.Z.; Zhou, Y.H.; Liang, H. (2004), Synthesis and characterization of 3-aminopropyltriethoxysilane-modified superparamagnetic magnetite nanoparticles. *Chem. Lett.*, 33: 1468-1469.

Shukla, A.; Degen, P. & Rehage, H. (2007), Synthesis and characterization of monodisperse poly(organosiloxane) nanocapsules with or without magnetic cores. *J. Am. Chem. Soc.*, 129: 8056-8057.

Silva, A.K.; da Silva, E.L.; Oliveira, E.E.; Nagashima, T.Jr.; Soares, L.A.; Medeiros, A.C.; Araújo, J.H.; Araújo, I.B.; Carriço, A.S. & Egito, E.S. (2007), Synthesis and characterization of xylan-coated magnetite microparticles. *Int. J. Pharm.*, 334: 42-47.

Sonvico, F.; Dubernet, C.; Colombo, P. & Couvreur, P. (2005), Metallic Colloid nanotechnology, applications in diagnosis and therapeutics. *Curr. Pharm. Des.* 11: 2091-2105.

Sugimoto, T. (2003), Formation of monodispersed nano- and micro-particles controlled in size, shape, and internal structure. *Chem. Eng. Technol.*, 26: 313-321.

Suzuki, M.; Shinkai, M.; Kamihira, M. & Kobayashi, T. (1995), Preparation and characteristics of magnetite-labelled antibody with the use of poly(ethylene glycol) derivatives, *Biotechnol. Appl. Biochem.*, 21: 335–345.

Suzuki, S.; Arai, K.; Koike, T. & Oguchi, K. (1990), Studies on liposomal ferromagnetic particles and a technique of high frequency inductive heating—in vivo studies of rabbits. *J. Japan. Soc. Cancer Therapy*, 25: 2649–2658.

Tanaka, T.; Takeda, H.; Ueki, F.; Obata, K.; Tajima, H.; Takeyama, H.; Goda, Y.; Fujimoto, S.; Tibbe, A.; de Grooth, B.; Greve, J.; Liberti, P.; Dolan, G. & Terstappen, L. (1999), Optical tracking and detection of immunomagnetically selected and aligned cells. *Nature Biotechnol.*, 17: 1210-1213.

Tanaka, T.; Takeda, H.; Ueki, F.; Obata, K.; Tajima, H.; Takeyama, H.; Goda, Y.; Fujimoto, S. & Matsunaga T. (2004), Rapid and sensitive detection of 17 β-estradiol in environmental water using automated immunoassay system with bacterial magnetic particles. *J. Biotechnol.*, 108: 153-159.

Tartaj, P.; Morales, M.P.; Veintemillas-Verdaguer, S.; Gonzalez-Carreno, T.; Serna, C.J. (2006), *Synthesis, properties and biomedical applications of magnetic nanoparticles.* Handbook of Magnetic Materials; Elsevier: Amsterdam, The Netherlands, p. 403.

Teo, B.; Chen, F.; Hatton, T.A.; Grieser F. & Ashokkumar, M. (2009), A novel one-pot synthesis of magnetite latex nanoparticles by ultrasound irradiation. *Langmuir*, 25: 2593-2595.

Tibbe, A.; de Grooth, B.; Greve, J.; Liberti, P.; Dolan, G. & Terstappen, L. (1999), Optical tracking and detection of immunomagnetically selected and aligned cells. *Nature Biotechnol.*, 17: 1210-1213.

Torchilin, V.P. (2000), Drug targeting. *Eur. J. Pharm. Sci.* 11: S81-S91.

Tronc, E.; Belleville, P.; Jolivet, J.P. & Livage, J. (1992), Transformation of ferric hydroxide into spinel by Fe(II) adsorption. *Langmuir*, 8: 313-319.

Vekas, L. (2008), Ferrofluids and magnetorheological fluids (review). *Advances in Science and Technology*, 54: 127-136.

Wang, L.; Yang, Z.; Gao, J.; Xu, K.; Gu, H.; Zhang, B.; Zhang, X.; Xu, B. (2006), A biocompatible method of decorporation: bisphosphonate-modified magnetite nanoparticles to remove uranyl ions from blood, *J. Am. Chem. Soc.*, 128: 13358-13359.

Watts, J.F. & Wolstenholme, J. (2003), An introduction to surface analysis by XPS and AES. John Wiley & Sons Ltd, England.

Wilson, K.S.; Goff, J.D.; Riffle, J.S.; Harris, L.A. & StPirre, T.G. (2005), Polydimethylsiloxane-magnetite nanoparticle complexes and dispersions in polysiloxane carrier fluids. *Polym. Adv. Technol.*, 16: 200-211.

Woo, K.; Lee, H.J.; Ahn, J.P.; Park, Y.S. (2003). Sol–Gel mediated synthesis of Fe_2O_3 nanorods. *Adv. Mater.*, 15: 1761-1764.

Wormuth, K. (2001), Superparamagnetic latex via inverse emulsion polymerization. *J. Colloid Interface Sci.*, 241: 366-377.

Yamamura, M.; Camilo, R.L.; Sampaio, L.C. ; Macedo, M.A. ; Nakamura, M. & Toma, H.E. (2004), Preparation and characterization of (3-aminopropyl) triethoxysilane-coated magnetite nanoparticles. *J. Magn. Magn.Mater.*, 279: 210-217.

Yang, H.H.; Zhang, S.Q.; Chen, X.L.; Zhuang, Z.X.; Xu, J.G. & Wang, X.R. (2004), Magnetite-containing spherical silica nanoparticles for biocatalysis and bioseparations. *Anal. Chem.*, 76: 1316-1321.

Ye, X.R.; Daraio, C.; Wang, C.; Talbot, J.B. & Jin, S. (2006), Room temperature solvent-free synthesis of monodisperse magnetite nanocrystals. *J. Nanosci. Nanotechnol.*, 6: 852-856.

Yoza, B.; Arakaki, A.; Maruyama, K.; Takeyama, H. & Matsunaga, T. (2003), Fully automated DNA extraction from blood using magnetic particles modified with hyperbranched polyamidoamine dendrimer. *J. Biosci. Bioeng.*, 95: 21-26.

Xu, X.Q.; Shen, H.; Xu, J.R. & Li, X.J. (2004), Aqueous-based magnetite fluids stabilized by surface small micelles of oleoylsarcosine. *App. Surf. Sci.*, 221: 430-436.

Zhang, L.; He, R. & Gu, H.C. (2006), Oleic acid coating on the monodisperse magnetite nanoparticles. *Appl. Surf. Sci.*, 253: 2611-2617.

Zhang, Y.; Kohler, N. & Zhang, M. (2002), Surface modification of superparamagnetic magnetite nanoparticles and their intracellular uptake. *Biomaterials*, 23: 1553-1561.

Zigeuner, R.E.; Riesenberg, R.; Pohla, H.; Hofstetter, A. & Oberneder, R. (2003), Isolation of circulating cancer cells from whole blood by immunomagnetic cell enrichment and unenriched immunocytochemistry in vitro. *J. Urol.*, 169: 701-705.

http://www.interactivearchitecture.org/ferrofluid-sculptures-by-sachiko-kodama.html

http://www.pricon.ro/ro/special.php

http://en.wikipedia.org/wiki/X-ray_photoelectron_spectroscopy

Part 3

Nanostructured Materials

8

Carbon Nanotubes – Imprinted Polymers: Hybrid Materials for Analytical Applications

Giuseppe Cirillo[1]*, Silke Hampel[2], Francesco Puoci[1], Diana Haase[2], Manfred Ritschel[2], Albrecht Leonhardt[2], Francesca Iemma[1] and Nevio Picci[1]
[1]*Department of Pharmaceutical Sciences, University of Calabria, Rende (CS),*
[2]*Leibniz Institute for Solid State and Materials Research Dresden, Dresden*
[1]*Italy*
[2]*Germany*

1. Introduction

Molecular imprinting is a recent new and rapidly evolving technique which allows the creation of synthetic receptors (MIPs) consisting of highly cross-linked porous-rich polymers with recognition properties comparable to the biological systems related to the presence of specific recognition sites complementary in shape, size and functional groups to a target molecule. It is a facile concept, which involves the construction of sites of specific recognition, commonly within synthetic polymers. The template of choice is entrapped within a pre-polymerization complex, consisting of functional monomers with good functionality, which chemically interacts with the template. Polymerization in the presence of crosslinker serves to freeze these template-monomer interactions and subsequent removal of the template results in the formation of a molecularly imprinted polymer matrix. Due to the advantages of MIPs such as low cost, stability, and easy preparation compared with natural molecular recognition products (e.g. antibody), Molecular imprinting is a well-developed tool in the analytical field, mainly for separating and quantifying very different substances, including drugs and bio-active molecules contained in relatively complex matrices. Despite the application of MIPs as sensor matrices or separation materials, they suffer from basic limitations associated with the limited concentration of imprinted sites, and the bulk volume of the polymer matrices that requires long diffusion paths of the imprinted host molecules. These limitations lead to inefficient sensing or separation processes. MIP nanomaterials are proposed as a pain reliever for headache by improving the accessibility and the homogeneity of the binding sites. In particular, with high strength, the extremely large surface area and unique chemical properties, Carbon nanotubes (CNTs) could serve as the reinforcing element or core in fabricating core–shell structural MIPs.

Since their discovery in 1991, CNTs have attracted great attention because of their unique properties (high electrical conductivity, chemical stability, mechanical strength, large specific surface area, and high thermal stability) indicating potential for various applications.

CNTs represent a new carbon material that has been widely recognized as the quintessential nanomaterial and, because the hexagonal arrays of carbon atoms of the CNTs surface have a strong interaction with other molecules or atoms, CNTs show a great analytical potential as a solid-phase extraction adsorbent. Additionally, CNTs' surfaces can be modified by introducing various organic functional groups, thus providing a strongly physic sorbing surface area, adjustable surface charge, and a source of protons for chemical ionization. It has been demonstrated that the surfaces of CNTs can be easily modified in numerous ways, either by covalent or non-covalent functionalization. All the facts mentioned before reveal that carbon nanotubes, and specially multiwalled carbon nanotubes, have great analytical potential as an effective solid-phase extraction adsorbent for chelates or ion pairs of metal ions, organic compounds, and organometallic compounds.

Based on these considerations, it could be summarized that MIPs-CNTs composites represent useful innovative materials for analytical determination of target analytes in complex matrices.

2. Molecularly Imprinted Polymers (MIPs): highly selective materials

Molecular imprinting is a very useful technique to incorporate specific substrate recognition sites into polymers (Puoci et al., 2010). In particular, by this polymer synthesis technique, the formation of macromolecular networks characterized by specific recognition sites for a desired target molecule, named template could be obtained (Byrne & Salian, 2008). The specific binding properties of MIP must be attributed to specific interactions between the template and the functional groups in the polymeric network, thus the choice of the functional monomers is of primary importance to obtain performing imprinted materials (Puoci et al., 2007; Cirillo et al, 2010, 2011a)

In the synthesis of these macromolecular system, many parameters involved can affect the information associated with the binding sites, such as functional monomers/polymers, crosslinkers and solvents/porogens. Thus, both the feasibility of imprinting and the proper preparation conditions need exploration for the preparation of efficient imprinted materials (Liu Z. et al., 2010). It is important to state that MIP can be obtained in different formats, depending on the preparation method followed. To date, the most common polymerizations for preparing MIPs involve conventional solution, suspension, precipitation, multi-step swelling and emulsion core-shell. There are also other methods, such as aerosol or surface rearrangement of latex particles, but they are not used routinely (Puoci et al., 2011).

MIPs are stable polymers with molecular recognition abilities and resistant to a wide range of conditions (pH, organic solvents, temperature, pressure), and were used for several different applications, such as chromatographic stationary phases (Hishiya et al., 2003), enantiomeric separation (Adbo & Nicholls, 2001), solid-phase extraction (SPE) (El-Sheikh et al., 2010), and catalysis (Anderson et al., 2005); they were also used as receptors (Haupt, 2003), antibodies (Svitel et al., 2001), enzyme mimics (Nicholls et al., 1996), affinity and sensing materials (Syu et al., 2006), and, in recent years, pharmaceutical applications, such as drug discovery, drug purification, or drug delivery (Mosbach, 2006; Yu Y. et al., 2002; Caldorera-Moore & Peppas, 2009; Hilt & Byrne, 2004).

MIPs can be synthesized following three different imprinting approaches named covalent, non-covalent and semi-covalent procedure, according to the kind of interaction between a template and functional groups during the synthesis and recognition phases (Caro et al., 2002).

In the non-covalent procedure (Figure 1), non-covalent interactions (hydrogen bonding, p-p interactions, Van der Waals forces, etc.) are involved in both the synthesis and the recognition step (Joshi et al., 1998). This method is still the most widely used method to prepare MIP because of the advantages that it offers from the point of view of synthesis. The covalent protocol requires the formation of covalent bonds between the template and the functional monomer prior to polymerization, as well as between template and functional group in the imprinted cavities during the re-binding process (Ikegami et al., 2004). Finally, the semi-covalent approach is a hybrid of the two previous methods. Specifically, covalent bonds are established between the template and the functional monomers before polymerization, while, once the template has been removed from the polymer matrix, the subsequent re-binding of the analyte to the MIP exploits non-covalent interactions (Curcio et al., 2010).

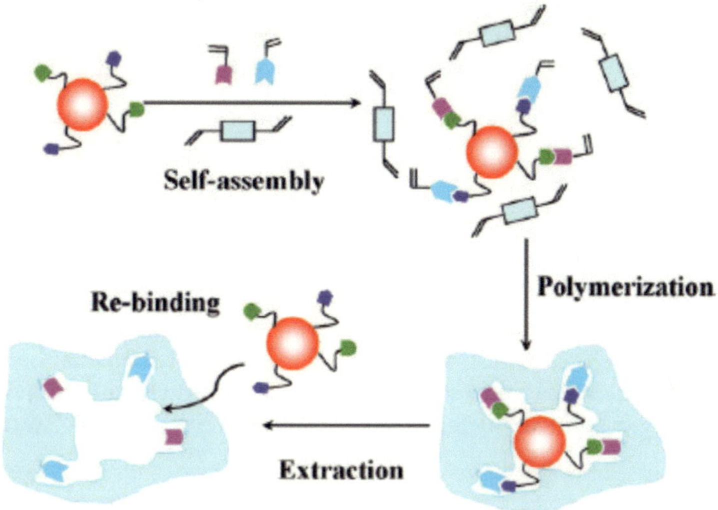

Fig. 1. Schematic representation of Non-Covalent Molecular Imprinting Process. Adapted from Liu Z. et al., 2010.

The binding sites obtained by molecular imprinting show different characteristics, depending on the type of imprinted approach. The average affinity of binding site prepared using bonding by non-covalent forces is generally weaker than those prepared using covalent methods because electrostatic, hydrogen bonding, π-π and hydrophobic interactions, between the template and the functional monomers, are used exclusively in forming the molecular assemblies (Hwang & Lee, 2002). However, when covalent bonds are established between the template and the functional monomer prior to polymerization, this gives rise to better defined and more homogeneous binding sites than the non-covalent approach, since the template-functional monomer interactions are far more stable and defined during the imprinting process.

It should also be mentioned that, as a control in each polymerization, a non-imprinted polymer (NIP) is also synthesised in the same way as the MIP but in absence of the template. To evaluate the imprinting effect, the selectivities of the NIP and MIP are then compared.

3. Carbon nanotubes: fascinating nano-objects

In the last decade, the dramatic development of nanotechnology in material science and engineering has led to the study and the development of innovative nanostructured materials (Xu T. et al., 2007; Niemeyer, 2001; Cui & Gao, 2003; Whitesides, 2003): certain materials with delicate structures of "small" sizes, falling in the 1–100 nm range, and specific properties and functions related to the "size effect" (Safarik & Safarikova, 2002; Laval et al., 1999). In particular, extensive researches have been focused on medicine and biomedical engineering for the investigation of the interactions between nanomaterials and biological systems (Foldvari & Bagonluri, 2008; Bianco et al., 2005; Desai, 2000). Potential products of bionanotechnology in the pharmaceutical and biomedical industry are refered as nanomedicines, including materials to be employed as clinical bio-analytical diagnostics (Bianco et al., 2005; Pantarotto et al., 2003), carriers to improve controlled and targeted drug release (Leary et al., 2006; Sinha & Yeow, 2005) and additives to improve solubility and bioavailability of poorly soluble drugs (Ajayan et al., 1993), novel tissue engineered scaffolds and devices (Bauer et al., 2004; Mazzola, 2003; Kikuchi et al., 2004).

Within the realm of bionanotechnology, carbon nanotubes (CNTs), a major class of carbon-based tubular nanostructures have been denoted great interest in the scientific community (Ke & Qiao, 2007; Klingeler & Sim, 2011).

The walls of CNTs are made up of a hexagonal lattice of carbon atoms analogous to the atomic planes of graphite (Dresselhaus et al., 2004). CNTs can be imaginatively produced by rolling up a graphene sheet (a single layer of graphite) forming a single-walled CNTs (SWCNTs), or by rolling up many graphene layers to form concentric cylinders (multi-walled CNTs; MWCNTs). The ends of CNT may be closed with halfspheres of fullerens. MWNTs are comprised of two up to several to tens of concentric carbon cylinders and thus generally have a larger outer diameter (2.5–100 nm) than SWNTs (0.6–2.4 nm) (Joselevich, 2004) (Figure 2).

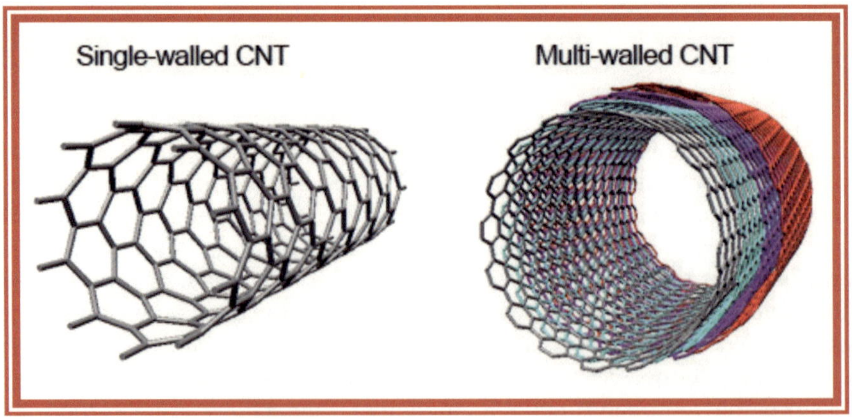

Fig. 2. Schematic representation of SWNTs and MWNTs.

Although there have been many interesting and successful attempts to grow CNTs by various methods (Klingeler et al., 2008; Vyalikh et al., 2008), the three most widely used techniques are: arc discharge, laser ablation, and chemical vapor deposition (CVD) (Ando et al., 2004).

The arc-discharge method is the one by which CNTs were first produced and recognized. In a model system, a dc arc voltage is applied between two graphite rods in the presence of an appropriate ambient gas. This method is useful for the production of both CNTs and fullerenes. In particular, when pure graphite rods are used, fullerenes are deposited in the form of soot in the chamber (Saito et al., 1992). However, a small part of the evaporated anode is deposited on the cathode, which includes CNTs (Iijima, 1991). As described by Ando et al., 2004, these CNTs represent MWNTs. They are found not only on the top surface of the cathode deposit (Ando, 1993) but also deep inside the deposit (Ando & Iijima, 1993). Large-scale synthesis of MWNTs by arc discharge has been achieved (Ebbesen & Ajayan, 1992; Colbert et al., 1994) in He gas. The same methodology could be applied to the synthesis of SWNTs if a graphite rod containing metal catalysts (Fe, Co, etc.) is used as the anode with a pure graphite cathode (Iijima & Ichihashi, 1993; Bethune et al., 1993). By using a dc pulsed arc discharge inside a furnace, homogeneous conditions in arc discharge are achieved and high-quality DWNTs are synthesized by a method called high-temperature pulsed arc discharge (Sugai et al., 1999, 2000, 2003; Shimada et al., 2004).

The laser ablation, based on the high energy density of lasers (typically a YAG or CO_2 laser) which is suitable for materials with a high boiling temperature such as carbon, was developed for fullerene and CNT production by Smalley's group (Guo T. et al., 1992; Thess et al., 1996). The laser has sufficiently high energy density not to cleave the target into graphite particles but to vaporize it at the molecular level converting the graphite vapor into amorphous carbon as the starting material of SWNTs (Puretzky et al., 2000; Sen et al., 2000; Kokai et al., 2000). The annealing conditions of the amorphous carbon in the laser ablation method are more homogeneous than those of the arc-discharge method, in which the electrodes and the convection flow disturb the homogeneity of the temperature and flow rate (Zhao X. et al., 2003; Kanai et al., 2001).

Chemical Vapor Deposition (CVD) is a simple and economic technique for synthesizing CNTs at low temperature and ambient pressure. Usually, a carbon feedstock is thermally decomposed in the presence of a metal catalyst (Cirillo et al., 2011b). The generated carbon dissolves in the catalyst particles and, after saturation, is deposited in the shape of CNTs.

To be distinguished from the many kinds of CVD used for various purposes, the method is also known as thermal or catalytic. Compared with arc-discharge and laser methods, CVD is more versatile because it offers better control over growth parameters. Furthermore, it harnesses a variety of hydrocarbons in any state (solid, liquid, or gas), enables the use of various substrates, and allows CNTs growth in a variety of forms, such as powder, thin or thick films, aligned or entangled, straight or coiled, or even a desired architecture of nanotubes at predefined sites on a patterned substrate. MWNTs were grown from benzene, ethylene, methane, and many other hydrocarbons (Endo et al., 1993; José-Yacamán et al., 1993; Satiskumar et al., 1999; Hernadi et al., 1996). SWNTs were first produced by Dai et al. from disproportionation of CO, and SWNTs were also produced from benzene, acetylene, ethylene, and methane using various catalysts. (Cheng et al., 1998; Satishkumar et al., 1998; Hafner et al., 1998; Kong J. et al., 1998; Flahaut et al., 1999). Due to lower synthesis temperatures CVD grown CNTs exhibit a lower crystallinity than tubes produced using the two alternative methods mentioned above.

When considering the whole of CNTs applications, the solubilization of pristine CNTs in aqueous solvents remains an obstacle to realizing their potential, due to the rather hydrophobic character of the graphene sidewalls. To successfully disperse CNTs the dispersing medium should be capable of both wetting the hydrophobic tube surfaces and modifying the tube surfaces to decrease tube aggregation. Four basic approaches have been used to obtain a dispersion: surfactant-assisted dispersion (Ham et al., 2005; Islam et al., 2003; Moore et al., 2003; Yurekli et al., 2004; Vaisman et al., 2006), solvent dispersion (Fu & Sun, 2003; Ausman et al., 2000; Kim D.S. et al., 2005), functionalization of CNT sidewalls (Dyke & Tour 2004; Fernando et al., 2004; Peng et al., 2003), and biomolecular dispersion (Gigliotti et al., 2006).

4. Chemical sensing

Before exploring the use of CNTs in chemical and electrochemical sensing it should be pointed out that the term sensor refers to a device that responds to a physical or chemical stimulus by producing a signal, usually an electrical one (Hillberg et al., 2005), while a biosensor is a sensor that uses biological selectivity to limit perception to the specific molecule of interest. A typical biosensor consists of two main components: the chemosensory materials (receptors) that can selectively bind target analytes and the efficient transducer that can transform the binding events into a readable signal output related to the analyte concentration in the sample (Eggins, 2002). The efficiency of chemosensors is largely dependent on the selectivity and sensitivity of the used sensory materials to a target species. In the traditional approaches a biological or biologically derived sensing element acting as a receptor is immobilized on the surface of a physical transducer to provide selective binding of analytes (Orellana & Moreno-Bondi, 2005; Jiang & Ju, 2007). As a sensing element, it is possible to use either biological macromolecules (e.g. antibodies, enzymes, receptors and ion channel proteins, nucleic acids, aptamers and peptide nucleic acids) or biological systems (e.g. ex vivo tissue, microorganisms, isolated whole cells and organelles). However, the small surface area and non-tunable surface properties of transducers greatly limit the efficiency of chemosensors, especially for the detection of ultratrace analytes. Recently, nanomaterials have found a wide range of applications as a material foundation of chemosensors, and have exhibited various degrees of success in the improvement of detection sensitivity and selectivity (Gao D. et al., 2007; Xie et al., 2006; Banholzer et al., 2008) due to their unique electrical, optical, catalytic or magnetic properties (Chen J.R. et al., 2004) and large surface-to-volume ratio (Xie et al., 2008).

4.1 CNTs in chemical sensing

Among nanomaterials, carbon surfaces represent very attractive materials for electrochemical studies, such as biosensor applications, due to their different allotropes (graphite, diamond and fullerenes/nanotubes). Carbon electrodes are well polarizable. However, their electrical conductivity strongly depends on the thermal treatment, microtexture, hybridization and content of heteroatoms. Additionally, the amphoteric character of carbon allows use of the rich electrochemical properties of this element from donor to acceptor state. Application of recently developed carbon materials include, environment-friendly and energy-saving automotive parts, high-purity components in semiconductor manufacturing equipment for highly integrated chips, internal walls of vacuum vessels in the nuclear fusion reactors expected to be an energy source in the 21st century, and negative electrodes of lithium ion secondary batteries, of which demand is rapidly increasing with the popularization of mobile electronic instruments (Kazua, 1999). During the last years a great interest has been focused on the application of carbons (Flavel et al., 2009; Ma G.P. et al., 2009) as electrode materials because of their accessibility, easy processability and relatively low cost. They are chemically stable in different solutions (from strongly acidic to basic) and able to show performance in a wide range of temperatures. Already well-established chemical and physical methods of activation allow for the production of materials with a developed surface area and a controlled distribution of pores that determine the electrode/electrolyte interface for electrochemical applications.

The possibility of using the activated carbon without binding substances (e.g. fibrous fabrics or felts) gives an additional advantage from the construction point of view. Unique properties of ultra-microelectrodes (UMEs) make them very attractive in electroanalytical measurements which require high spatial and temporal resolution (Wigthman, 1988). At ultra-microelectrodes, at typical slow scan rates of voltammetry, signal to noise ratios (S/N) can increase up to one decade due to efficient mass transportation to the electrode resulting from edge effects. In addition, because of the small dimensions and low IR drop, fast scan voltammetric measurements (Garreau et al., 1990; Hsueh et al., 1993) and measurements in highly resistive media are possible at UMEs (Howell & Wightman, 1984) while high spatial resolution and minimal tissue damage (Musallam et al., 2007). can be achieved in vivo analysis. Microelectrodes are useful for measurements for changes in easily oxidized neurotransmitter concentrations in extracellular brain fluid (Kawagoe et al., 1993) as they provide a way to observe the rapid chemical changes associated with the release of neurotransmitters from neurons and their subsequent removal from the extracellular fluid (Atesa & Sarac, 2009).

The nanodimensions of CNTs guarantee a very large active surface area and are especially suited for the conception of miniaturized sensors. In addition to the high porosity and reactivity, this makes them ideal candidates for the storage of neutral species as well as electron donors, when used as electrodes in electrochemical reactions, for which they may also be called "nanoelectrodes" (Basabe-Desmonts et al., 2007; Daniel et al., 2007; Guo et al., 2004). Hence, it is not surprising that gas sensors made from individual nanotubes show good sensitivities at room temperature (J. Kong et al., 2001; Vasanth Kumar et al., 2010, 2011) in comparison to commercially available classical semiconductor sensors, which in general operate above 200°C. However, a necessary prerequisite is that the molecules to be detected must have a distinct electron donating or accepting ability, which is fulfilled, for example, by ammonia (NH_3) as a donor and nitrogen dioxide (NO_2) as an acceptor. The adsorption of these molecules on the nanotubes is associated with a partial charge transfer, which alters the charge-carrier concentration or, alternatively, the adsorbed molecules may affect the potential barriers present at the tube–electrode contacts. In any event, the resulting change in the electrical resistance of the nanotube is utilized as a sensor signal. However, for the detection of molecules that are only weakly adsorbed (e.g., carbon monoxide and hydrogen), the change in resistance is often too small. A possible method to overcome this drawback is accomplished by the modification of the nanotube sidewalls with nanoparticles made of a suitable metal. For instance, sensitive hydrogen sensors operating at room temperature can be obtained via the deposition of palladium nanoparticles either by direct evaporation (Kong J. et al., 2000) or through electrodeposition (Geng et al., 2004; Teles et al., 2008). Electrodeposition offers the specific advantage of site-selectivity, since the metal decoration is restricted to the current-carrying tubes, so that the remaining substrate surface is unaffected. An example is a semiconducting SWCNT with electrodeposited Pd particles upon exposure to hydrogen. The operation mechanism is largely analogous to that of a palladium gate field-effect transistor realized within classical silicon technology (Lundstroem et al., 1975). Specifically, molecular hydrogen is split on the surface of a Pd particle into atomic hydrogen, which diffuses to the Pd/SWCNT interface. At this interface, a dipole layer is formed, which acts like a microscopic gate electrode that locally changes the charge-carrier concentration. It should be mentioned that the recovery of this type of room-temperature-operated hydrogen sensor requires a supply of oxygen to remove the hydrogen atoms in the form of water (Balasubramanian & Burghard, 2005). Carbon nanotubes exhibit high electron transfer rates for different redox couples in various media (Balasubramanian et al., 2004) which has stimulated an increasing amount of research into CNT based amperometric sensors for the detection of specific analytes in solution. The length scales of CNTs are similar to that of typical biological molecules, which gives CNTs an edge over other materials in functioning as effective electrodes in bioelectrochemical sensing (Guiseppi-Elie et al., 2002). In particular, their high aspect ratio and their diameter in the nanometer range make CNTs particularly well suited for direct electrochemical communication with the redox site of a protein, without requiring any mediator. When properly arranged, a nano- tube should have the capability to act as a 1D channel that guides electrons towards the redox center. These materials can be used to preconcentrate analytes and for the magnetic separation and molecular identification of biomolecules, and organic and inorganic species (AguilarArteaga et al., 2010).

The application of CNTs as absorption and/or transducing materials was shown in several different works. Recently, a critical review dealing with the adsorption mechanism of analytes of different nature on carbon nanotubes was published (Cruz et al., 2010; Fontanals et al., 2007; See et al., 2010; Petrovic et al., 2010; Zeng J. et al., 2009; Augusto et al., 2010; Lucena et al., 2011;). The authors stated that this procedure cannot be explained by a single mechanism but through a combination of mechanisms. Indeed, the introduction of functional groups on the surface can facilitate the selective interaction with a given family of compounds. However, this fact has not yet been exploited for the analysis of urine samples. MWCNTs are proposed as a sorbent in an off-line SPE method for the determination of antidepressants using LC and UV detection (Bakker & Qin, 2006). MWCNTs have been applied as adsorbents for preconcentration of sulfonylurea herbicides (Lemos et al., 2008; Zhou Q. et al., 2006a, 2007), bisphenol A, 4-n-nonylphenol, 4-tert-octylphenol (Cai et al., 2003a), dichlorodiphenyltrichloroethane and its metabolites (Zhou Q. et al., 2006b); atrazine, simazine (Zhou Q. et al., 2006c), and trihalomethanes (Lu et al., 2005) in environmental water samples, barbiturates determination in pork (Zhao H. et al., 2007), for simultaneous determination of 10 sulfonamides in eggs and pork (Fang et al., 2006), for trapping volatile organic compounds in a purge-and trap GC system (Li Q. et al., 2004), to extract di-ethylphthalate, di-n-propyl-phthalate, di-iso-butyl-phthalate, and di-cyclohexylphthalate from aqueous solutions (Cai et al., 2003b).

Direct electron transfer has been achieved with various types of CNT electrodes for cytochrome c (Wang J. et al., 2002), horseradish peroxidase (Zhao Y. et al., 2002), myoglobin

(Zhao G.C. et al., 2003), as well as glucose oxidase (Guiseppi-Elie et al., 2003). It is noteworthy that in the latter case, the redox-active center is deeply embedded within the protein. In some cases, an oxidative pretreatment that introduces negatively charged surface groups on the CNTs was necessary to achieve high electron transfer rates. In a strategy to optimize the accessibility of the redox center, aligned CNT arrays have been fabricated using self-assembly techniques, followed by the covalent attachment of microperoxidase to the tube ends (Gooding et al., 2003). On this basis, various types of amperometric biosensors have been fabricated (Rubianes et al., 2003; Wang J. & Musameh, 2003; Wang J. et al., 2003). A glucose sensor, for example, is obtained by immobilizing glucose oxidase onto SWCNTs deposited on a glassy carbon surface (Davis et al., 2003). Hrapovic et al., 2004 combined platinum nanoparticles with carbon nanotubes and developed a new approach for the electrochemical detection of glucose (Aponte et al., 2006). Effective deposition of Pt nanoparticles on SWCNT was achieved by using a negatively charged polymer, Nafion, to dissolve and disperse SWCNT. Pt nanoparticles can easily be deposited on the Nafion-modified CNT due to charge interactions. Compared to the sensors containing only Pt or CNT, the integration of Pt and CNT lead to improved sensitivity and detection limit.

Zhang and Gorski used carbon nanotubes and redox mediators dispersed into a polymeric matrix for NADH detection and found a reduction of the overpotential (by 0.3 V), a higher sensitivity, and faster response times (Zhang M. & Gorski, 2005). Wang J. & Musameh, 2004 reported on the use of carbon nanotubemodified glassy carbon electrodes in order to accelerate the electrooxidation of insulin, which resulted in a lowering of the required detection potential and a 14 nM detection limit as studied by flow injection analysis. Wang K. et al., 2005 proposed a potentiometric sensor for ascorbic acid on the basis of cobalt phthalocyanine nanoparticles directly coated on a glassy carbon electrode by drop coating. The authors explained the selective and rapid response to ascorbic acid on the basis of the redox potential, with cobalt acting as the mediator.

Kerman et al., 2004 also described the label-free electrochemical detection of DNA based on the direct attachment of adenine probes to the sidewall and end of functionalized MWCNTs. The MWCNTs were attached onto the carbon paste electrode surface modified with thymine probes by hybridization between adenine and thymine. The combination of sidewall and end functionalization of MWNT showed enhancement of the guanine oxidation signal in the direct measurement compared to the ones from only end-modified MWCNT.

The attachment of numerous enzymes on a single carbon nanotube label provided for enormous signal enhancement. Wang J. et al., 2004 reported on a novel method to dramatically amplify enzyme-based DNA sensing by using carbon nanotubes. In the new sensing scheme, the DNA duplex acted as connectors between microbeads and CNTs. The CNT loaded with enzymes linked to the magnetic microbeads through the interaction of complementary oligonucleotides. Without the recognition event, the CNT cannot attach to the particles and is removed during the magnetic separation. The CNTs function as carriers for enzyme tags, and the enzyme reaction products were also accumulated in the CNTs. The coupling of enzyme-tagged CNTs provided a better detection limit than single-enzyme labels and a conventional glassy carbon transducer.

4.2 MIPs in chemical sensing

In the sensors and biosensors technology, often there is the overriding complication that it is unusual, in "real" samples, for there to be a single species present. More commonly an analyte of interest is accompanied by a number of different species, all present at different concentrations and all adding to the complexity of the analytical problem.

All over the world, billions of dollars are spent annually on chemical/biological detections related to medical diagnosis, environmental monitoring, public security and food safety because lab analysis using expensive equipment is usually cumbersome and time-consuming. Therefore, there has been a pressing societal need for the development of chemo/biosensors for the detection of various analytes in solution and atmosphere, which are both less expensive and simpler to construct and operate. Although considerable progress was made in the past several decades, the chemo/biosensor field remains underdeveloped and at a low level of commercialization because of the lack of alternative strategies and multidisciplinary approaches (Guan et al., 2008).

The standard approach to the analytical analysis of complex matrices is the separation of the different components. Typically, therefore, before a sensor can be used to perceive and quantify one component in a mixed solution, the various components of the complex mixture must be separated, usually by a chromatographic process, so that some form of non-selective sensor, e.g. UV absorbance measurements, can be used to detect and quantify each individual component. In order to improve the performance of chemical sensors, an improvement of their selectivity is required, so that a particular chemical species can be detected and assayed without the need for a possibly lengthy separation stage. In this direction, a technological approach is the development of the biosensor (Updike & Hicks, 1967).

Molecular imprinting is one of the most efficient strategies that offer a synthetic route to artificial recognition systems by a template polymerization technique (Ye & Mosbach, 2001; Spivak, 2005; Zhang H.Q. et al., 2006). While most research in this direction is targeted to the design of chromatographic stationary phases, their use in electrochemical sensors is expanding for electroactive analytes. In convincing work, the group of Mandler explored sol-gel polymers imprinted with the organophosphate pesticide parathion and performed gasphase and liquid-phase partitioning experiments as well as cyclic voltammetric studies (Marx et al., 2004). Imprinted films showed a more than 10-fold increased equilibrium binding over nonimprinted polymers and discriminated well against a range of other structurally similar organophosphates.

To date molecularly imprinted polymers have been successfully used with most types of transduction platforms and a range of methods have been used to bring about close integration of the platform with the polymer (Adhikari & Majumdar 2004). By using MIP it is possible to overcome the commonly founded limitations in the traditional biosensor approach (Orellana & Moreno-Bondi, 2005; Jiang & Ju, 2007), based on the use of biological macromolecules (e.g. antibodies, enzymes, receptors and ion channel proteins, nucleic acids, aptamers and peptide nucleic acids) or biological systems (e.g. ex vivo tissue, microorganisms, isolated whole cells and organelles) as sensing element. In the last years, indeed, although biological receptors have specific molecular affinity and have been widely used in diagnostic bioassays and chemo/biosensors, great affords have been made in synthesizing artificial recognition receptors to overcome the limit in the efficiency of chemosensors, especially for the detection of ultratrace analytes due to the small surface area and non-tunable surface properties of transducers. Another limitation of traditional biosensors is that often natural receptors for many detected analytes do not exist;

furthermore, they are produced via complex protocols with a high cost and require specific handling conditions because of their poor stability (Whitcombe et al., 2000; Wulff, 2002; Haupt & Mosbach, 2000; Ye & Haupt, 2004).

During the past ten years, the literatures on the development of MIP-based sensors, in particularly electrochemical (Riskin et al., 2008;) and optical (McDonagh et al., 2008; Basabe-Desmonts et al., 2007; Li J. et al., 2007a; Feng et al., 2008) sensors, have been dramatically growing (Nilsson et al., 2007; Ramanavicius et al., 2006). It was found that the manufacture of composites consisting of molecularly imprinted conducting polymers results in obtaining materials that exhibit both predetermined selective molecular recognition and electrical conductivity (Deore et al, 2000). This type of materials, mostly frequented based on overoxidized polypyrrole, is of special interest for use in the field of sensor technology (Shiigi et al., 2003).

MIP-based electrochemical sensors were first reported in the early 1990s by Mosbach's group (Andersson et al., 1990), and to date, remarkable progress in MIP-based electrochemical sensors have been achieved in the use and the performance of conductometric/potentiometric MIP nanomaterials (Zhou Y.X. et al., 2003), which were used to detect many different analytes (Augisto et al., 2010) such as barbituric acid (Mirsky et al., 1999), amino acid derivatives (Panasyuk et al., 1999), morphine (Kriz et al., 1995), atrazine (Kim **Y.** et al., 2007), benzyltriphenylphosphonium chloride (Kriz & Mosbach, 1995), thiophenol (Kröger et al., 1999), glutamic acid (Ouyang et al., 2007), folic acid (Prasad et al., 2010a; Prasad et al., 2010c), tolazoline (Zhang Z. et al., 2010a), tryptophan (Prasad et al., 2010d; Kong Y. et al., 2010), clindamycin (Zhang J. et al., 2010), 2,4-dichlorophenoxy acetic acid (Xie et al, 2010), histamine (Bongaers et al., 2010), caffeine (Alizadeh et al., 2010; Vinjamuri et al., 2008), theophylline (Kan et al., 2010) uracil and 5-fluorouracil (Prasad et al., 2009), salicylic acid (Kang et al., 2009), uric acid (Patel et al., 2009), resveratrol (Xiang & Li, 2009), hydroquinone (Kan et al., 2009; Kan et al., 2008a), bisphenol (Kolarz & Jakubiak, 2008), and dopamine (Kan et al., 2008b).

Despite the application of MIPs as sensor matrices or separation materials, they suffer from basic limitations associated with the limited concentration of imprinted sites, and the bulk volume of the polymer matrices that requires long diffusion paths of the imprinted host molecules. These limitations lead to inefficient sensing or separation processes (Shi et al., 2007). In order to increase the efficiency of MISPE, a method is increasing the surface area of polymer. Higher surface area can contain more functional groups, thus the interaction between polymer and bioactive compounds will be increased accordingly (Tian et al., 2011). A useful approach is the fabrication of robust multilayer structures by the photo-crosslinking of the layers (Sun et al., 1998, 2000). This process transformed the electrostatic interlayer stabilizing interactions into covalent bonds. This achievement, together with the stated limitations of MIPs, suggested that the surface imprinting of LbL nanostructured films might be a viable technique to fabricate effective MIP matrices.

5. CNTs-MIPs composites: innovative materials for analytical determinations

Among the different applications of CNTs in nanotechnology, polymer composites, consisting of additives and polymer matrices, including thermoplastics, thermosets and elastomers, are considered to be an important group of relatively inexpensive materials for many engineering applications (Ma et al., 2010; Yang et al., 2007). The technology implications are significant to many fields, from semiconductor device manufacturing to emerging areas of nanobiotechnology, nanofluidics, and chemistry where the ability to mold structures with molecular dimensions might open up new pathways to molecular recognition, drug discovery, catalysis, and molecule specific chemio-biosensing (Hua et al., 2004).

The main approaches for the fabrication of these materials can be divided into "grafting to" and "grafting from" approaches (Liu M. et al., 2009; Baskaran et al., 2004).

The "grafting to" approach involves pre-formed polymer chains reacting with the surface of either pristine or pre-functionalized carbon nanotubes. (See Figure 3). The main approaches exploited in this functionalization strategy are radical or carbanion additions as well as cycloaddition reactions to the CNT double bonds. Since the curvature of the carbon nanostructures imparts a significant strain upon the sp^2 hybridized carbon atoms that make up their framework, the energy barrier required to convert these atoms to sp^3 hybridization is lower than that of the flat graphene sheets, making them susceptible to various addition reactions. Therefore, to exploit this chemistry, it is only necessary to produce a polymer centered transient in the presence of CNT material. Alternatively, defect sites on the surface of oxidized CNTs, as openended nanostructures with terminal carboxylic acid groups, allow covalent linkages of oligomer or polymer chains. The "grafting to" method onto CNT defect sites means that the readymade polymers with reactive end groups can react with the functional groups on the nanotube surfaces. An advantage of the "grafting to" method is that preformed commercial polymers of controlled molecular weight and polydispersity can be used. The main limitation of the technique is that initial binding of polymer chains sterically hinders diffusion of additional macromolecules to the CNT surface, leading to a low grafting density. Also, only polymers containing reactive functional groups can be used.

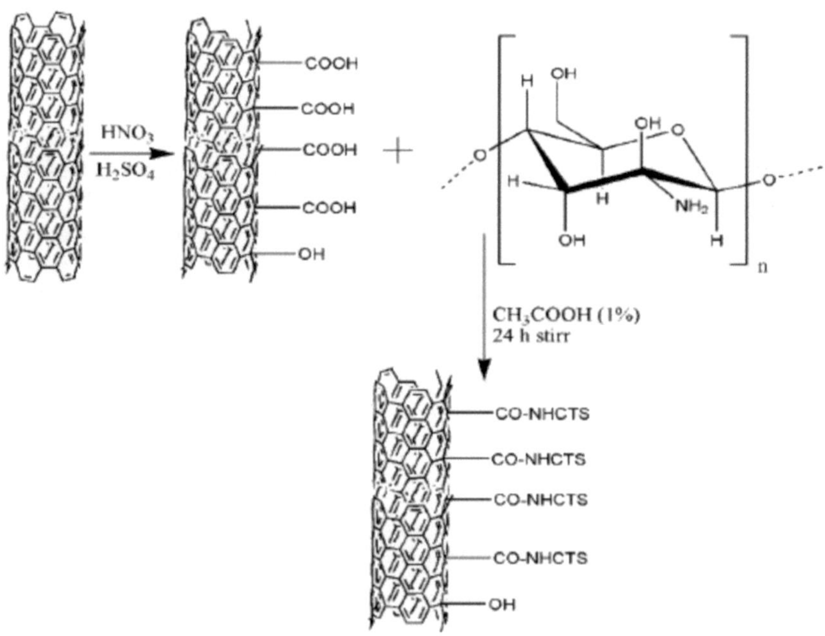

Fig. 3. Synthesis of CNTs-polymer composites by "Grafting to" approach as reported by Venkatesan & Kim, 2010.

The "grafting from" approach involves the polymerization of monomers from surface-derived initiators on CNTs. These initiators are covalently attached using the various functionalization reactions developed for small molecules, including acid-defect group chemistry and side-wall functionalization of CNTs. (See Figure 4). The advantage of "grafting from" approach is that the polymer growth is not limited by steric hindrance, allowing high molecular weight polymers to be efficiently grafted. In addition, nanotube-polymer composites with quite high grafting density can be prepared. However, this method requires strict control of the amounts of initiator and substrate as well as accurate control of conditions required for the polymerization reaction. Moreover, the continuous π-electronic properties of CNTs would be destructed by the acid oxidation, even worse, CNTs may be destroyed to several hundred nanometers in length. As a result, compared with the "grafting from", the "grafting to" has much less alteration of the structure of CNTs (Yan & Yang, 2009; Mylvaganam & Zhang, 2004; Zhang M.N. et al., 2004).

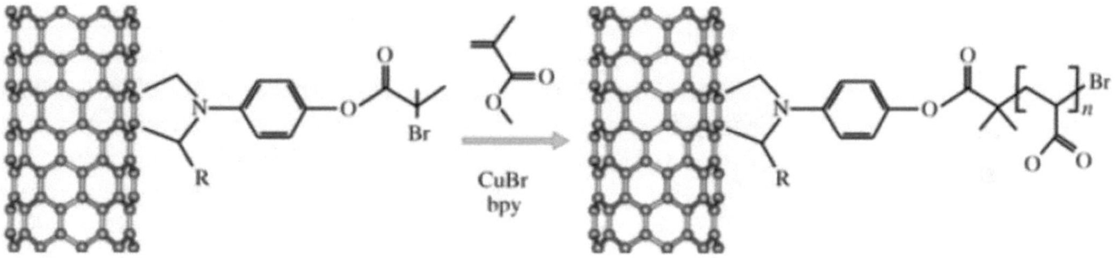

Fig. 4. Synthesis of CNTs-polymer composites by "Grafting from" approach as reported by Balasubramanian & Burghard, 2005.

Many techniques including esterification (Gao C. et al., 2007; Sano et al., 2001; Kahn et al., 2002; Huang W. et al., 2003; Lin et al., 2003), "click" chemistry (Li H.M. et al., 2005), layer-by-layer self-assembly (Kong H. et al., 2005; He & Bayachou, 2005; Qin et al., 2005; Artyukhin et al., 2004), pyrene moiety adsorption (Bahun et al., 2006; Martin et al., 2004; Qu et al., 2002; Petrov et al., 2003; Gomez et al., 2003), radical coupling (Liu Y.Q. et al., 2005; Lou et al., 2004), anionic coupling (Huang H.M. et al.,2004), radical polymerization (Qin et al., 2004a; Shaffer & Koziol, 2002), supercritical CO_2-solubilized polymerization or coating (Wang J.W. et al., 2006; Dai et al., 2004), γ-ray irradiation (Xu H.X. et al., 2006), cathodic electrochemical grafting (Petrov et al., 2004) polycondensation (Zeng H.L. et al., 2006a; Gao C. et al., 2005; Nogales et al., 2004), reversible addition fragmentation chain-transfer (RAFT) polymerization (Xu G.Y. et al., 2006; You et al., 2006; Cui H. et al., 2004; Hong et al., 2005, 2006), anionic polymerization (Chen S.M. et al. 2006; Liu I.C. et al., 2004), ring-opening polymerization (Zeng H.L. et al., 2006b; Qu et al., 2005; Buffa et al., 2005; Gao J.B. et al., 2005) and atom transfer radical polymerization (ATRP) (Kong H. et al., 2004; Qin et al., 2004b; Yao et al., 2003) have been employed to functionalize CNTs with polymers.

CNTs exhibit a high aspect ratio and high conductivity, which makes CNTs excellent candidates for conducting composites (Sahoo et al., 2010) (Figure 5).

The observation of an enhancement of electrical conductivity by several orders of magnitude of CNTs in polymer matrices without compromising other performance aspects of the polymers such as their low weight, optical clarity, low melt viscosities, etc., has

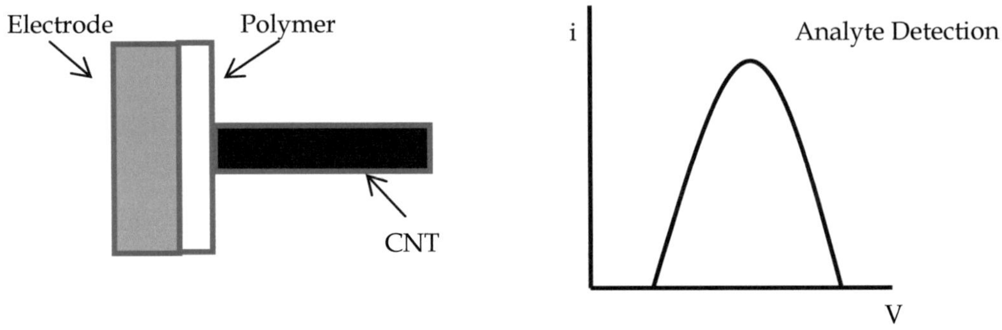

Fig. 5. Schematic representation of electrochemical detection by CNTs composite electrode. Adpted from Teles & Fonseca, 2008.

triggered an enormous activity worldwide in this scientific area (Spitalsky et al., 2010). Nanotube-filled polymers could potentially, among the others, be used for transparent conductive coatings, electrostatic dissipation, electrostatic painting and electromagnetic interference shielding applications (Bauhofer & Kovacs, 2009; Breuer & Sundararaj, 2004; Moniruzzaman & Winey, 2006; Winey et al., 2007). The electrical conductivity in CNTs-polymer nanocomposites depends on dispersion (Sandler et al., 2003; Li J. et al., 2007b; Song & Youn, 2005), alignment (Choi et al., 2003; Fangming et al., 2003), aspect ratio (Bai & Allaoui, 2003; Bryning et al., 2005), degree of surface modification (Georgakilas et al., 2002) of CNTs, polymer types (Ramasubramaniam et al., 2003) and composite processing methods (Li J. et al., 2007b). Based on their characteristic, CNTs-polymer composites can behave as conductors, semiconductors or insulators (Maruccio et al., 2004).

CNTs-polymer composites were successfully synthesized and employed for different applications, both in biomedical and engineering fields. One of the most representative example is a study regarding an ammonia gas sensors based on single-walled carbon nanotubes functionalized with covalently attached poly(m-aminobenzenesulfonic acid). The sensor was operated as a chemiresistor, with the carbon nanotubes forming a random network between interdigitated electrodes, and improved response and recovery times were observed (on the order of 15 min) (Bekyarova et al., 2004). In another work, a carbon nanotube/poly(ethylene-co-vinyl acetate) composite electrode was developed for amperometric detection in Capillary Electrophoresis (Frost et al., 2010). The new electrode also generated improved S/N, decreased fouling, and resulted in better long-term stability (Chen Z. et al., 2009). Zhou D. et al., 2008 describes a novel sieving matrix composed of both a quasi-interpenetrating polymer network (IPN) and PDMA functionalized MWNTs. Atom transfer radical polymerization was used to graft PDMA on MWNTs. The functionalized MWNTs were compatible with the quasi-IPN network. The rigid structure of MWNTs increased the stability and sieving ability of the matrix. Results showed that this novel matrix was advantageous in terms of resolution, speed, and reproducibility. Carbon nanotubes have been used to improve the efficiency of $Ru(bpy)_3^{2+}$ modified polyacrylamide electrode because of their high conductivity (Xing & Yin, 2009). Carbon nanotubes modified with polypyrrole-silica nanocomposites seem very promising for electrochemical DNA sensor design (Ramanacius et al., 2006). MWCNTs were also grafted with poly(acrylamide) (PAAM) and with poly(N,Ndimethylacrylamide) (PDMA) at same grafting percentage by using N_2 plasma technique and used in the removal of Pb^{2+} from aqueous solution under ambient conditions (Shao et al., 2010). Authors found that the grafted PAAM and PDMA improved MWCNT adsorption capacity in the removal of Pb^{2+} from large volumes of aqueous solutions. Furthermore, MWCNTgPAAM had much higher adsorption capacity than MWCNT-g-PDMA, which was attributed to higher amide group content in acrylamide than that in N,N-dimethylacrylamide.

A particular kind of CNTs-polymer composites is represented by CNTs-MIPs composites, in which the polymer part is a molecularly imprinted polymer (Chang et al., 2011; Walcarius et al., 2005). CNTs impart electrical conductivity to MIPs, while molecular imprinting on these one-dimensional nanostructures will endow the nanotubes with molecular recognition functions, further expanding their application fields (Guan et al., 2008). Several example of using these materials are reported in literature for biomedical, pharmaceutical and environmental applications (Figure 6).

In pharmaceutical fields, in Zhang Z. et al., 2010b, a novel sensitive and selective imprinted electrochemical sensor was constructed for the direct detection of L-histidine by combination of a molecular imprinting film and MWNTs. The sensor was fabricated onto an indium tin oxide electrode via stepwise modification of MWNTs and a thin film of MIPs via sol–gel technology. The introduced MWNTs exhibited noticeable enhancement on the sensitivity of the MIPs sensor, meanwhile, the molecularly imprinted film displayed high sensitivity and excellent selectivity for the target molecule. H. Y. Lee & Kim, 2009 reports of the synthesis of CNTs-MIP composite to be potentially applied to probe materials in biosensor system for theophylline recognition based on CNT field effect. Hydroxyl-

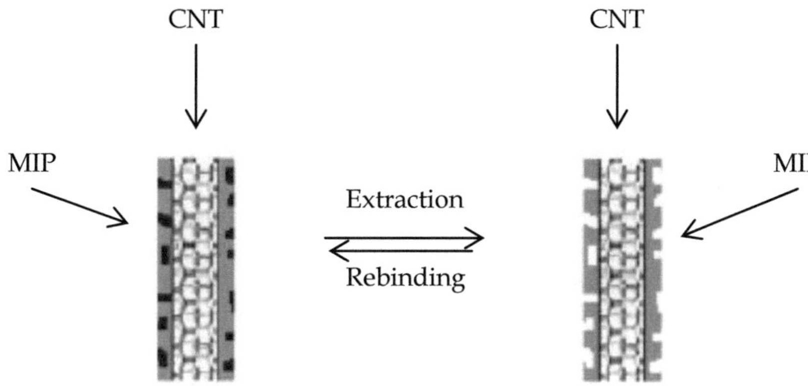

Fig. 6. Schematic representation of CNTs-MIPs recognition process. Adapted from Z. Zhang et al., 2010d.

functionalized CNT was modified by silanisation with 3-chloropropyl trimethoxysilane. The iniferter groups were then introduced by reacting the CNT-bound chloropropyl groups with sodium *N,N*-diethyldithiocarbamate. UV light-initiated copolymerization of ethylene glycol dimethacrylate (crosslinking agent) and methacrylic acid (functional monomer) resulted in grafting of MIP on CNT for theophylline as a model template. The theophylline-imprinted polymer on CNT showed higher binding capacity for theophylline than non-imprinted polymer on CNT and selectivity for theophylline over caffeine and theobromine (similar structure molecules). Another theophylline sensor based on CNTs-MIPs composites is shown in Lee E. et al., 2008. In this study acrylated Tween 20 was used as a linking molecule of MIPs to CNT. MIPs were formed for theophylline as a model template on the surface of CNT with methacrylic acid (functional monomer) and ethylene glycol dimethacrylate (crosslinking agent) using a photografting polymerization technique. The adsorbed layer of 2,2-dimethoxy-2-phenylacetophenone initiated a radical polymerization near the surface by UV-light irradiation. The theophylline-imprinted polymer on CNT showed higher binding capacity for theophylline than non-imprinted polymer (NIP) on CNTand selectivity for theophylline over caffeine (similar structure molecules).

In another work (Prasad et al., 2010a), MIP–carbon composite is prepared via in situ free radical polymerization of a synthetic monomer and subsequent crosslinkage with ethylene glycol dimethacrylate, in the presence of carbon powder and folic acid as template . The detection of folic acid with the MIP fiber sensor was found to be specific and quantitative in aqueous, blood serum and pharmaceutical samples, without any problem of nonspecific false positive contribution and crossreactivity.

An insulin imprinted polymer (Prasad et al., 2010b) was synthesized over the surface of vinyl group functionalized MWCNTs using phosphotidylcholine containing functional monomer and crosslinker. Phosphotidylcholine is a major component of all biological membrane; its incorporation in polymer backbone assures water compatibility, biocompatibility and specificity to molecularly imprinted nanomaterials, without any crossreactivity or interferences from biological sample matrices. An electrochemical sensor fabricated by modifying multiwalled carbon nanotubes molecularly imprinted polymer onto the pencil graphite electrode, was used for trace level detection of insulin in aqueous, blood serum, and pharmaceutical samples by differential pulse anodic stripping voltammetry.

A sensitive molecularly imprinted electrochemical sensor has been developed in Huang J. et al., 2011a for the selective detection of tyramine by combination of MWCNTs-gold nanoparticle composites and chitosan. Chitosan acts as a bridge for the imprinted layer and the nanocomposites. The molecularly imprinted polymer (MIP) was synthesized using tyramine as the template molecule, silicic acid tetracthyl ester and triethoxyphenylsilane as the functional monomers. The molecularly imprinted film displayed excellent selectivity towards tyramine. A thymidine sensor (Zhang Z.H. et al., 2010) was developed by casting thin film of molecularly imprinted sol–gel polymers with specific binding sites for thymidine on carbon electrode by electrochemical deposition. The excellent performance of the imprinted sol–gel/MWCNTs electrode towards thymidine can be ascribed to the MWCNTs functional layer with electrochemical catalytic activities and the porous imprinted film with plentiful selective binding sites. Under the optimized analytical conditions, the peak current was linear to thymidine concentration from 2 to 22 μmol^{-1} with the detection limit of $1.6\ 10^{-9}$ mol L^{-1}.

A different MWNTs-MIPs composites was prepared by using dopamine as a template molecule by the selective copolymerization of methacrylic acid and trimethylolpropane trimethacrylate in the presence of the template and vinyl group functionalized MWNT surface (Kan et al., 2008b). In this work, for grafting MIPs on MWNTs, the vinyl group was first introduced on the surface of MWNTs, which directed the selective polymerization of functional monomers and cross linkers in the presence of DA on the MWNTs surface. The thickness of the MIPs can be adjusted by changing the concentration of prepolymerization monomers. The resulting MWNTs-MIPs possessed a faster adsorption dynamics, higher selectivity for the template. The modified electrode fabricated by modifying the MWNTs-MIPs on the glassy carbon electrode can recognize dopamine with a linear range from 5.0 10^{-7} to 2.0 10^{-4} mol/L.

A sensor for the detection of Uric acid was proposed in (Yu J.C.C. & Lai, 2006) by polymerization of polymethacrylic acid on the surface of vinyl-functionalized CNTs in the presence of the template. The MIP adsorbs more Uric acid than NIP and the imprinting efficiency was found to be about 4.41. The MIP modified MWCNTs can be deposited on the MWCNTs electrode surface and used for the electrochemical detection for the template. The differences of adsorption amounts between MIP and NIP electrodes were determined by CVs with different adsorption times. The adsorption reached saturation after 5min of adsorption and the result was close to the rebinding experiments. The sensitivities of the MIP and NIP modified electrodes were about 11.03 and 5.39 mAM^{-1} cm^{-2}, respectively, and the difference mainly came from the affinity cavities which were created by the imprinted template.

A novel protein molecularly imprinted membrane (PMIM) was synthesized in Zhang M. et al., 2010 on the surface of MWNTs through a surface molecular imprinting technique by using bovine serum albumin as the template molecule, acrylamide as the functional monomer, N,N'methylenebisacrylamide as the crosslinker. The selectivity adsorption experiments showed that the PMIM/MWNTs also had higher adsorption capacities for BSA than for such molecules, as HSA, HB, pepsin and HRP. The PMIM/MWNTs displayed a 2.6 fold increase in affinity to BSA compared to the nPMIM/MWNTs. The PMIM/MWNTs, on the other hand, did not exhibit any significant change in affinity to other molecules compared to the nPMIM/MWNTs.

An estrone sensor was developed in Gao R. et al., 2011 by using a semicovalent imprinting strategy, which employed a thermally reversible covalent bond at the surface of silica coated CNTs. The synthesis of the nanocomposites involves silica shell deposition on the surface of CNTs, MIPs functionalized onto the silica surface, and final extraction of estrone by thermal reaction and generation of the recognition site. Authors state that the core-shell CNTs@MIPs nanocomposites developed in this work can also be applied as a selective coating for electrochemical or quartz crystal microbalance sensors to monitor for estrone residue in environmental water

MIP-CNTs nanocomposites are also proposed as innovative drug delivery devices (Yin et al., 2010). MIP nanotubes were fabricated by atom transfer radical polymerization (ATRP) and applied in enantioselective drug delivery and controlled release. Authors found that S-propranolol imprinted nanotubes provided differential release of enantiomers, whereby the release of the more therapeutically active S-propranolol (eutomer) is greatly promoted whilst the release of the less active R-enantiomer (distomer) is retarded.

In environmental field, a recent work (Zhang Z. et al., 2010c) a novel MIPs with core-shell structure is fabricated by using MWNTs as the core material. Prior to polymerization, the silicon-oxygen group was grafted onto the MWNTs surface, then silicon-oxygen groups grafted onto the MWNTs surface could copolymerize directly with functional monomers and crosslinkers in the presence of the template molecules by hydrolysis and condensation, which can lead to the formation of MIPs on the surface of MWNTs. The imprinted material, which showed a good selective recognition to Sudan IV with Qmax of 63.2mmol g^{-1}, was applied as sorbets for the enrichment and determination of trace Sudan IV in real samples by online SPE–HPLC.

CNTs were used successfully to enhance the binding capacity of a molecularly imprinted polypyrrole modified stainless steel frit for determination of ochratoxin A in red wines (Yu J.C.C. & Lai, 2007; Wei et al., 2007). In a different work (Li Y. et al., 2010), a molecularly imprinted polymer–graphene oxide hybrid material was synthesized by reversible addition and fragmentation chain transfer (RAFT) polymerization for the selective detection of 2,4-dichlorophenol in aqueous solution with an appreciable selectivity over structurally related compounds.

Diaz-Diaz et al., 2011 describes an electrochemical sensors based on a catalytic 2,4,6-trichlorophenol molecularly imprinted microgel that mimics the dehalogenative function of the natural enzyme chloroperoxidase for p-halophenols. Two strategies were explored: a carbon paste modified with the polymer and the drop-coating of screen-printed electrodes with powder suspensions of the polymer and carbon nanotubes. With this last design, 2,4,6-trichlorophenol concentrations above 25 mM could be detected.

In similar studies, an electrochemical imprinted sensor for sensitive and convenient determination of Bisphenol A (Huang J. et al., 2011b) and clindamycin (Zhang Z. et al., 2010a) were developed. In both the cases, MWCNTs and gold nanoparticles were introduced for the enhancement of electronic transmission and sensitivity, while thin film of molecularly imprinted sol-gel polymers with specific binding sites for the templates were cast on gold electrode by electrochemical deposition. The resulting composites displayed excellent selectivity towards.

A simple method was developed (Gao R. et al., 2010) to synthesize core–shell molecularly imprinted polymers for the extraction of triclosan with fast kinetics, high capacity and favorable selectivity by combining a surface molecular imprinting technique with a sol–gel process based on carbon nanotubes coated with silica.

By a surface imprinting technique, in Zhang H. et al., 2011, a composite imprinted material, on the basis of a MWCNTs-incorporated layer using melamine as a template, methacrylic acid as a functional monomer, and ethylene glycol dimethacrylate as a cross-linker, was synthesized. In this work, the poly(acrylic-acid)-functionalized CNTs were synthesized to increase the diameter of CNTs. Then, the vinyl group was introduced to the surface of poly(acrylic-acid)-functionalized CNTs by an amidation. Using Melamine as a template molecule, imprinted CNT composite material was fabricated by a thermal polymerization. Applied as a sorbent, the imprinted materials were used for the determination of Melamine in the spiked sample by online SPE combined with HPLC.

By the same approach, Ga(III)imprinted-CNTs sorbent was prepared in Zhang Z. et al., 2010d by using Ga(III) ion-8hydroxyquinoline complex as a template molecule. The imprinted sorbent was applied successfully for extraction of Ga(III) ion from fly ash lixivium followed by FAAS detection. Authors state that compared with the others literature methods for Ga determination, their method is sufficiently accurate and precise to be used for Ga(III) ion analysis in fly ash samples, and performed better characteristics such as selectivity and cleanliness of the extracts.

6. Acknowledgements

Authors are solely responsible for this work. Financial support of Regional Operative Program (ROP) Calabria ESF 2007/2013 - IV Axis Human Capital - Operative Objective M2 - Action D.5 is gratefully acknowledged.

7. References

Adbo, K. & Nicholls, I.A. (2001). Enantioselective solid-phase extraction using Troger's base molecularly imprinted polymers. *Analytica Chimica Acta*, 435, 115-120.

Adhikari, B. & Majumdar, S. (2004) Polymers in sensor applications. *Progress in Polymer Science*, 29, 699- 766.

AguilarArteaga, K.; Rodriguez, J.A. & Barrado, E. (2010). Magnetic solids in analytical chemistry: A review. *Analytica Chimica Acta*, 674, 157–165.

Ajayan, P.M.; Lambert, J.M.; Bernier, P.; Barbedette, L.; Colliex, C. & Planeix, J.M. (1993). Growth morphologies during cobalt-catalyzed single-shell carbon nanotube synthesis. *Chemical Physics Letters*, 215, 509-517.

Alizadeh, T.; Ganjali, M.R.; Zare, M. & Norouzi, P. (2010). Development of a voltammetric sensor based on a molecularly imprinted polymer (MIP) for caffeine measurement. *Electrochimica Acta*, 55, 1568-1574.

Anderson, C.D.; Shea, K.J. & Rychnovsky, S.D. (2005). Strategies for the generation of molecularly imprinted polymeric nitroxide catalysts. *Organic Letters*, 7, 4879-4882.

Andersson, L.I.; Miyabayashi, A.; O'Shannessy, D.J. & Mosbach, K. (1990). Enantiomeric resolution of amino acid derivatives on molecularly imprinted polymers as monitored by potentiometric measurements. *Journal of Chromatography*, 516, 323-331.

Ando, Y. & Iijima, S. (1993). Preparation of Carbon Nanotubes by Arc-Discharge Evaporation. *Japanese Journal of Applied Physics*, 32, L107-L109.

Ando, Y. (1993). Carbon Nanotubes at As-Grown Top Surface of Columnar Carbon Deposit. *Japanese Journal of Applied Physics*, 32, L1342-L1345.

Ando, Y.; Zhao, X.; Sugai, T. & Kumar, M. (2004). Growing carbon nanotubes. *Materials Today*, 7, 22-29.

Aponte, V.M.; Finch, D.S. & Klaus, D.M. (2006). Considerations for non-invasive in-flight monitoring of astronaut immune status with potential use of MEMS and NEMS devices. *Life Sciences*, 79, 1317–1333.

Artyukhin, A.B.; Bakajin, O.; Stroeve, P. & Noy, A. (2004). Layer-by-Layer Electrostatic Self-Assembly of Polyelectrolyte Nanoshells on Individual Carbon Nanotube Templates. *Langmuir*, 20, 1442-1448.

Atesa, M. & Sarac A.S. (2009). Conducting polymer coated carbon surfaces and biosensor applications. *Progress in Organic Coatings*, 66, 337–358.

Augusto, F.; Carasek, E.; Costa Silva, R.G.; Rivellino, S.R.; Batista, A.D. & Martendal E. (2010). New sorbents for extraction and microextraction techniques. *Journal of Chromatography A*, 1217, 2533–2542.

Ausman, K.; Piner, R., Lourie, O. & Ruoff, R. (2000). Organic solvent dispersion of single-walled carbon nanotubes: towards solutions of pristine nanotubes. *Journal of Physical Chemistry B*, 104, 8911-8915.

Bahun, G.J.; Wang, C. & Adronov, A. (2006). Solubilizing single-walled carbon nanotubes with pyrene-functionalized block copolymers. *Journal of Polymer Science, Part A: Polymer Chemistry*, 44, 1941-1951.

Bai, J.B. & Allaoui, A. (2003). Effect of the length and the aggregate size of MWNTs on the improvement efficiency of the mechanical and electrical properties of nanocomposites experimental investigation. *Composites Part A*, 34, 689–694.

Bakker, E. & Qin, Y. (2006). Electrochemical Sensors. *Analytical Chemistry*, 78, 3965-3983.

Balasubramanian, K. & Burghard, M. (2005). Chemically Functionalized Carbon Nanotubes. *Small*, 1 180 -192.

Balasubramanian, K.; Burghard, M. & Kern, K. (2004). In: *Dekker Encyclopedia of Nanoscience and Nanotechnology*, J.A. Schwarz, C.I. Contescu & K. Putyera, 507-517, Marcel Dekker, New York.

Banholzer, M.J.; Millstone, J.E.; Qin, L.D. & Mirkin, C.A. (2008). Rationally designed nanostructures for surface-enhanced raman spectroscopy. *Chemical Society Reviews*, 37, 885-897.

Basabe-Desmonts, L.; Reinhoudt, D.N. & Crego-Calama, M. (2007). Design of fluorescent materials for chemical sensing. *Chemical Society Reviews*, 36, 993–1017.

Baskaran, D., Mays, J.W. & Bratcher, M.S. (2004). Polymer-Grafted Multiwalled Carbon Nanotubes through Surface-Initiated Polymerization. *Angewandte Chemie International Edition in English*, 43, 2138–2142.

Bauer, L.A.; Birenbaum, N.S. & Meyer, G.J. (2004). Biological applications of high aspect ratio nanoparticles. *Journal of Materials Chemistry*, 14, 517-526.

Bauhofer, W. & Kovacs, J.Z. (2009). A review and analysis of electrical percolation in carbon nanotube polymer composites. *Composites Science and Technology*, 69, 1486–1498.

Bekyarova, E.; Davis, M.; Burch, T.; Itkis, M. E.; Zhao, B.; Sunshine, S. & Haddon, R.C. (2004). Chemically Functionalized Single-Walled Carbon Nanotubes as Ammonia Sensors. *Journal of Physical Chemistry B*, 108, 19717-19720.

Bethune, D.S.; Johnson, R.D.; Salem, J.R.; de Vries, M.S. & Yannoni, C.S. (1993). Atoms in carbon cages: the structure and properties of endohedral fullerenes. *Nature*, 366, 123-128.

Bianco, A., Kostarelos. K.; Partidos, C.D. & Prato, M. (2005). Biomedical applications of functionalised carbon nanotubes. *Chemical Communications*, 5, 571-577.

Bianco, A.; Kostarelos, K. & Prato, M. (2005). Applications of carbon nanotubes in drug delivery. *Current Opinion in Chemical Biology*, 9, 674-679.

Bongaers, E.; Alenus, J.; Horemans, F.; Weustenraed, A.; Lutsen, L.; Vanderzande, D.; Cleij, T.J.; Troost, G.J.; Brummer, R.-J. & Wagner, P. (2010). A MIP-based biomimetic sensor for the impedimetric detection of histamine in different pH environments. *Physica Status Solidi (A) Applications and Materials*, 207, 837-843.

Breuer, O. & Sundararaj, U. (2004). Big returns from small fibers: a review of polymer/carbon nanotube composites. *Polymer Composites*, 25, 630–645.

Bryning, M.B.; Islam, M.F.; Kikkawa, J.M. & Yodh, A.G. (2005). Very low conductivity threshold in bulk isotropic single walled carbon nanotube epoxy composites. *Advanced Materials*, 17, 1186–1191.

Buffa, F.; Hu, H. & Resasco, D.E. (2005). Side-Wall Functionalization of Single-Walled Carbon Nanotubes with 4-Hydroxymethylaniline Followed by Polymerization of ε-Caprolactone. *Macromolecules*, 38, 8258-8263.

Byrne, M.E. & Salian, V. (2008). Molecular imprinting within hydrogels II: progress and analysis of the field. *International Journal of Pharmaceutics*, 364, 188-212.

Cai, Y.; Jiang, G.; Liu, J. & Zhou, Q. (2003a). Multi walled carbon nanotubes as a solid-phase extraction adsorbent for the determination of bisphenol a, 4-n-nonylphenol, and 4-tert-octylphenol. *Analytical Chemistry*, 75, 2517–2521.

Cai, Y.; Jiang, G.; Liu, J. & Zhou, Q. (2003b). Multi-walled carbon nanotubes packed cartridge for the solid-phase extraction of several phthalate esters from water samples and their determination by high performance liquid chromatography. *Analytica Chimica Acta*, 494, 149–156.

Caldorera-Moore M. & Peppas, N.A. (2009). Micro- and nanotechnologies for intelligent and responsive biomaterial-based medical systems. *Advanced Drug Delivery Reviews*, 61, 1391-1401.

Caro, E.; Masquè, N.; Marcè, R.M.; Borrull, F.; Cormack, P.A.G. & Sherrington, D.C. (2002). Non-covalent and semi-covalent molecularly imprinted polymers for selective on-line solid-phase extraction of 4-nitrophenol from water samples. *Journal of Chromatography A*, 963, 169-178.

Chang, L.; Wu, S.; Chen, S. & Li, X. (2011). Preparation of graphene oxide–molecularly imprinted polymer composites via atom transfer radical polymerization. *Journal of Materials Science*, 46, 2024–2029.

Chen, J.R.; Miao, Y.Q.; He, N.Y.; Wu, X.H. & Li, S.J. (2004). Nanotechnology and biosensors. *Biotechnology Advances*, 22, 505-518.

Chen, S.M.; Chen, D.Y. & Wu, G.Z. (2006). Grafting of Poly(tBA) and PtBA-b-PMMA onto the Surface of SWNTs Using Carbanions as the Initiator. *Macromolecular Rapid Communications*, 27, 882-887.

Chen, Z.; Zhang, L.Y. & Chen, G. (2009). Carbon nanotube/poly(ethylene-co-vinyl acetate) composite electrode for capillary electrophoretic determination of esculin and esculetin in Cortex Fraxini. *Electrophoresis*, 30, 3419–3426.

Cheng, H.M.; Li, F.; Sun, X.; Brown, S.D.M.; Pimenta, M.A.; Marucci, A.; Dresselhaus, G. & Dresselhaus, M.S. (1998). Bulk morphology and diameter distribution of single-walled carbon nanotubes synthesized by catalytic decomposition of hydrocarbons. *Chemical Physics Letters*, 289, 602-610.

Choi, E.S.; Brooks, J.S.; Eaton, D.L.; Al Haik, M.S.; Hussaini, M.Y.; Garmestani, H.; Li, D. & Dahmen, K. (2003). Enhancement of thermal and electrical properties of carbon nanotube polymer composites by magnetic field processing. *Journal of Appled Physics*, 94, 6034–6039.

Cirillo, G.; Curcio, M.; Parisi, O.I.; Puoci, F.; Iemma, F.; Spizzirri, U.G.; Restuccia, D. & Picci, N. (2011a). Molecularly Imprinted Polymers for the Selective Extraction of Glycyrrhizic acid from Liquorice Roots. *Food Chemistry*, 125, 1058-1063.

Cirillo, G.; Hampel, S.; Klingeler, R.; Puoci, F.; Iemma, F.; Curcio, M.; Parisi, O.I.; Spizzirri, U.G.; Picci, N.; Leonhardt, A.; Ritschel, M. & Büchner, B. (2011b). Antioxidant Multi-Walled Carbon Nanotubes by Free Radical Grafting of Gallic acid: new materials for biomedical applications. *Journal of Pharmacy and Pharmacology*. 63, 179-188.

Cirillo, G.; Parisi, O.I.; Curcio, M.; Puoci, F.; Iemma, F.; Spizzirri, U.G. & Picci, N. (2010). Molecularly Imprinted Polymers as Drug Delivery Systems for the Sustained Release of Glycyrrhizic acid. *Journal of Pharmacy and Pharmacology*, 62, 577-582.

Colbert, D.T.; Zhang, J.; McClure, S.M.; Nikolaev, P.; Chen, Z.; Hafner, J.H.; Owens, D.W.; Kotula, P.G.; Carter, C.B.; Weaver, J.H.; Rinzler, A.G. & Smalley, R.E. (1994). Growth and Sintering of Fullerene Nanotubes. *Science*, 266, 1218-1222.

Cruz-Vera, M.; Lucena, R.; Cárdenas, S. & Valcárcel, M. (2010). Highly selective and non-conventional sorbents for the determination of biomarkers in urine by liquid chromatography. *Analytical and Bioanalytical Chemistry*, 397, 1029-1038.

Cui, D.X. & Gao, H.J. (2003). Advance and prospect of bionanomaterials. *Biotechnology Progress*, 19, 683-692.

Cui, H.; Wang, W.P.; You, Y.Z.; Liu, C.H. & Wang, P.H. (2004). Functionalization of multiwalled carbon nanotubes by reversible addition fragmentation chain-transfer polymerization. *Polymer*, 45, 8717-8721.

Curcio, M.; Puoci, F.; Cirillo, G.; Iemma, F.; Spizzirri, U.G. & Picci, N. (2010). Selective Determination of Melamine in Aqueous Medium by Molecularly Imprinted Solid Phase Extraction. *Journal of Agricultural and Food Chemistry*, 58, 11883-11887.

Dai, X.H.; Liu, Z.M.; Han, B.X.; Sun, Z.Y.; Wang, Y.; Xu, J.; Guo, X.L.; Zhao, N. & Chen, J. (2004). Carbon nanotube/poly(2,4-hexadiyne-1,6-diol) nanocomposites prepared with the aid of supercritical CO_2. *Chemical Communications*, 19, 2190-2191.

Daniel, S.; Rao, T.P.; Rao, K.S.; Rani, S.U.; Naidu, G.R.K.; Lee, H.Y. & Kawai, T. (2007). A review of DNA functionalized/grafted carbon nanotubes and their characterization. *Sensors and Actuators, B*, 122, 672-682.

Davis, J.J.; Coleman, K.S.; Azamian, B.R.; Bagshaw, C.B. & Green, M.L.H. (2003). Chemical and Biochemical Sensing with Modified Single Walled Carbon Nanotubes. *Chemistry A European Journal*, 9, 3732-3739.

Deore, B.; Chen, Z.D. & Nagaoka, T. (2000). Potential-Induced Enantioselective Uptake of Amino Acid into Molecularly Imprinted Overoxidized Polypyrrole. *Analytical Chemistry*, 72, 3989-3994.

Desai, T.A. (2000). Micro- and nanoscale structures for tissue engineering constructs. *Medical Engineering and Physics*, 22, 595-606.

Diaz-Diaz, G.; Blanco-Lopez, M.C.; Lobo-Castanon, M.J.; Miranda-Ordieres, A.J. & Tunon-Blanco, P. (2011). Preparation and Characterization of a Molecularly Imprinted Microgel for Electrochemical Sensing of 2,4,6-Trichlorophenol. *Electroanalysis*, 23, 201-208.

Dresselhaus, M.S.; Dresselhaus, G.; Charlier, J.C. & Hernandez, E. (2004). Electronic, thermal and mechanical properties of carbon nanotubes. *Philosophical Transactions of the Royal Society A: Mathematical, Physical and Engineering Sciences*, 362, 2065-2098.

Dyke, C.A. & Tour, J.M. (2004). Overcoming the insolubility of carbon nanotubes through high degrees of sidewall functionalization. *Chemistry*, 10, 812-817.

Ebbesen, T.W. & Ajayan, P.M. (1992). Large-scale synthesis of carbon nanotubes. *Nature*, 358, 220-222.

Eggins, B.R. (2002). *Chemical sensors and biosensors*. John Wiley, Chichester, UK.

El-Sheikh, A.H.; Al-Quse, R.W.; El-Barghouthi, M.I. & Al-Masri, F.S. (2010). Derivatization of 2-chlorophenol with 4-amino-anti-pyrine: A novel method for improving the selectivity of molecularly imprinted solid phase extraction of 2-chlorophenol from water. *Talanta,* 83, 667-673.

Endo, M.; Takeuchi, K.; Igarashi, S.; Kobori, K.; Shiraishi, M. & Kroto, H.W. (1993). The production and structure of pyrolytic carbon nanotubes (PCNTs). *Journal of Physics and Chemistry of Solids,* 54, 1841-1848.

Fang, G.; He, J. & Wang, S. (2006). Multiwalled carbon nanotubes as sorbent for on-line coupling of solid-phase extraction to high-performance liquid chromatography for simultaneous determination of 10 sulfonamides in eggs and pork. *Journal of Chromatography A,* 1127, 12–17.

Fangming, D.; Fisher, J.E. & Winey, K.I. (2003). Coagulation method for preparing single walled carbon nanotube/poly(methyl methacrylate) composites and their modulus, electrical conductivity, and thermal stability. *Journal of Polymer Science Part B, Polymer Physics,* 41, 3333–3338.

Feng, F.D.; He, F.; An, L.L.; Wang, S.; Li, Y.H. & Zhu, D.B. (2008). Fluorescent conjugated polyelectrolytes for biomacromolecule detection. *Advanced Materials,* 20, 2959-2964.

Fernando, K.A.; Lin, Y. & Sun, Y.P. (2004). High aqueous solubility of functionalized single-walled carbon nanotubes. *Langmuir,* 20, 4777-4778.

Flahaut, E.; Govindaraj, A.; Peigney, A.; Laurent, Ch.; Rousset, A. & Rao, C.N.R. (1999). Synthesis of single-walled carbon nanotubes using binary (Fe, Co, Ni) alloy nanoparticles prepared in situ by the reduction of oxide solid solutions. *Chemical Physics Letters,* 300, 236-242.

Flavel, B.S.; Yu, J.X.; Shapter, J.G. & Quinton, J.S. (2009). Patterned polyaniline & carbon nanotube–polyaniline composites on silicon. *Soft Matter,* 5, 164-172.

Foldvari, M. & Bagonluri, M. (2008). Carbon nanotubes as functional excipients for nanomedicines: I. pharmaceutical properties. *Nanomedicine: Nanotechnology, Biology, and Medicine,* 4, 173–182.

Fontanals, N.; Marcé, R.M. & Borrull, F. (2007). New materials in sorptive extraction techniques for polar compounds *Journal of Chromatography A,* 1152, 14–31.

Frost, N.W.; Jing, M. & Bowser, M.T. (2010). Capillary Electrophoresis. *Analytical Chemistry,* 82, 4682–4698.

Fu, K. & Sun, Y.P. (2003). Dispersion and solubilization of carbon nanotubes. *Journal of Nanoscience and Nanotechnology,* 3, 351-364.

Gao, C.; Jin, Y.Z.; Kong, H.; Whitby, R.L.D.; Acquah, S.F.A.; Chen, G.Y.; Qian, H.H.; Hartschuh, A.; Silva, S.R.P.; Henley, S.; Fearon, P.; Kroto, H.W. & Walton, D.R.M. (2005). Polyurea-Functionalized Multiwalled Carbon Nanotubes: Synthesis, Morphology, and Raman Spectroscopy. *Journal of Physical Chemistry B,* 109, 11925-11932.

Gao, C.; Muthukrishnan, S.; Li, W.; Yuan, J.; Xu, Y. & Mu1ller, A.H.E. (2007). Linear and Hyperbranched Glycopolymer-Functionalized Carbon Nanotubes: Synthesis, Kinetics, and Characterization. *Macromolecules,* 40, 1803-1815.

Gao, D.; Zhang, Z.; Wu, M.; Xie, C.; Guan, G. & Wang, D. (2007). A surface functional monomer-directing strategy for highly dense imprinting of TNT at surface of silica nanoparticles. *Journal of the American Chemical Society,* 129, 7859-7866.

Gao, J.B.; Itkis, M.E.; Yu, A.P.; Bekyarova, E.; Zhao, B. & Haddon, R.C. (2005). Continuous Spinning of a Single-Walled Carbon Nanotube−Nylon Composite Fiber. *Journal of the American Chemical Society*, 127, 3847-3854.

Gao, R.; Kong, X.; Su, F.; He, X.; Chen, L. & Zhang, Y. (2010). Synthesis and evaluation of molecularly imprinted core–shell carbon nanotubes for the determination of triclosan in environmental water samples. *Journal of Chromatography A*, 1217, 8095–8102.

Gao, R.; Su, X.; He, X.; Chen, L. & Zhang, Y. (2011). Preparation and characterisation of core–shell CNTs@MIPs nanocomposites and selective removal of estrone from water samples. *Talanta*,83, 757–764.

Garreau, D.; Hapiot, P. & Saveant, J.M. (1990). Fast cyclic voltammetry at ultramicroelectrodes: Current measurement and ohmic drop positive positive feedback compensation by means of current feedback operational amplifiers. *Journal of Electroanalytical Chemistry and Interfacial Electrochemistry*, 281, 73-83.

Geng, H.Z.; Zhang, X.B.; Mao, S.H.; Kleinhammes, A.; Shimoda, H.; Wu, Y. & Zhou, O. (2004). Opening and closing of single-wall carbon nanotubes. *Chemical Physics Letters*, 399, 109-113.

Georgakilas, V.; Kordatos, K.; Prato, M.; Guldi, D.M.; Holzingger, M. & Hirsch, A. (2002). Organic functionalization of carbon nanotubes. *Journal of the American Chemical Society*, 124, 760–761.

Gigliotti, B.; Sakizzie, B.; Bethune, D.S.; Shelby, R.M. & Cha, J.N. (2006). Sequence independent helical wrapping of single-walled carbon nanotubes by long genomic DNA. *Nano Letters*, 6, 159-164.

Gomez, F.J.; Chen, R.J.; Wang, D.W.; Waymouth, R.M. & Dai, H.J. (2003). Ring opening metathesis polymerization on non-covalently functionalized single-walled carbon Nanotubes. *Chemical Communications*, 2, 190-191.

Gooding, J.J.; R. Wibowo, Liu, J.; Yang, W.; Losic, D.; Orbons, S.; Mearns, F.J.; Shapter, J.G. & Hibbert, D.B. (2003). Protein Electrochemistry Using Aligned Carbon Nanotube Arrays. *Journal of the American Chemical Society*, 125, 9006-9007.

Guan, G.; Liu, B.; Wang, Z. & Zhang, Z. (2008). Imprinting of Molecular Recognition Sites on Nanostructures and its Applications in Chemosensors. *Sensors*, 8, 8291-8320.

Guiseppi-Elie, A.; Lei, C. & Baughman, R.H. (2002). Direct electron transfer of glucose oxidase on carbon nanotubes. *Nanotechnology*, 13, 559-564.

Guo, M.; Chen, J.; Liu, D.; Nie, L. & Yao, S. (2004). Electrochemical characteristics of the immobilization of calf thymus DNA molecules on multi-walled. *Bioelectrochemistry*, 62, 29-35.

Guo, T.; Diener, M.D.; Chai, Y.; Alford, M.J.; Haufler, R.E.; McClure, S.M.; Ohno, T.; Weaver, J.H.; Scuseria, G.E. & Smalley, R.E. (1992). Uranium Stabilization of C28: A Tetravalent Fullerene. *Science*, 257, 1661-1164.

Hafner, J.H.; Bronikowski, M.J.; Azamian, B.R.; Nikolaev, P.; Rinzler, A.G.; Colbert, D.T.; Smith, K.A. & Smalley, R.E. (1998). Catalytic growth of single-wall carbon nanotubes from metal particles. *Chemical Physics Letters*, 296, 195-202.

Ham, H.T.; Choi, Y.S.; Chung, I.J. (2005). An explanation of dispersion states of single-walled carbon nanotubes in solvents and aqueous surfactant solutions using solubility parameters. *Journal of Colloid and Interface Science*, 286, 216-223.

Haupt K. (2003). Imprinted polymers - Tailor-made mimics of antibodies and receptors. *Chemical Communications*, 2, 171-178.

Haupt, K. & Mosbach, K. (2000). Molecularly imprinted polymers and their use in biomimetic sensors. *Chemical Revies* 100, 2495-2504.

He, P.G. & Bayachou, M. (2005). Layer-by-Layer Fabrication and Characterization of DNA-Wrapped Single-Walled Carbon Nanotube Particles. *Langmuir*, 21, 6086-6092.

Hernadi, K.; Fonseca, A.; Nagy, J.B.; Bernaerts, D. & Lucas, A.A. (1996). Fe-catalyzed carbon nanotube formation. *Carbon*, 34, 1249-1257.

Hillberg, A.L.; Brain, K.R. & Allender, C.J. (2005). Molecular imprinted polymer sensors: Implications for therapeutics. *Advanced Drug Delivery Reviews*, 57, 1875–1889.

Hilt, J.Z. & Byrne, M.E. (2004). Configurational biomimesis in drug delivery: molecular imprinting of biologically significant molecules. *Advanced Drug Delivery Reviews*, 56, 1599-1620.

Hishiya, T.; Asanuma, H. & Komiyama, M. (2003). Molecularly imprinted cyclodextrin polymers as stationary phases of high performance liquid chromatography. *Polymer Journal*, 35, 440-445.

Hong, C.Y.; You, Y.Z. & Pan, C.Y. (2005). Synthesis of Water-Soluble Multiwalled Carbon Nanotubes with Grafted Temperature-Responsive Shells by Surface RAFT Polymerization. *Chemistry of Materials*, 17, 2247- 2254.

Hong, C.Y.; You, Y.Z. & Pan, C.Y. (2006). Functionalized multi-walled carbon nanotubes with poly(N-(2-hydroxypropyl)methacrylamide) by RAFT polymerization. *Journal of Polymer Science, Part A: Polymer Chemistry*, 44, 2419-2427.

Howell, J.O. & Wightman, R.M. (1984). Ultrafast voltammetry and voltammetry in highly resistive solutions with microvoltammetric electrodes. *Analytical Chemistry*, 56, 524-529.

Hrapovic, S.; Liu, Y.; Male, K.B. & Luong, J.H.T. (2004). Electrochemical Biosensing Platforms Using Platinum Nanoparticles and Carbon Nanotubes *Analytical Chemistry*, 76, 1083-1088.

Hsueh, C.C. & Brajter-Toth, A. (1993). Fast-scan voltammetry in aqueous solutions at carbon fiber ultramicroelectrodes with on-line iR compensation. *Analytical Chemistry*, 65, 1570-1575.

Hua, G.; Sun, Y.; Gaur, A.; Meitl, M.A.; Bilhaut, L.; Rotkina, L.; Wang, J.; Geil, P.; Shim, M.; Rogers, J.A. & Shim, A. (2004). Polymer Imprint Lithography with Molecular-Scale Resolution. *Nano Letters*, 4, 2467-2471.

Huang, H.M.; Liu, I.C.; Chang, C.Y.; Tsai, H.C.; Hsu, C.H. & Tsiang, R.C.C. (2004). Preparing a polystyrene-functionalized multiple-walled carbon nanotubes via covalently linking acyl chloride functionalities with living polystyryllithium. *Journal of Polymer Science, Part A: Polymer Chemistry*, 42, 5802-5810.

Huang, J.; Xing, X.; Zhang, X.; He, X.; Lin, Q.; Lian, W. & Zhu, H. (2011a).A molecularly imprinted electrochemical sensor based on multi walled carbon nanotube-gold nanoparticle composites and chitosan for the detection of tyramine. *Food Research International*, 44, 276–281.

Huang, J.; Zhang, X.; Lin, Q.; He, X.; Xing, X.; Huai, H.; Lian, W. & Zhu, H. (2011b). Electrochemical sensor based on imprinted solegel and nanomaterials. *Food Control*, 22, 786-791.

Huang, W.; Fernando, S.; Allard, L.F. & Sun, Y.P. (2003). Solubilization of Single-Walled Carbon Nanotubes with Diamine-Terminated Oligomeric Poly(ethylene Glycol) in Different Functionalization Reactions. *Nano Letters*, 3, 565-568.

Hwang, C.C. & Lee, W.C. (2002). Chromatographic characteristics of cholesterol-imprinted polymers prepared by covalent and non-covalent imprinting methods. *Journal of Chromatography A*, 962, 69-78.

Iijima, S. & Ichihashi, T. (1993). Single-shell carbon nanotubes of 1nm diameter. *Nature*, 363, 603-605.

Iijima, S. (1991). Helical microtubules of graphitic carbon. *Nature*, 354, 56-58.

Ikegami, T.; Mukawa, T.; Nariai, H. & Takeuchi, T. (2004). Bisphenol A-recognition polymers prepared by covalent molecular imprinting. *Analytica Chimica Acta*, 504, 131-135.

Islam, M.; Rojas, R.; Bergey, D.; Johnson, A. & Yodh, A. (2003). High weight fraction surfactant solubilization of single-wall carbon nanotubes in water. *Nano Letters*, 3, 269-273.

Jiang, H. & Ju, H.X. (2007). Enzyme-quantum dots architecture for highly sensitive electrochemiluminescenece biosensing of oxidase substrates. *Chemical Communications*, 4, 404-406.

Joselevich E. (2004). Electronic structure and chemical reactivity of carbon nanotubes: a chemist's view. *Chemphyschem*, 5, 619-624.

José-Yacamán, M.; Miki-Yoshida, M.; Rendón, L. & Santiesteban, J. G. (1993). Catalytic growth of carbon microtubules with fullerene structure. *Applied Physics Letters*, 62, 657-659.

Joshi, V.P.; Karode, S.K.; Kulkarni, M.G. & Mashelkar, R.A. (1998). Novel separation strategies based on molecularly imprinted adsorbents. *Chemical Engineering Science*, 53, 2271-2284.

Kahn, M.G.C.; Banerjee, S. & Wong, S.S. (2002). Solubilization of Oxidized Single-Walled Carbon Nanotubes in Organic and Aqueous Solvents through Organic Derivatization. *Nano Letters*, 2, 1215-1218.

Kan, X.; Geng, Z.; Wang, Z. & Zhu, J.-J. (2009). Core-Shell molecularly imprinted polymer nanospheres for the recognition and determination of hydroquinone. *Journal of Nanoscience and Nanotechnology*, 9, 2008-2013.

Kan, X.; Liu, T.; Zhou, H.; Li, C. & Fang, B. (2010). Molecular imprinting polymer electrosensor based on gold nanoparticles for theophylline recognition and determination. *Microchimica Acta*, 171, 423–429.

Kan, X.; Zhao, Q.; Zhang, Z.; Wang, Z. & Zhu, J.-J. (2008a). Molecularly imprinted polymers microsphere prepared by precipitation polymerization for hydroquinone recognition. *Talanta*, 75, 22-26.

Kan, X.; Zhao, Y.; Geng, Z.; Wang, Z. & Zhu, J.J. (2008b). Composites of Multiwalled Carbon Nanotubes and Molecularly Imprinted Polymers for Dopamine Recognition. *Journal of Physical Chemistry C*, 112, 4849-4854.

Kanai, M., Koshio, A.; Shinohara, H.; Mieno, T.; Kasuya, A.; Ando, Y. & Zhao, X. (2001). High-yield synthesis of single-walled carbon nanotubes by gravity-free arc discharge. *Applied Physics Letters*, 79, 2967-2969.

Kang, J.; Zhang, H.; Wang, Z.; Wu, G. & Lu, X. (2009). A novel amperometric sensor for salicylic acid based on molecularly imprinted polymer-modified electrodes. *Polymer - Plastics Technology and Engineering,* 48, 639-645.

Kawagoe, K.T.; Zimmerman, J.B. & Wightman, R.M. (1993). Principles of voltammetry and microelectrode surface states. *Journal of Neuroscience Methods,* 48, 225-240.

Kazua, Y. (1999). *Hitachi Chemical Technical Report,* 32, 7.

Ke, P.C. & Qiao, R. (2007). Carbon nanomaterials in biological systems. *Journal of Physics Condensed Matter,*19, 373101.

Kerman, K.; Morita, Y.; Takamura, Y.; Ozsoz, M. & Tamiya, E. (2004). DNA-Directed Attachment of Carbon Nanotubes for Enhanced Label-Free Electrochemical Detection of DNA Hybridization. *Electroanalysis,* 16, 1667-1672.

Kikuchi, M.; Ikoma, T.; Itoh, S.; Matsumoto, H.N.; Koyama, Y.; Takakuda, K.; Shinomiya, K. & Tanaka, J. (2004). Biomimetic synthesis of bone-like nanocomposites using the self-organization mechanism of hydroxyapatite and collagen. *Composites Science and Technology,* 64, 819-825.

Kim, D.S., Nepal, D. & Geckeler, K.E. (2005). Individualization of single walled carbon nanotubes: is the solvent important? *Small,* 1, 1117-1124.

Kim, S.N.; Rusling, J.F. & Papadimitrakopoulos, F. (2007). Carbon nanotubes for electronic and electrochemical detection of biomolecules. *Advanced Materials,* 19, 3214-3228.

Klingeler, R. & Sim, R.B. (2011). *Carbon Nanotubes for Biomedical Applications.* Springer, Germany.

Klingeler, R.; Hampel, S. & Büchner, B. (2008). Carbon nanotube based biomedical agents for heating, temperature sensoring and drug delivery. *International Journal of Hyperthermia,* 24, 496-505.

Kokai, F.; Takahashi, K; Yudasaka, M. & Iijima, S. (2000). Laser Ablation of Graphite−Co/Ni and Growth of Single-Wall Carbon Nanotubes in Vortexes Formed in an Ar Atmosphere. *Journal of Physical Chemistry B,* 104, 6777-6784.

Kolarz, B.N. & Jakubiak, A. (2008). Catalytic activity of molecular imprinted vinylpyridine/acrylonitrile/ divinylbenzene terpolymers with guanidyl ligands-Cu(II) inside the active centres. *Polimery/Polymers,* 53, 848-853.

Kong, H.; Gao, C. & Yan, D. (2004). Controlled Functionalization of Multiwalled Carbon Nanotubes by in Situ Atom Transfer Radical Polymerization. *Journal of the American Chemical Society,* 126, 412-413.

Kong, H.; Luo, P.; Gao, C. & Yan, D. (2005). Polyelectrolyte-functionalized multiwalled carbon nanotubes: preparation, characterization and layer-by-layer self-assembly. *Polymer,* 46, 2472-2485.

Kong, J.; Cassell, A.M. & Dai, H. (1998). Chemical vapor deposition of methane for single-walled carbon nanotubes. *Chemical Physics Letters,* 292, 567-574.

Kong, J.; Chapline, M.G. & Dai, H. (2001). Functionalized Carbon Nanotubes for Molecular Hydrogen Sensors. *Advanced Materials,* 13, 1384-1386.

Kong, J.; Franklin, N.R.; Zhou, C.; Chapline, M.G.; Peng, S.; Cho, K. & Dai, H. (2000). Nanotube Molecular Wires as Chemical Sensors. *Science,* 287, 622-625.

Kong, Y.; Zhao, W.; Yao, S.; Xu, J.; Wang, W. & Chen, Z.(2010). Molecularly imprinted polypyrrole prepared by electrodeposition for the selective recognition of tryptophan enantiomers. *Journal of Applied Polymer Science,* 115, 1952-1957.

Kriz, D. & Mosbach, K. (1995). Competitive amperometric morphine sensor-based on an agarose immobilized molecularly imprinted polymer. *Analytica Chimica Acta*, 300, 71- 75.

Kriz, D.; Ramström, O.; Svensson, A. & Mosbach, K. (1995). Introducing biomimetic sensors based on molecularly imprinted polymers as recognition elements. *Analytical Chemistry*, 67, 2142–2144.

Kröger, S.; Turner, A.P.F.; Mosbach, K. & Haupt, K. (1999). Imprinted polymer based sensor system for herbicides using differential pulse voltammetry on screen printed electrodes. *Analytical Chemistry*, 71, 3698- 3702.

Laval, J.M.; Mazeran, P-E. & Thomas, D. (1999). Nanobiotechnology and its role in the development of new analytical devices. *Analyst*, 125, 29-33.

Leary, S.P.; Liu, C.Y. & Apuzzo, M.L. (2006). Toward the emergence of nanoneurosurgery: part III— nanomedicine: targeted nanotherapy, nanosurgery, and progress toward the realization of nanoneurosurgery. *Neurosurgery*, 58, 1009-1026.

Lee, E.; Park, D.W.; Lee, J.O.; Kim, D.S.; Lee, B.H. & Kim, B.S. (2008). Molecularly imprinted polymers immobilized on carbon nanotube. *Colloids and Surfaces A: Physicochemical and Engineering Aspects*, 313–314, 202-206.

Lee, H.Y. & Kim, B.S. (2009). Grafting of molecularly imprinted polymers on iniferter-modified carbon Nanotube. *Biosensors and Bioelectronics*, 25, 587–591.

Lemos, V.A.; Sena Gomes Teixeira, L.; Almeida Bezerra, M.; Spınola Costa, A.C.; Teixeira Castro, J.; Cardoso, L.A.M.; Santiago de Jesus, D.; Souza Santos, E.; Xavier Baliza, P. & Novaes Santos, L. (2008). New Materials for Solid-Phase Extraction of Trace Elements. *Applied Spectroscopy Reviews*, 43, 303–334.

Li, H.M.; Cheng, F.O.; Duft, A.M. & Adronov, A. (2005). Functionalization of Single-Walled Carbon Nanotubes with Well-Defined Polystyrene by "Click" Coupling. *Journal of the American Chemical Society*, 127, 14518-14524.

Li, J.; Kendig, C.E. & Nesterov, E.E. (2007a). Chemosensory performance of molecularly imprinted fluorescent conjugated polymer materials. *Journal of the American Chemical Society*, 129, 15911-15918.

Li, J.; Ma, P.C.; Chow, W.S.; To, C.K.; Tang, B.Z. & Kim, J.K. (2007b). Correlations between percolation threshold, dispersion state, and aspect ratio of carbon nanotubes. *Advanced Functional Materials*, 17, 3207–3215.

Li, Q.; Yuan, D. & Lin, Q. (2004). Evaluation of multi-walled carbon nanotubes as an adsorbent for trapping volatile organic compounds from environmental samples. *Journal of Chromatography A*, 1026, 283–288.

Li, Y.; Li, X.; Dong, C.; Qi, J. & Han, X. (2010). Graphene oxide-based molecularly imprinted polymer platform for detecting endocrine disrupting chemicals. *Carbon*, 48, 3427-3433.

Lin, Y.; Zhou, B.; Fernando, K.A.S.; Allard, L.F. & Sun, Y.P. (2003). Polymeric Carbon Nanocomposites from Carbon Nanotubes Functionalized with Matrix Polymer. *Macromolecules*, 36, 7199-7204.

Liu, I.C.; Huang, H.M.; Chang, C.Y.; Tsai, H.C.; Hsu, C.H. & Tsiang, R.C.C. (2004). Preparing a Styrenic Polymer Composite Containing Well-Dispersed Carbon Nanotubes: Anionic Polymerization of a Nanotube-Bound p-Methylstyrene. *Macromolecules*, 37, 283-287.

Liu, M.; Zhu, T.; Li, Z. & Liu, Z. (2009). One-Step in Situ Synthesis of Poly(methyl methacrylate)-Grafted Single-Walled Carbon Nanotube Composites. *Journal of Physical Chemistry C*, 113, 9670–9675.

Liu, Y.Q.; Yao, Z.L. & Adronov, A. (2005). Functionalization of Single-Walled Carbon Nanotubes with Well-Defined Polymers by Radical Coupling. *Macromolecules*, 38, 1172-1179.

Liu, Z.; Bucknall, D.G. &Allen, M.G. (2010). Absorption performance of iodixanol-imprinted polymers in aqueous and blood plasma media. *Acta Biomaterialia*, 6, 2003-2012.

Lou, X.D.; Detrembleur, C.; Pagnoulle, C.; Jerome, R.; Bocharova, V.; Kiriy, A. & Stamm, M. (2004). Surface Modification of Multiwalled Carbon Nanotubes by Poly(2-vinylpyridine): Dispersion, Selective Deposition, and Decoration of the Nanotubes. *Advanced. Materials*, 16, 2123-2127.

Lu, C.; Chung, Y. & Chang, K. (2005). Adsorption of trihalomethanes from water with carbon nanotubes. *Water Research*, 39, 1183–1189.

Lucena, R.; Simonet, B.M.; Cárdenas, S. & Valcárcel, M. (2011). Potential of nanoparticles in sample preparation. *Journal of Chromatography A*, 1218, 620–637.

Lundstroem, I.; Shivaraman, S.; Svensson, C. & Lundkvist, L. (1975). A hydrogen−sensitive MOS field−effect transistor. *Applied Physics Letters*, 26, 55-57.

Ma, G.P.; Yang, D.Z. & Nie, J. (2009). Preparation of porous ultrafine polyacrylonitrile (PAN) fibers by electrospinning. *Polymers for Advanced Technologies*, 20, 147-150.

Ma, P.C.; Siddiqui, N.A.; Marom, G. & Kim, J.K. (2010). Dispersion and functionalization of carbon nanotubes for polymer-based nanocomposites: A review. *Composites A*, 41, 1345–1367.

Martin, R.B.; Qu, L.W.; Lin, Y.; Harruff, B.A.; Bunker, C.E.; Gord, J.R.; Allard, L.F. & Sun, Y P. (2004). Functionalized Carbon Nanotubes with Tethered Pyrenes: Synthesis and Photophysical Properties. *Journal of Physical Chemistry B*, 108, 11447-11453.

Maruccio, G.; Cingolani, R. & Rinaldi, R. (2004). Projecting the nanoworld: Concepts, results and perspectives of molecular electronics. *Journal of Materials Chemistry*, 14, 542-554.

Marx, S.; Zaltsman, A.; Turyan, I. & Mandler, D. (2004). Parathion Sensor Based on Molecularly Imprinted Sol−Gel Films. *Analytical Chemistry*, 76, 120-126.

Mazzola, L. (2003). Commercializing nanotechnology. *Nature Biotechnology*, 21, 1137-1143.

McDonagh, C.; Burke, C.S. & MacCraith, B.D. (2008). Optical chemical sensors. *Chemical Reviews*, 108, 400-422.

Mirsky, V.M.; Hirsch, T.; Piletsky, S.A. & Wolfbeis, O.S. A spreader-bar approach to architecture: formation of stable artificial chemoreceptors. *Angewandte Chemie - International Edition*, 38, 1108-1110.

Moniruzzaman, M. & Winey, K,I. (2006). Polymer nanocomposites containing carbon nanotubes. *Macromolecules*, 39, 5194–5205.

Moore, V.; Strano, M.; Haroz, E.; Hauge, R. E. Smalley R. (2003). Individually suspended single-walled carbon nanotubes in various surfactants. *Nano Letters*, 3, 1379-1382.

Mosbach, K. (2006). The promise of molecular imprinting. *Scientific American*, 295, 86-91.

Musallam, S.; Bak, M.J.; Troyk, P.R. & Anderson, R.A.(2007). A floating metal microelectrode array for chronic implantation. *Journal of Neuroscience Methods*, 160, 122-127.

Mylvaganam, K. & Zhang, L.C. (2004). Nanotube Functionalization and Polymer Grafting: An ab Initio Study. *Journal of Physical Chemistry B*, 108, 15009–15012.

Nicholls, I.A.; Matsui, J.; Krook, M. & Mosbach, K. (1996). Some recent developments in the preparation of novel recognition systems: A recognition site for the selective catalysis of an aldol condensation using molecular imprinting and specific affinity motifs for alpha-chymotrypsin using a phage display peptide library. *Journal of Molecular Recognition*, 9, 652-657.

Niemeyer, C.M. (2001). Nanoparticles, Proteins, and Nucleic Acids: Biotechnology Meets Materials Science. *Angewandte Chemie International Edition in English*, 40, 4128-4158.

Nilsson, C.; Birnbaum, S. & Nilsson, S. (2007). Use of nanoparticles in capillary and microchip electrochromatography. *Journal of Chromatography A*, 1168, 212-224.

Nogales, A.; Broza, G.; Roslaniec, Z.; Schulte, K.; Sics, I.; Hsiao, B.S.; Sanz, A.; Garcia-Gutierrez, M.C.; Rueda, D.R.; Domingo, C. & Ezquerra, T.A. (2004). Low Percolation Threshold in Nanocomposites Based on Oxidized Single Wall Carbon Nanotubes and Poly (butylene terephthalate). *Macromolecules*, 37, 7669-7672.

Orellana, G. & Moreno-Bondi, M.C. (2005). *Frontiers in chemical sensors: novel principles and techniques*. Springer-Verlag, Berlin.

Ouyang, R.; Lei, J.; Ju, H. & Xue, Y. (2007). A molecularly imprinted copolymer designed for enantioselective recognition of glutamic acid. *Advanced Functional Materials* 17, 3223-3230.

Panasyuk, T.L.; Mirsky, V.M.; Piletsky, S.A. & Wolfbeis, O.S. (1999). Electropolymerized molecularly imprinted polymers as receptor layers in a capacitive chemical sensors. *Analytical Chemistry*, 71, 4609-4613.

Pantarotto, D.; Partidos, C.D.; Graff, R.; Hoebeke, J.; Briand, J.P.; Prato, M. & Bianco, A. (2003). Synthesis, structural characterization, and immunological properties of carbon nanotubes functionalized with peptides. *Journal of the American Chemical Society*, 125, 6160-6164.

Patel, A.K.; Sharma, P.S. & Prasad, B.B. (2009). Electrochemical sensor for uric acid based on a molecularly imprinted polymer brush grafted to tetraethoxysilane derived sol-gel thin film graphite electrode. *Materials Science and Engineering C*, 29, 1545-1553.

Peng, H.; Alemany, L.B.; Margrave, J.L. & Khabashesku, V.N. (2003). Sidewall carboxylic acid functionalization of single-walled carbon nanotubes. *Journal of the American Chemical Society*, 125, 15174-15182.

Petrov, P.; Lou, X.D.; Pagnoulle, C.; Jerome, C.; Calberg, C. & Jerome, R. (2004). Functionalization of Multi-Walled Carbon Nanotubes by Electrografting of Polyacrylonitrile. *Macromolecular Rapid Communications*, 25, 987-990.

Petrov, P.; Stassin, F.; Pagnoulle, C. & Jerome, R. (2003). Noncovalent functionalization of multi-walled carbon nanotubes by pyrene containing polymers. *Chemical Communications*, 23, 2904-2905.

Petrovic, M.; Farré, M.; Lopez de Alda, M.; Perez, S.; Postigo, C.; Köck, M.; Radjenovic, J.; Gros, M. & Barcelo, D. (2010). Recent trends in the liquid chromatography–mass spectrometry analysis of organic contaminants in environmental samples. *Journal of Chromatography A*, 1217, 4004–4017.

Prasad, B.B.; Madhuri, R.; Tiwari, M.P. & Sharma, P.S. (2010a). Imprinted polymer–carbon consolidated composite fiber sensor for substrate selective electrochemical sensing of folic acid. *Biosensors and Bioelectronics*, 25, 2140–2148.

Prasad, B.B.; Madhuri, R.; Tiwari, M.P. & Sharma, P.S. (2010b). Imprinting molecular recognition sites on multiwalled carbon nanotubes surface for electrochemical detection of insulin in real samples. *Electrochimica Acta*, 55, 9146–9156.

Prasad, B.B.; Madhuri, R.; Tiwari, M.P. & Sharma, P.S. (2010c). Electrochemical sensor for folic acid based on a hyperbranched molecularly imprinted polymer-immobilized sol-gel-modified pencil graphite electrode. *Sensors and Actuators, B: Chemical,* 146, 321-330.

Prasad, B.B.; Madhuri, R.; Tiwari, M.P. & Sharma, P.S. (2010d). Enantioselective recognition of d- and l-tryptophan by imprinted polymer-carbon composite fiber sensor. *Talanta,* 81, 187-196.

Prasad, B.B.; Srivastava, S.; Tiwari, K. & Sharma, P.S. (2009). Development of uracil and 5-fluorouracil sensors based on molecularly imprinted polymer-modified hanging mercury drop electrode. *Sensors and Materials,* 21, 291-306.

Puoci, F.; Cirillo, G.; Curcio, M.; Iemma, F.; Parisi, O.I.; Spizzirri, U.G. & Picci, N. (2010). Molecularly Imprinted Polymers (MIPs) in Biomedical Applications. In: *Biopolymers,* Magdy M. Elnashar, 547-574, Sciyo, Rijeka, Croatia.

Puoci, F.; Cirillo, G.; Curcio, M.; Iemma, F.; Spizzirri, U.G. & Picci, N. (2007). Molecularly Imprinted Solid Phase Extraction for the selective HPLC determination of α-tocopherol in Bay Leaves. *Analytica Chimica Acta,* 2007, 593, 164-170.

Puoci, F.; Cirillo, G.; Curcio, M.; Parisi, O.I.; Iemma, F. & Picci, N. (2011). Molecularly Imprinted Polymers in Drug Delivery: state of art and future perspectives. *Expert Opinion on Drug Delivery*, DOI 10.1517/17425247.2011.609166.

Puretzky, A.A. Geohegan, D.B.; Fan, X. & Pennycook, S.J. (2000). Dynamics of single-wall carbon nanotube synthesis by laser vaporization. *Applied Physics A Materials Science Processing,* 70, 153-160.

Qin, S.H.; Qin, D.Q.; Ford, W.T.; Herrera, J.E. & Resasco, D.E. (2004a). Grafting of Poly(4-vinylpyridine) to Single-Walled Carbon Nanotubes and Assembly of Multilayer Films. *Macromolecules,* 37, 9963-9967.

Qin, S.H.; Qin, D.; Ford, W.T.; Resasco, D.E. & Herrera, J.E. (2004b). Polymer Brushes on Single-Walled Carbon Nanotubes by Atom Transfer Radical Polymerization of n-Butyl Methacrylate. *Journal of the American Chemical Society,* 126, 170-176.

Qin, S.H.; Qin, D.Q.; Ford, W.T.; Zhang, Y.J. & Kotov, N.A. (2005). Covalent Cross-Linked Polymer/Single-Wall Carbon Nanotube Multilayer Films. *Chemistry of Materials,* 17, 2131-2135.

Qu, L.W.; Martin, R.B.; Huang, W.J.; Fu, K.F.; Zweifel, D.; Lin, Y.; Sun, Y.P.; Bunker, C.E.; Harruff, B.A.; Gord, J.R. & Allard, L.F. (2002). Interactions of functionalized carbon nanotubes with tethered pyrenes in solution. *Journal of Chemical Physics,* 117, 8089-8094.

Qu, L.W.; Veca, L.M.; Lin, Y.; Kitaygorodskiy, A.; Chen, B.L.; McCall, A.M.; Connell, J.W. & Sun, Y.P. (2005). Soluble Nylon-Functionalized Carbon Nanotubes from Anionic Ring-Opening Polymerization from Nanotube Surface. *Macromolecules,* 38, 10328-10331.

Ramanavicius, A.; Ramanaviciene, A. & Malinauskas, A. (2006). Electrochemical sensors based on conducting polymer—polypyrrole. *Electrochimica Acta,* 51, 6025–6037.

Ramasubramaniam, R.; Chen, J. & Liu, H. (2003). Homogeneous carbon nanotubepolymer composites for electrical applications. *Applield Physics Letters*, 83, 2928–2930.

Riskin, M.; Tel-Vered, R.; Bourenko, T.; Granot, E. & Willner, I. (2008). Imprinting of molecular recognition sites through electropolymerization of functionalized Au nanoparticles: development of an electrochemical TNT sensor based on π-donor-acceptor interactions. *Journal of the American Chemical Sociery*, 130, 15911-15918.

Rubianes, M.D. & Rivas, G.A. (2003). Carbon nanotubes paste electrode. *Electrochemistry Communications*, 5, 689-694.

Safarik, I. & Safarikova, M. (2002). Magnetic nanoparticles and biosciences. *Monatshefte Fur Chemie*, 133, 737-759.

Sahoo, N.G.; Rana, S.; Cho, J.W.; Li, L. & Chan, S.H. (2010). Polymer nanocomposites based on functionalized carbon nanotubes. *Progress in Polymer Science*, 35, 837–867.

Saito, Y.; Inagaki, M.; Shinohara, H.; Nagashima, H.; Ohkohchi, M. & Ando, M. (1992). Yield of fullerenes generated by contact arc method under He and Ar: dependence on gas pressure. *Chemical Physics Letters*, 200, 643-648.

Sandler, J.K.W.; Kirk, J.E.; Kinloch, I,A.; Shaffer, M.S.P. & Windle, A.H. (2003). Ultralow electrical percolation threshold in carbon nanotube epoxy composites. *Polymer*, 44, 5893–5899.

Sano, M.; Kamino, A.; Okamura, J. & Shinkai, S. (2001). Self-Organization of PEO-graft-Single-Walled Carbon Nanotubes in Solutions and Langmuir−Blodgett Films. *Langmuir*, 17, 5125-5128.

Satishkumar, B.C.; Govindaraj, A.; Sen, R. & Rao, C.N.R. (1998). Single-walled nanotubes by the pyrolysis of acetylene-organometallic mixtures. *Chemical Physics Letters*, 293, 47-52.

Satiskumar, B. C.; Govindaraj, A. & Rao, C.N.R. (1999). Bundles of aligned carbon nanotubes obtained by the pyrolysis of ferrocene–hydrocarbon mixtures: role of the metal nanoparticles produced in situ. *Chemical Physics Letters*, 307, 158-162.

See, H.H.; Marsin Sanagi, M.; Ibrahim, W.A. & Naim, A.A. (2010). Determination of triazine herbicides using membraneprotected carbon nanotubes solid phase membrane tip extraction prior to microliquid chromatography. *Journal of Chromatography A*, 1217, 1767–1772.

Sen, R.; Ohtsuka, Y.; Ishigaki, T.; Kasuya, D.; Suzuki, S.; Kataura, H. & Achiba, Y. (2000). Time period for the growth of single-wall carbon nanotubes in the laser ablation process: evidence from gas dynamic studies and time resolved imaging. *Chemical Physics Letters*, 332, 467-473.

Shaffer, M.S.P. & Koziol, K. (2002). Polystyrene grafted multi-walled carbon nanotubes. *Chemical Communications*, 18, 2074-2075.

Shao, D.; Ren, X.; Hu, J.; Chen, Y. & Wang, X. (2010). Preconcentration of Pb^{2+} from aqueous solution using poly(acrylamide) and poly(N,Ndimethylacrylamide) grafted multiwalled carbon nanotubes. *Colloids and Surfaces A: Physicochemical Engineering Aspects*, 360, 74–84.

Shi, F.; Liu, Z.; Wu, G.; Zhang, M.; Chen, H.; Wang, Z.; Zhang, X. & Willner, I. (2007). Surface Imprinting in Layer-by-Layer Nanostructured Films. *Advanced Functional Materials*, 17, 1821–1827.

Shiigi, H.; Okamura, K.; Kijima, D.; Deore, B.; Sree, U. & Nagaoka, T. (2003). An Overoxidized Polypyrrole/Dodecylsulfonate Micelle Composite Film for Amperometric Serotonin Sensing. *Journal of The Electrochemical Society*, 150, H119-H123.

Shimada, T.; Sugau, T.; Ohno, Y.; Kishimoto, S.; Mizutani, T.; Yoshida, H.; Okazaki, T. & Shinohara, H. (2004). Double-wall carbon nanotube field-effect transistors: Ambipolar transport Characteristics. *Applied Physics Letters*, 84, 2412-2414.

Sinha, N. & Yeow, J. (2005). Carbon nanotubes for biomedical applications. *IEEE Transactions on NanoBioscience*, 4, 180-195.

Song, Y.S. & Youn, J.R. (2005). Influence of dispersion states of carbon nanotubes on physical properties of epoxy nanocomposites. *Carbon*, 43, 1378–1385.

Spitalsky, Z.; Tasis, D.; Papagelis, K. & Galiotis, C. (2010). Carbon nanotube–polymer composites: Chemistry, processing, mechanical and electrical properties. *Progress in Polymer Science*, 35, 357–401.

Spivak, D.A. (2005). Optimization, evaluation, and characterization of molecularly imprinted polymers. *Advanced Drug Delivery Reviws*, 57, 1779-1794.

Sugai, T.; Omote, H.; Bandow, S.; Tanaka, N. & Shinohara, H. (1999). Production of Single-Wall Nanotubes by High-Temperature Pulsed Arc Discharge: Mechanisms of their Production. *Japanese Journal of Applied Physics*, (1999) 38, L477-L479.

Sugai, T.; Omote, H.; Bandow, S.; Tanaka, N. & Shinohara, H. (2000). Production of fullerenes and single-wall carbon nanotubes by high-temperature pulsed arc discharge. *Journal of Chemical Physics*, 112, 6000-6005.

Sugai, T.; Yoshida, H.; Shimada, T.; Okazaki, T. & Shinohara, H. (2003). New Synthesis of High-Quality Double-Walled Carbon Nanotubes by High-Temperature Pulsed Arc Discharge. *Nano Letters*, 3, 769-773.

Sun, J.Q.; Wu, T.; Liu, F.; Wang, Z.Q.; Zhang, X. & Shen, J.C. (2000). Covalently Attached Multilayer Assemblies by Sequential Adsorption of Polycationic Diazo-Resins and Polyanionic Poly(acrylic acid). *Langmuir*, 16, 4620-4624.

Sun, J.Q.; Wu, T.; Sun, Y.P.; Wang, Z.Q.; Zhang, X.; Shen, J.C. & Cao, W.X. (1998). Fabrication of a covalently attached multilayer via photolysis of layer-by-layer self-assembled films containing diazo-resins. *Chemical Communications*, 17, 1853-1854.

Svitel, J.; Surugiu, I.; Dzgoev, A.; Ramanathan, K. & Danielsson, B. (2001). Functionalized surfaces for optical biosensors: Applications to in vitro pesticide residual analysis. *Journal of Materials Science: Materials in Medicine*, 12, 1075-1078.

Syu, M.J.; Chiu, T.C.; Lai, C.Y. & Chang Y.S. (2006). Amperometric detection of bilirubin from a micro-sensing electrode with a synthetic bilirubin imprinted poly(MAA-co-EGDMA) film. *Biosensors and Bioeletronics*, 22, 550-557.

Teles, F.R.R. & Fonseca, L.P. (2008). Applications of polymers for biomolecule immobilization in electrochemical biosensors. *Materials Science and Engineering C*, 28, 1530–1543.

Thess, A.; Lee, R.; Nikolaev, P.; Dai, H.; Petit, P.; Robert, J.; Xu, C.; Lee, Y.H.; Kim, S.G.; Rinzler, A.G.; Colbert, D.T.; Scuseria, G.E.; Tománek, D.; Fischer, J.E. & Smalley, R.E. (1996). Crystalline Ropes of Metallic Carbon Nanotubes. *Science*, 273, 483-487.

Tian, M.; Bi, W. & Row, K.H. (2011). Molecular imprinting in ionic liquid-modified porous polymer for recognitive separation of three tanshinones from Salvia miltiorrhiza Bunge. *Analytical Bioanalytical Chemistry*, 399, 2495–2502.

Updike, S.J & Hicks, G.P. (1967). The enzyme electrode. *Nature*, 214, 986–988.

Vaisman, L.,Wagner, H.D. & Marom, G. (2006). The role of surfactants in dispersion of carbon nanotubes. *Advances in Colloid and Interface Science*, 128-130, 37-46.

Vasanth Kumar, K.; Monteiro de Castro, M.; Martinez-Escandell, M.; Molina-Sabio, M. & Rodriguez-Reinoso, F. (2010). A Continuous Binding Site Affinity Distribution Function from the Freundlich Isotherm for the Supercritical Adsorption of Hydrogen on Activated Carbon. *Journal of Physical Chemistry C*, 114, 13759–13765.

Vasanth Kumar, K.; Monteiro de Castro, M.; Martinez-Escandell, M.; Molina-Sabio, M. & Rodriguez-Reinoso, F. (2011). A site energy distribution function from Toth isotherm for adsorption of gases on heterogeneous surfaces. *Physical Chemistry Chemical Physics*, 13, 5753–5759.

Venkatesan, J. & Kim S.K. (2010). Chitosan Composites for Bone Tissue Engineering - An overview. *Marine Drugs*, 8, 2252-2266.

Vinjamuri, A.K.; Burns, S.C. & Dahl, D.B. (2008). Caffeine and Theobromine Selectivity Using Molecularly Imprinted Polypyrrole Modified Electrodes. *ECS Transactions*, 13, 9-20.

Vyalikh, A.; Wolter, A.U.; Hampel, S.; Haase, D.; Ritschel, M.; Leonhardt, A.; Grafe, H.J.; Taylor, A., Krämer. K.; Büchner, B. & Klingeler, R. (2008). *Nanomedicine*, 3: 175-182.

Walcarius, A.; Mandler, D.; Cox, J.A.; Collinson, M. & Lev, O. (2005). Exciting new directions in the intersection of functionalized sol–gel materials with electrochemistry *Journal of Material Chemistry*, 15, 3663–3689.

Wang, J. & Musameh, M. (2003). Carbon Nanotube/Teflon Composite Electrochemical Sensors and Biosensors. *Analytical Chemistry*, 75, 2075-2079.

Wang, J. & Musameh, M. (2004). Electrochemical detection of trace insulin at carbon-nanotube-modified electrodes. *Analytica Chimica Acta*, 511, 33-36.

Wang, J.; Li, M.; Shi, Z.; Li, N. & Gu, Z. (2002). Direct Electrochemistry of Cytochrome c at a Glassy Carbon Electrode Modified with Single-Wall Carbon Nanotubes. *Analytical Chemistry*, 74, 1993-1997.

Wang, J.; Liu, G. & Jan, M.R. (2004). Ultrasensitive Electrical Biosensing of Proteins and DNA: Carbon-Nanotube Derived Amplification of the Recognition and Transduction Events. *Journal of the American Chemical Society*, 126, 3010-3011.

Wang, J.; Musameh, M. & Lin, Y. (2003). Solubilization of Carbon Nanotubes by Nafion toward the Preparation of Amperometric Biosensors. *Journal of the American Chemical Society*, 125, 2408-2409.

Wang, J.W.; Khlobystov, A.N.; Wang, W.X.; Howdle, S.M. & Poliakoff, M. (2006). Coating carbon nanotubes with polymer in supercritical carbon dioxide. *Chemical Communications*, 15, 1670-1672.

Wang, K.; Xu, J.J.; Tang, K.S. & Chen, H.Y. (2005). Solid-contact potentiometric sensor for ascorbic acid based on cobalt phthalocyanine nanoparticles as ionophore. *Talanta*, 67, 798-805.

Wei, Y.; Qiu, L.; Yu, J.C.C. & Lai, E.P.C.(2007). Molecularly Imprinted Solid Phase Extraction in a Syringe Needle Packed with Polypyrrole-encapsulated Carbon Nanotubes for

Determination of Ochratoxin A in Red Wine. *Food Science and Technology International*, 13, 375–380.

Whitcombe, M.J.; Alexander, C. & Vulfson, E.N. (2000). Imprinted polymers: Versatile new tools in synthesis. *Synlett*, 911-923.

Whitesides, G.M. (2003). The right size in nanobiotechnology. *Nature Biotechnology*, 21, 1161-1165.

Wigthman, R.M. (1988). Voltammetry with Microscopic Electrodes in New Domains. *Science*, 240, 415-420.

Winey, K.I.; Kasiwagi, T. & Mu, M. (2007). Improving electrical conductivity and thermal properties of polymers by addition of carbon nanotubes as fillers. *MRS Bulletin*, 32, 348–353.

Wulff, G. (2002). Enzyme-like catalysis by molecularly imprinted polymers. *Chemical Reviews*, 102, 1-27.

Xiang, H.Y. & Li, W.G. (2009). Electrochemical sensor for trans-resveratrol determination based on indium tin oxide electrode modified with molecularly imprinted self-assembled films. *Electroanalysis*, 21, 1207-1210.

Xie, C.; Gao, S.; Guo, Q. & Xu, K. (2010) Electrochemical sensor for 2,4-dichlorophenoxy acetic acid using molecularly imprinted polypyrrole membrane as recognition element. *Microchimica Acta*, 169, 145-152.

Xie, C.; Liu, B.; Wang, Z.; Gao, D.; Guan, G. & Zhang, Z. (2008). Molecular imprinting at walls of silica nanotubes for TNT recognition. *Analytical Chemistry*, 80, 437-443.

Xie, C.; Zhang, Z.; Wang, D.; Guan, G.; Gao, D. & Liu, J. (2006). Surface molecular self-assembly strategy for TNT imprinting of polymer nanowire/nanotube arrays. *Analytical Chemistry*, 78, 8339-8346.

Xing, B. & Yin, X.B. (2009). Electrochemiluminescence from hydrophilic thin film $Ru(bpy)_3^{2+}$ modified electrode prepared using natural halloysite nanotubes and polyacrylamide gel. *Biosensors and Bioelectronics*, 24, 2939–2942.

Xu, G.Y.; Wu, W.T.; Wang, Y.S.; Pang, W.M.; Wang, P.H.; Zhu, G.R. & Lu, F. (2006). Synthesis and characterization of water-soluble multiwalled carbon nanotubes grafted by a thermoresponsive polymer. *Nanotechnology*, 17, 2458-2465.

Xu, H.X.; Wang, X.B.; Zhang, Y.F. & Liu, S.Y. (2006). Single-Step in Situ Preparation of Polymer-Grafted Multi-Walled Carbon Nanotube Composites under ^{60}Co γ-Ray Irradiation. *Chemistry of Materials*, 18, 2929-2934.

Xu, T.; Zhang, N.; Nichols, H.L.; Shi, D. & Wen, X. (2007). Modification of nanostructured materials for biomedical applications. *Materials Science and Engineering C*, 27, 579–594.

Yan, D. & Yang, G. (2009). A novel approach of in situ grafting polyamide 6 to the surface of multi-walled carbon nanotubes. *Materials Letters*, 63, 298–300.

Yang, B.X.; Pramoda, K.P.; Xu, G.Q. & Goh, S.H. (2007). Mechanical reinforcement of polyethylene using polyethylene-grafted multiwalled carbon nanotubes. *Advanced Functional Materials*, 17, 2062–2069.

Yao, Z.; Braidy, N.; Botton, G.A. & Adronov, A. (2003). Polymerization from the Surface of Single-Walled Carbon Nanotubes – Preparation and Characterization of Nanocomposites. *Journal of the American Chemical Society*, 125, 16015-16024.

Ye, L. & Haupt, K. (2004). Molecularly imprinted polymers as antibody and receptor mimics for assays, sensors and drug discovery. *Analytical and Bioanalytical Chemistry*, 378, 1887-1897.

Ye, L. & Mosbach, K. (2001). Molecularly imprinted microspheres as antibody binding mimics. *Reactive and Functional Polymers*, 48, 149-157.

Yin, J.; Cui, Y.; Yang, G. & Wang H. (2010). Molecularly imprinted nanotubes for enantioselective drug delivery and controlled release. *Chemical Communications*, 46, 7688–7690.

You, Y.Z.; Hong, C.Y. & Pan, C.Y. (2006). Directly growing ionic polymers on multi-walled carbon nanotubes via surface RAFT polymerization. *Nanotechnology*, 17, 2350- 2354.

Yu, J.C.C. & Lai, E.P.C. (2006). Molecularly imprinted polypyrrole modified carbon nanotubes on stainless steel frit for selective micro solid phase pre-concentration of ochratoxin A. *Reactive and Functional Polymers*, 66, 702-711.

Yu, J.C.C. & Lai, E.P.C. (2007). Determination of ochratoxin A in red wines by multiple pulsed elutions from molecularly imprinted polypyrrole. *Food Chemistry*, 105, 301–310.

Yu, Y.; Ye, L.; Haupt, K. & Mosbach, K. (2002). Formation of a class of enzyme inhibitors (drugs), including a chiral compound, by using imprinted polymers or biomolecules as molecular-scale reaction vessels. *Angewandte Chemie International Edition in English*, 114, 4640-4643.

Yurekli, K., Mitchell, C.A. & Krishnamoorti R. (2004). Small-angle neutron scattering from surfactant-assisted aqueous dispersions of carbon nanotubes. *Journal of the American Chemical Society*, 126, 9902-9903.

Zeng, H.L.; Gao, C.; Wang, Y.P.; Watts, P.C.P.; Kong, H.; Cui, X.W.; Yan, D.Y. (2006a). In situ polymerization approach to multiwalled carbon nanotubes-reinforced nylon 1010 composites: Mechanical properties and crystallization behavior. *Polymer*, 47, 113-122.

Zeng, H.L.; Gao, C. & Yan, D.Y. (2006b). Poly(ϵ-caprolactone)-Functionalized Carbon Nanotubes and Their Biodegradation Properties. *Advanced Functional Materials*, 16, 812- 818.

Zeng, J.; Chen, J.; Chen, W.; Huang, X.; Chen, L. & Chen, X. (2009). Recent Development of Laboratory-made Solid-phase Microextraction Fibers on the Application of Food Safety Analysis. *Food Science and Biotechnology*, 18, 579-585.

Zhang, H.; Zhang, Z.; Hu, Y.; Yang, X. & Yao, S. (2011). Synthesis of a Novel Composite Imprinted Material Based on Multiwalled Carbon Nanotubes as a Selective Melamine Absorbent. *Journal of Agricultural and Food Chemsitry*, 59, 1063-1071.

Zhang, H.Q.; Ye, L & Mosbach, K. (2006). Non-covalent molecular imprinting with emphasis on its application in separation and drug development. *Journal of Molecular Recognition*, 19, 248-259.

Zhang, J.; Wang, Y.; Lv, R. & Xu, L. (2010). Electrochemical tolazoline sensor based on gold nanoparticles and imprinted poly-o-aminothiophenol film. *Electrochimica Acta*, 55, 4039-4044.

Zhang, M. & Gorski, W. (2005). Electrochemical Sensing Based on Redox Mediation at Carbon Nanotubes *Analytical Chemistry*, 77, 3960-3965.

Zhang, M.; Huang, J.; Yu, P. & Chen, X. (2010). Preparation and characteristics of protein molecularly imprinted membranes on the surface of multiwalled carbon nanotubes. *Talanta*, 81, 162–166.

Zhang, M.N.; Yan, Y.M.; Gong, K.P.; Mao, L.Q.; Guo, Z.X. & Chen Y. (2004). Electrostatic Layer-by-Layer Assembled Carbon Nanotube Multilayer Film and Its Electrocatalytic Activity for O_2 Reduction. *Langmuir*, 20, 8781–8785.

Zhang, Z.; Hu, Y.; Zhang, H. & Yao, S. (2010a). Novel layer-by-layer assembly molecularly imprinted sol–gel sensor for selective recognition of clindamycin based on Au electrode decorated by multi-wall carbon nanotube. *Journal of Colloid and Interface Science*, 344, 158–164.

Zhang, Z.; Hu, Y.; Zhang, H.; Luo, L. & Yao, S. (2010b). Layer by layer assembly sensitive electrochemical sensor for selectively probing l-histidine based on molecular imprinting sol–gel at functionalized indium tin oxide electrode. *Biosensors and Bioelectronics*, 26, 696–702.

Zhang, Z.; Zhang, H.; Hu, Y. & Yao, S. (2010c). Synthesis and application of multiwalled carbon nanotubes–molecularly imprinted sol–gel composite material for online solid phase extraction and high performance liquid chromatography determination of trace Sudan IV. *Analytica Chimica Acta*, 661, 173–180.

Zhang, Z.; Zhang, H.; Hu, Y.; Yang X & Yao, S. (2010d). Novel surface molecularly imprinted material modified multi walled carbon nanotubes as solid phase extraction sorbent for selective extraction gallium ion from fly ash. *Talanta*, 82, 304–311.

Zhang, Z.H.; Hu, Y.F.; Zhang, H.B.; Luo, L.J. & Yao, S.Z. (2010). Electrochemical layer-by-layer modified imprinted sensor based on multi-walled carbon nanotubes and sol-gel materials for sensitive determination of thymidine. *Journal of Electroanalytical Chemistry*, 644, 7–12.

Zhao, G.C.; Zhang, L.; Wei, X.W. & Yang, Z.S. (2003). Myoglobin on multi-walled carbon nanotubes modified electrode: direct electrochemistry and electrocatalysis. *Electrochemistry Communications*, 5, 825-829.

Zhao, H.; Wang, L.; Qiu, Y.; Zhou, Z.; Zhong, W. & Li, X. (2007). Multiwalled carbon nanotubes as a solid-phase extraction adsorbent for the determination of three barbiturates in pork by ion trap gas chromatography–tandem mass spectrometry (GC/MS/MS) following microwave assisted derivatization. *Analytica Chimica Acta*, 586, 399–406.

Zhao, X.; Inoue, S.; Jinno, M.; Suzuki, T. & Ando, Y. (2003). Macroscopic oriented web of single-wall carbon nanotubes. *Chemical Physics Letters*, 373, 266-271.

Zhao, Y.; Zhang, W.; Chen, H.; Luo, Q. & Li, S.F.Y. (2002). Direct electrochemistry of horseradish peroxidase at carbon nanotube powder microelectrode. *Sensors and Actuators B*, 87, 168-172.

Zhou, D.; Yang, L.P.; Yang, R.M.; Song, W.H.; Peng, S.H. & Wang, Y.M. (2008). Novel quasi-interpenetrating network/ functionalized multi-walled carbon nanotubes double-network composite matrices for DNA sequencing by CE. *Electrophoresis*, 29, 4637–4645.

Zhou, Q.; Wang, W. & Xiao, J. (2006a). Preconcentration and determination of nicosulfuron, thifensulfuron-methyl and metsulfuron-methyl in water samples using carbon nanotubes packed cartridge in combination with high performance liquid chromatography. *Analytica Chimica Acta*, 559, 200–206.

Zhou, Q.; Xiao, J. & Wang, W. (2006b). Using multi-walled carbon nanotubes as solid phase extraction adsorbents to determine dichlorodiphenyltrichloroethane and its metabolites at trace level in water samples by high performance liquid chromatography with UV detection. *Journal of Chromatography A*, 1125, 152–158.

Zhou, Q.; Xiao, J.; Wang, W.; Liu, G.; Shi, Q. & Wang, J. (2006c). Determination of atrazine and simazine in environmental water samples using multiwalled carbon nanotubes as the adsorbents for preconcentration prior to high performance liquid chromatography with diode array detector. *Talanta*, 68, 1309–1315.

Zhou, Q.; Xiao, J. & Wang, W. (2007). Comparison of multiwalled carbon nanotubes and a conventional absorbent on the enrichment of sulfonylurea herbicides in water samples. *Analytical Science*, 23, 189–192.

Zhou, Y.X.; Yu, B. & Levon, K. (2003). Potentiometric sensing of chiral amino acids. *Chemistry of Materials*, 15, 2774-2779.

9

Mechanism of Nano-Machining and Mechanical Behavior of Nanostructure

Jiaxuan Chen[1], Na Gong[1] and Yulan Tang[2]
[1]*Harbin Institute of Technology,*
[2]*Shenyang Jianzhu University*
China

1. Introduction

As the development of observing technology (such as the atomic force microscopy, AFM) and the increasing need of micro mechanical system in modern manufacturing industry, Nano-machining with machine tools and position-control techniques aim to produce high quality surfaces in terms to form accuracy, surface finish, surface integrity for optical mechanical and electronic components, has been a fundamental researching subject in ultra-precision machining area. Lots of researchers have shown extreme interest on the properties of nanoscale materials and the forming mechanism of nano-machining. Nowadays, although the AFM developed by Binnig et al [1] has been an effective instrument which is able to obtain directly the three-dimentional surface topography and can also been used to carry out the nanoindentation [2, 3] and nanocutting experiments[4, 5] in order to research on the mechanical properties and plastic deformation of materials. The details of plastic deformation, evolvement of defects and other important phenomenon in nano-machining processing are difficult to be observed clearly just by AFM. However, the molecular dynamics (MD) methods provides us a practical and effective way to research from the atomic perspective the removal mechanism, states of the stress and strain, the subsurface damaged layer, dislocation nucleation and propagation, surface friction and tool wear and so on [6-9]. In this chapter, we will introduce the method of applying MD simulation on the observing and analyzing of the nano-machining process to the readers in detail based on our researching results which have been published previously.

Meanwhile, as the basic components of micro mechanical system, nanostructures loaded show different mechanical response compared with macrostructures. Due to size effects, surface effects, and interface effects of nanostructures, properties of nanomaterials are enhanced, and nanoscale research has been an area of active research over the past decades. Many researchers use MD numerical simulation to investigate the physical mechanism of nanostructures by atomic motion in detail and have a rapid progress in recent years [10-21]. Most of those studies mainly concentrated on materials with free defects or artificial defects, however, as a matter of fact, a variety of defects can be generated in nano components during nanomachining process. Therefore, it is greatly important to have a suitable description of the material properties of nano-machined components. In this chapter, in order to find a better way to predict the material properties of microstructures, we established the model of real nanostructure with defect, and conduct the integrated MD simulation of scratching and tension or scratching and shearing with the same specimen, respectively. So that, we will eventually find the regulation of real microstructures under different loading conditions and direct the design of nano-scale device.

2. Model and theory

With the development of the electronic industry, copper has been one of the important materials in many fields such as electrical interconnects, and in the past decade, which has been of significant interest to many researchers [22-27].In our simulations of nano-machining and nanostructures, we also considered the monocrystalline copper and pyramid rigid diamond as the workpiece and tool, respectively, as shown in Fig.1. The initial atomic configuration of the workpiece material is created from the face-centred cubic (FCC) copper lattice and the tool is tri-pyramid diamond tip with a rake angle of -60°. The atoms of workpiece are divided two parts. The upper part of workpiece is made of the 36 layers newtonial atoms and the atoms of surfaces are free, while the bottom of workpiece is made of the 4 layers boundary atoms and atoms are fixed in space at the process of scratching. The sizes of slab at x[1 0 0], y[0 1 0] and z[0 0 1] direction are 30, 17 and $18a_0$, where a_0 is the equilibrium lattice constant (for FCC Cu, a_0 = 362 nm).

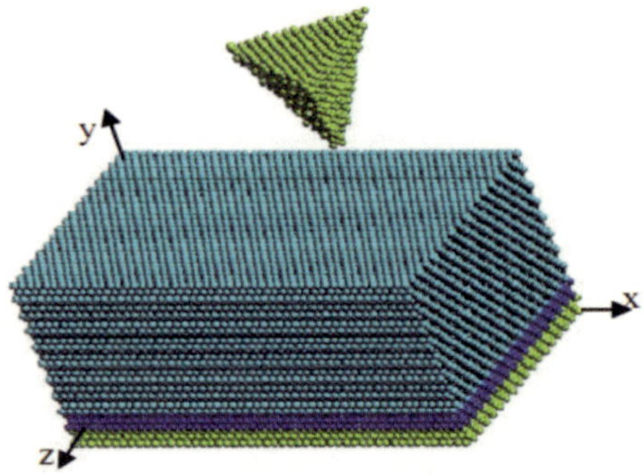

Fig. 1. MD simulation model of nano cutting on single crystal copper surface.

For the Cu-C interactions between workpiece atoms and tool atoms, the Morse potential is used. The Morse potential model [28-31] is adopted for the diamond atoms and the Cu atoms of the calculation process. The Morse potential is written as (1):

$$\mu r_{ij} = D[\exp(-2A(r_{ij} - r_0) - 2\exp(-A(r_{ij} - r_0)))] \quad (1)$$

where μr_{ij} is a pair potential energy function, D , A , r_0 and r_{ij} correspond to the cohesion energy, the elastic modulus, the atomic distance at equilibrium and the atomic distance between atom i and atom j, respectively.

The Cu-Cu interactions between workpiece atoms are described by the established embedded atom method (EAM) potential [32-36]. For the EAM potential, the total atomic potential energy of a system E is expressed as (2):

$$E = \sum_{i}^{N}[F(\rho_i) + \sum_{j>i}^{N} u(r_{ij})] \quad (2)$$

where E is the sum of the energy, $F(\rho_i)$ and $u(r_{ij})$ correspond to the sum of the embedding energy, the total potential energy. ρ_i is the host electron density at site induced by all other atoms in the system, which is given by (3):

$$\rho_i = \sum_j \phi(r_{ij}) \tag{3}$$

where $\phi(r_{ij})$ is the embedding function the electron density around an atom.

Usually, it is necessary to calculate stress of every atom in order to analyze the relations between the stress states of atoms and the onset of defects. The stress in the atomistic simulation mn on plane m and in the n-direction can be given by

$$\sigma_{mn} = \frac{1}{N_S}\sum_i [\frac{m_i v_i^m v_i^n}{v_i} + \frac{1}{2V_i}\sum_j F_{ij}\frac{r_{ij}^m r_{ij}^n}{r_{ij}}] \tag{4}$$

Where m_i is the mass of atom i, v_i is the volume assigned to atom i, N_S is the number of particles contained in region S, where S is defined as the region of atomic interaction, r_{ij} is the distance between atoms i and j, and r_{ij}^m and r_{ij}^n are two components of the vector from atom i to j.

The temperature of the entire system was maintained at 293 K with a Nos'e–Hoover thermostat [37, 38], and the half step frog leap algorithm was used for time integration of the atomic coordinates. Through preset the different simulation conditions, such as the cutting face and direction, the depth of nano-cutting, scratching speed, rank angle and so on, we are able to study in depth the removing mechanism of material, state of stress and strain, forming and evolvement of defects, fluctuating of cutting force and other important and interesting processing phenomenon, during the course of nano-machining or nanostructure loading.

3. Mechanism of nano-metric cutting

The workpiece is relaxed to minimum energy before scratched. At the initial stage of scratching process, some atoms of workpiece leave workpiece to be absorbed on the fronter surface of tool under the cutting and attracting action of tool, material ahead and beneath the cutting tool is deformed during scratching, which is similar to the conventional cutting process. With the pas- sage of tool, more and more atoms of workpiece begin to climb along the frontier surface of tool under the shearing and attracting force of tool, and stack the frontier surface on of tool, which is the primeval stage of chip forming as shown in Figure 2 below [39]. In the whole process of scratching, the lattice of the workpiece is deformed as by buckling and compressing due to plowing of the tool, and those atoms near the tool, especially that ahead the tool are affected by the stress from the motion of tool and obtain the higher energy .The atoms with high migration energy leave its initial position, and accomplish the workpiece removed by the attraction and compression of the tool.

Due to shearing and compressing action of tool, atoms of workpiece are applied forceful shear and nor-mal stress. When those stresses get to some extents, dislocations begin to

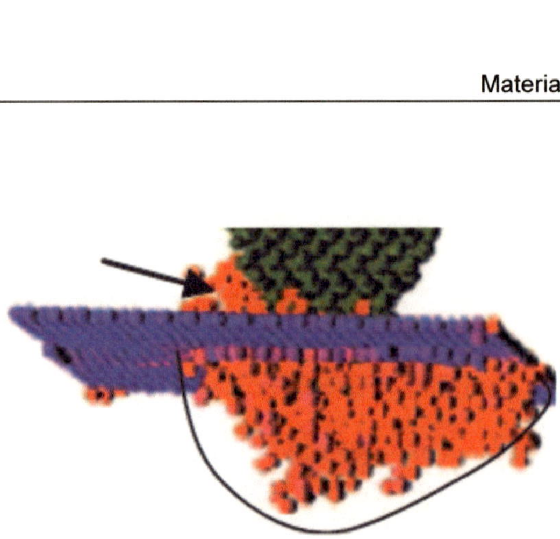

Fig. 2. The chip deforming of single crystal copper scratched on (010) plane by triangular-based pyramid tool with -60°rank angle. Arrow shows chip deforming.

appear. At the initial scratching stage dislocations are driven and propagate only in surface and subsurface of workpiece, and expand to another free surface along the [−101], [10−1] direction on the surface. With progress of the cutting edge, the numbers of pile-up atoms on surface of tool increase obviously, the numbers of atoms which play the cutting role increase accordingly, atoms of workpiece are applied to the more forceful stress and the range of atoms affected is wider, some dislocations are driven by shearing stress and normal stress, and begin to propagate downwards on {111} plane in workpiece. The deformed regions beneath machined surface of workpiece are smaller than that beneath surface being machined, and show the elastic restore to some extents, as shown in Figure 3.

Fig. 3. Triangular-based pyramid tool with -45°rake angle scratching monocrystal copper {010} plane, dislocations propogate downwards on {111} plane in workpiece.

Compared with the condition of -60°rake angle, we can see that the deformed region beneath tool is broadened, and the deformed region beneath machined surface of workpiece is also broadened with the increasing scratching depths and rake angle. As a result, the numbers of defects in workpiece increase under the big scratching depths, while the order degree of latticedecreases, especially for the distribution of short-range order, the atomic numbers of regular position decrease, which are also shown in Figure 4.

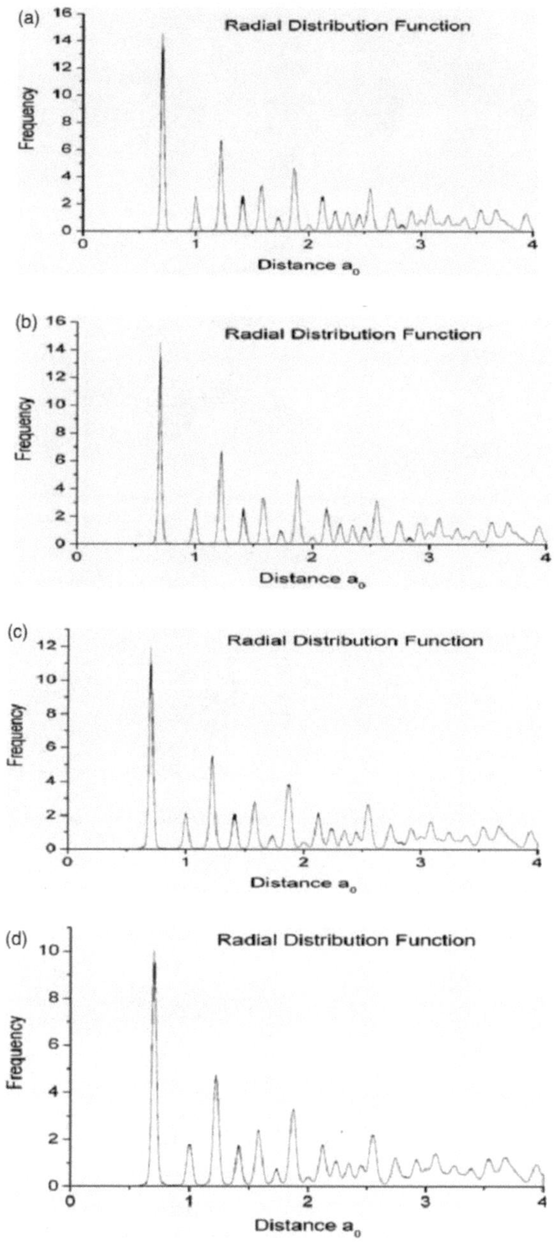

Fig. 4. The order degree of atoms in workpiece after triangular-based pyramid tool with -45°rake angle scratch monocrystal copper {010} plane with scratching depths 0.5, 1, 2, 3 a_0.

In the scratching process, with increasing depths of scratching, the numbers of atoms what play the cutting role are augmented, and the number of removal atoms also increase, which need larger cutting force, the cutting force for tool with rake angle -30° is larger than that for tool with rake angle -45°at the same scratching depths, the curves of scratching force for different scratching depths and rake angle are shown in Figure 5 below.

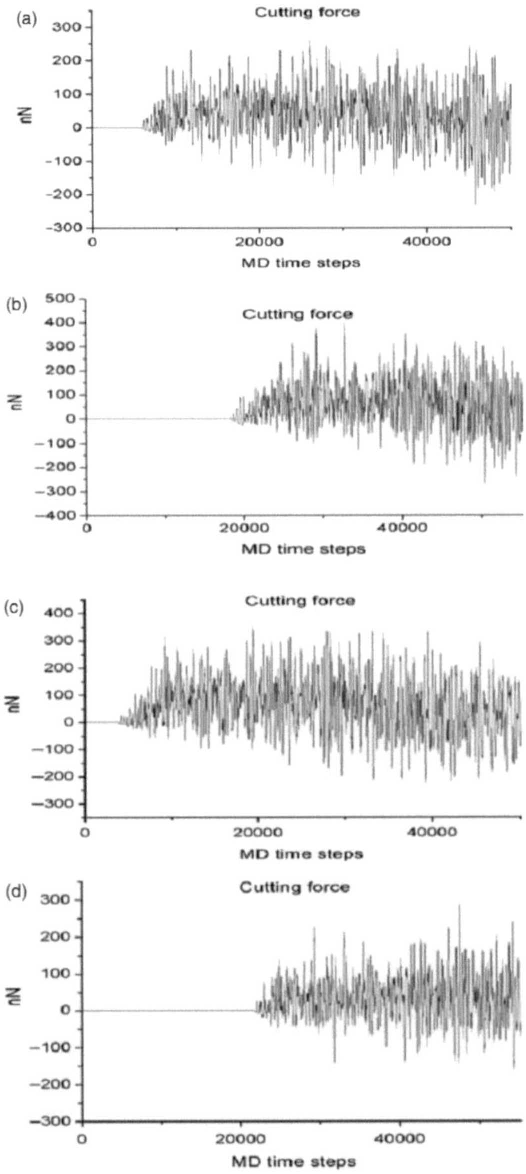

Fig. 5. The scratching force of triangular-based pyramid tool scratching monocrystal copper {010} plane under scratching depths1, 3 a_0: (a–b) −30°rake angle, (c–d) −45°rake angle.

With increasing scratching depths, atoms of workpiece receive the more forceful action of shearing and compression, the effected range of atoms in workpiece are also broaden, though some effected atoms restore their usual position, there are still more defects' atoms to exist in workpiece after tool scratch the workpiece. At the same time, there exists the higher residual stress in workpiece after tool withdraws the workpiece as shown in Figure 6.Though residual stress can decrease by anneal process,the level of residual stress in subsurface of workpiece with the increasing scratching depths enhance accordingly, the level of residual stress in subsurface of workpiece that scratched by tool with rake angle -45° is higher than that rake angle -30°.

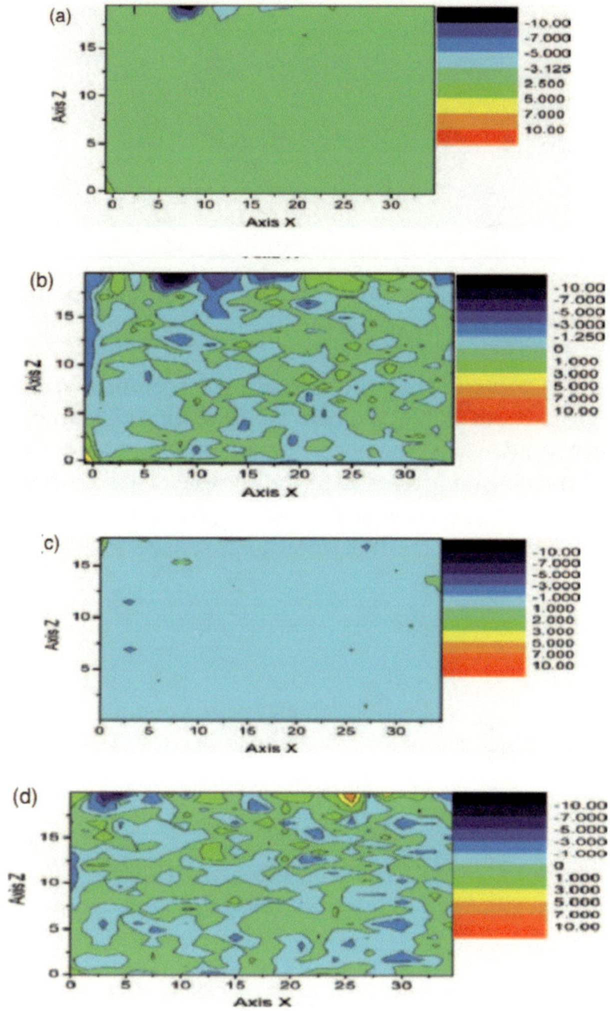

Fig. 6. The shear stress distribution of atoms in subsurface with differ-ent rake angles and scratching depths after relaxation: (a–b)-scratching depths 0.5, 3 a_0 for -45°rake angle; (c–d)-scratching depths 0.5, 3 a_0 for -30°rake angle.

The workpiece crystal orientations also have important effects on defects evolvement, as shown in Fig.7. When the AFM diamond tip scratches the surface of single crystal copper as shown in Fig. 7(a), it can be seen that dislocation is prone to nucleate ahead the tool, and those dislocations glide in the [101] direction as the effects of the edge of the AFM diamond tip. The Fig. 7(b) shows the case of (110) plane of single crystal copper. From the Fig. 7(b), it is found that dislocations glide in the [-101] direction. For the case of (111) plane of single crystal copper, the dislocations seem to propagate the dislocations propagate at 30o to the cutting direction at the top surface, and dislocations glide in the [1-10] direction and [01-1] direction as shown in Fig. 7(c) [40].

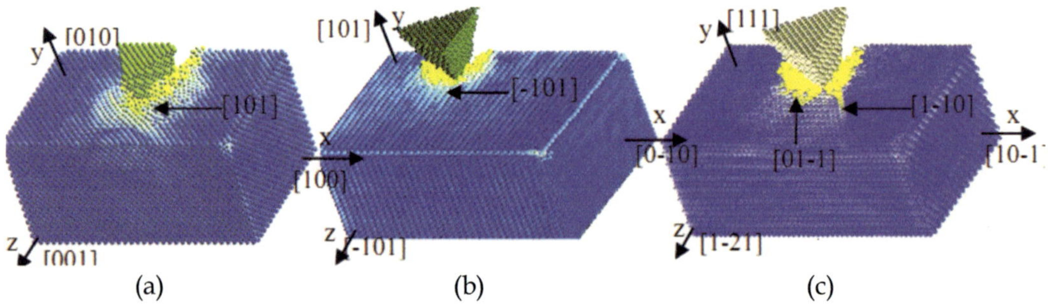

Fig. 7. The atom snaps of nano-cutting single copper on different orientation: (a) (100) orientation, (b) (110) orientation and (111) orientation.

Fig.8 shows the variation curve of fluctuating cutting force under different orientation and the same cutting depths. From Fig.8, we can see that the nucleation and propagation of dislocations results in the release of the accumulated cutting energy, which corresponds to the temporary drop of the cutting force due to the complex local motions of the dislocations. It also can be seen from the curves that the cutting force of the steady state from (111) orientation is the biggest, that from (110) orientation is the lowest.

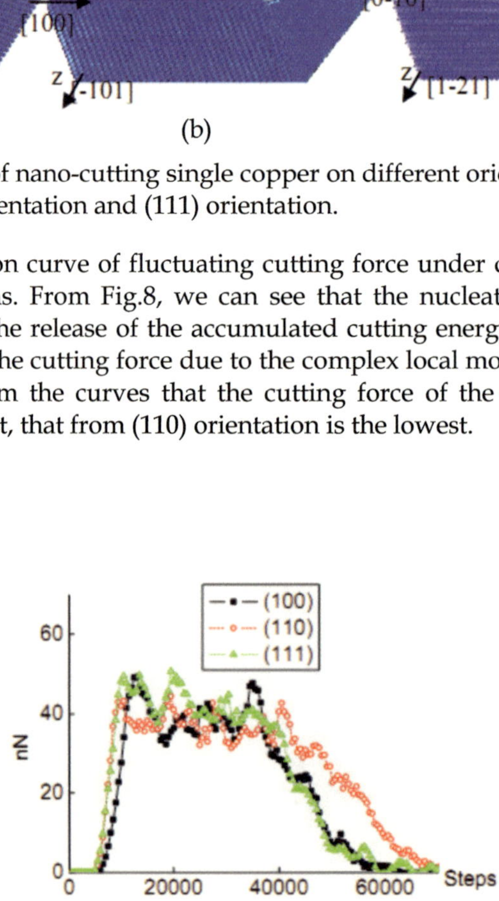

Fig. 8. Effect of crystal orientation on cutting force.

We also find that the scratching speed has important effects on dislocations evolvement. Fig.9 shows the relationship of the numbers of defects and the scratching speed. From Fig.9 we can see that dislocations can propagate succeeding under the low scratching speed. We also can see that the order of the atoms of workpiece under high scratching speed is better than low scratching; the numbers of defects inside workpiece with high scratching speed are low corresponding to the low scratching.

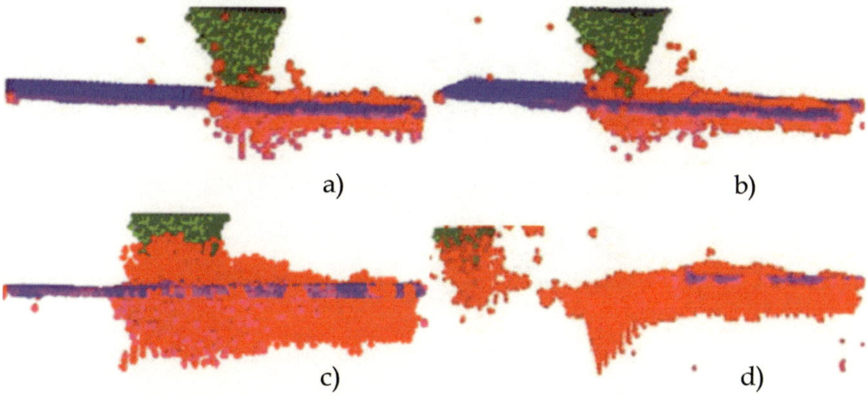

Fig. 9. The picture a1 and a2 indicate the effects of dislocations evolvement under the scratching speed 360m/s. The picture b1 and b2 under the scratching speed 180m/s.

4. Mechanical behavior of nanostructure

The nanostructure with no defects or artificial defects are not suitable to simulate the material properties of practical nanostructure. in this part, we introduce an integrated MD simulation method, in order to study the mechanical properties of nanostructure with practical processing defects. We consider the nanostructure scratched by diamond tip under the machining conditions of different scratching depth of different scratching lattice plane or direction, just like we have described above, as the practical processing defects, then, apply tension or shearing force on such nanostructure to observe their physical and mechanical behavior.

4.1 Scratching and tension

The models we established are shown in figure 10, we also introduce the ideal nanostructure with no defects for comparing. For ideal copper specimen, as shown in Fig.10 (a), the sizes of slab at X[100], Y [010], and Z[001] direction are 12a_0, 17a_0, and 20 a_0, respectively, where a0 is the equilibrium lattice constant (for fcc Cu, a_0=0.362 nm); For machined specimen, as shown in Fig 10 (b) and (c), which can be obtained as followed: a tri-pyramid diamond tool scratches an ideal fcc copper slab on the (010) plane along [001] and [100] directions, Newton atoms layers are and the states of these atoms should be kept, the two ends of Newton atoms are picked up 2a_0 length along axis Z [001] and Z [00-1] direction to become the boundary layer atoms respectively [41].

Before simulation, the ideal specimen is relaxed for 10000 MD steps for minimizing energy. After that, double tensile loads are applied to the two ends of specimen. At the initial stage, though the atoms of specimen begin to deviate from their positions, the configurations of lattice don't change, as shown in Fig 11 (a) and 11(b), which is corresponding to elastic stage in stress-strain curve, as shown in Fig 12 (a). During tensile process, when normal stress σ_{zz} is up to the first peak point, defects such as dislocations begin to occur and propagate along different {111} planes under tensile loads due to the smallest Burgers vector existing in the [110] close-packed directions for fcc crystal structures, making it most energetically favorable to reconstruct along this plane. Because of the motion of dislocations, the higher

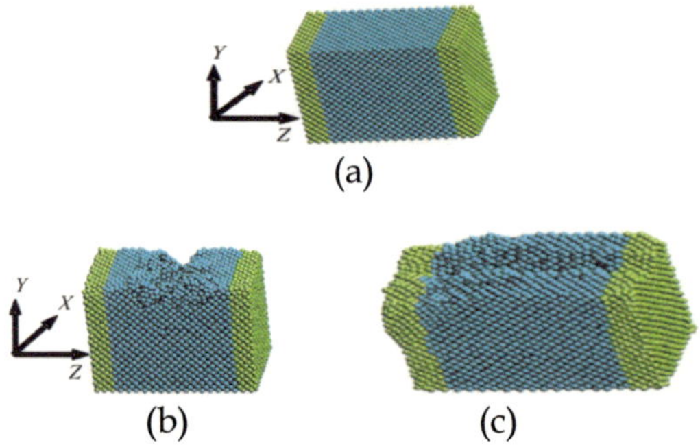

Fig. 10. (a) Model for ideal specimen; (b) Model for mahined specimen along [001] direction; (c) Model for machined specimen along [100] direction.

stress is released and normal stress σ_{zz} begins to decrease. As the increasing of defects, which hinder the motion of dislocation freely, and result in work-harden formation. It is necessary to obtain forceful force to drive the motion of dislocations, which exhibits that normal stress σ_{zz} increase secondly in stress-strain curve. After the second yield, ideal specimen begin to neck under tensile loads, and corresponding to the stress-strain curve begin to decrease quickly, and occur to rupture at the neck region firstly, as shown in Fig.13 (a)and(b). So the deformation of nanostructure experiences next process: elastic stage, first yielding and work-harden stage, second yielding and full plastic stage, rupture [42].

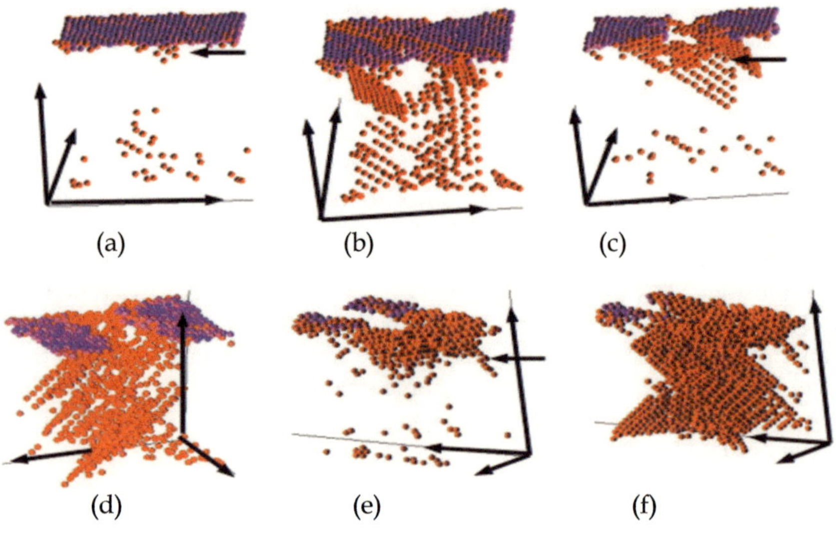

Fig. 11. (a) and (b) ideal specimen at strain 0.066 and 0.089,and atoms are shown for slip index 0.20, respectively, and tensile speed is 7.6 m/s; (c) and (d) scratched specimen with groove for depth $1a_0$ along $[\bar{1}00]$ direction at strain 0.0615 and 0.075 respectively, and atoms are shown for slip index >0.20, and tensile speed is 3.6 m/s; (e) and (f) scratched specimen with groove for $2a_0$ depth along $[00\bar{1}]$ direction at strain 0.0486 and 0.077 respectively, and atoms are shown for slip index >0.20, and tensile speed is 3.6 m/s arrows denote dislocations.

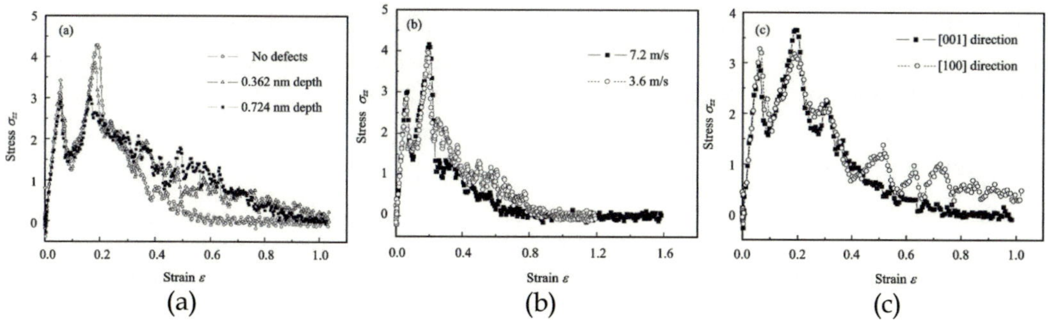

Fig. 12. (a) Specimen with scratched groove along $[00\bar{1}]$ direction for tensile speed 3.6 m/s; (b) Specimen with scratched groove at $1a_0$ depth along $[00\bar{1}]$ direction; (c) Spec-imen with scratched groove at $2a_0$ depth.

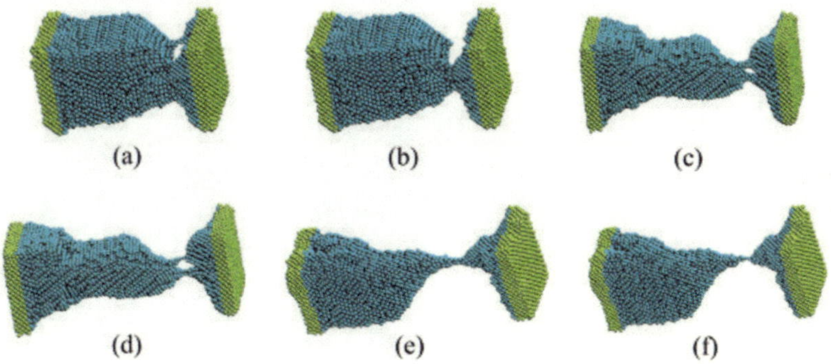

Fig. 13. (a) and (b) ideal specimen at strain 0.437 and 0.495 for tensile speed 3.6 m/s; (c) and (d) specimen with scratched groove along $[00\bar{1}]$ direction for $2a_0$ depth, strain 1.068 and 1.071; specimen with scratched groove along $[00\bar{1}]$ direction for $1a_0$ depth, strain 1.23 and 1.28, tensile speed 3.6 m/s.

For scratched specimen case, the process of relaxation and force applying are in the same way with the ideal specimen case. The stress-strain curves are shown in Fig.12 (a)-(c). From Fig.12 (a)-(c), we can see that specimens machined also exhibit the initial stage of nearly elastic response, then two peaks, work-hardening stage and rapture at last, which is also the same as the specimen with on defects. From Fig.12(a), it can be seen that the first yielding and second yielding normal stress decrease as the depths of scratched groove on surface of specimen increase. From Fig.12 (b), it can be seen the numerical value of the first and the second yielding stress for specimen with scratched groove of $1a_0$ depth under higher tensile speed are higher than lower tensile speed. It shows that yielding strength of specimen increase with increasing tensile speed. From Fig.12(c), we can see that when the direction of scratched groove is consistent with the extension direction, under the same groove of depth and tensile speed, the second yielding stress is higher than the direction of groove is in perpendicular direction of extension.

For machined specimen, we also notice that dislocations prone to nucleate not only from surface and corner of specimen just as the ideal specimen case, but also from groove, in particular from the position of ends of groove, and then dislocations begin to propagate along different (111) planes, as shown in Fig.11 (c)-(g). For the case that the direction of scratched groove is in the perpendicular direction of extension, dislocations prone to nucleate from the position that tool withdraw specimen firstly.

4.2 Scratching and shearing

Another usual loading condition is shearing for nanostructure. Here, following the method we have described above, we begin to study the mechanical behavior of practical nanostructure under the shearing load. The model established is shown as Fig.14. Through setting the different simulation conditions, such as the model size and the depth of groove , we can research in depth the affecting regulation of these elements on the mechanical and physical behavior.

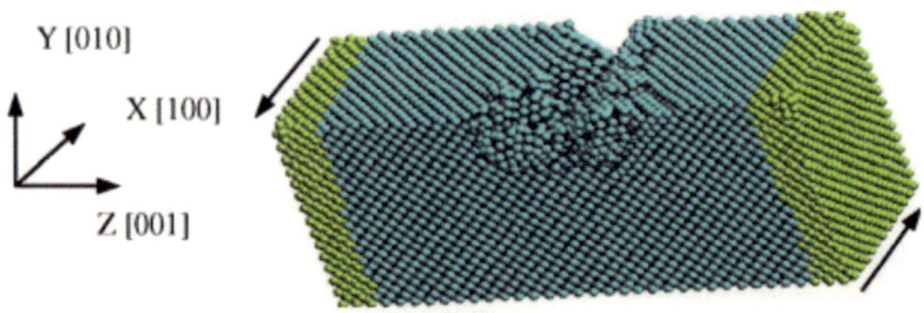

Fig. 14. Shearing simulation model for scratched specimen with arrows denoting directions of loading.

The stress-strain curves generated from the simulations of shearing with specimens without defect and specimen with groove at scratch depths of 0.5, 1, 2, 3, $5a_0$ are as shown in Fig. 15. From Fig.15 we can see that the deformation of specimen experience an elastic response at the initial stages, and then elastic-plastic stage near PSS (peak shear stress), and total plastic stage after PSS. From the moment when plastic yielding occurs in the specimen to the moment when the specimen plastically fails under continuous shear loads, shear stress τ_{xz} drops firstly, and then remains steady for a certain length of time before another decline starts, and the stress-strain curves decrease again, and are shown in zigzag changes, and a total failure occurs at the end of the second decline. Normal stress σ_{xx} fluctuates around zero firstly, and then it begins to gradually increase when the nanostructures begin to yield. When normal stress σ_{xx} goes up to the same level as shear stress τ_{xz}, normal stress σ_{xx} begins to decrease with shear stress τ_{xz}. It can be seen from the stress-strain curves shown in Fig. 15a and b that shear stress τ_{xz} plays a leading role in the initial stage of shearing process. The contribution of normal stress σ_{xx} begins to increase when the yielding starts, and normal stress σ_{xx} and shear stress τ_{xz} play the same role for plastic deformation of specimen.

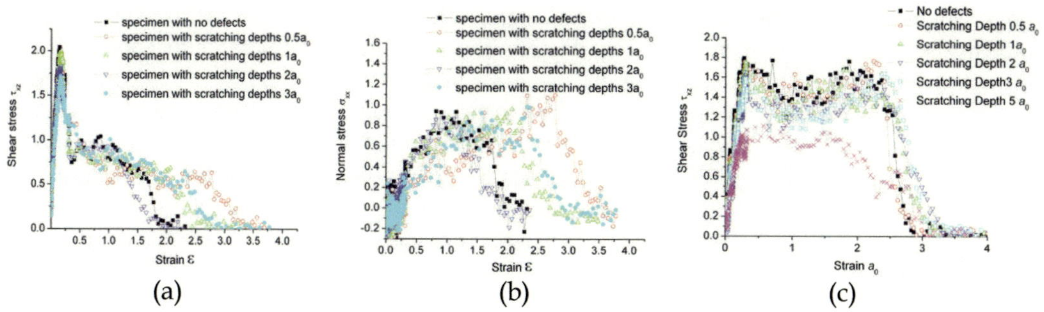

Fig. 15. (a) And (b) Normal stress σ_{xx} and shear stress τ_{xz} stress–strain curves for specimen at size 12-15-20a$_0$ without defects and scratching depths of 0.5, 1, 2 and 3a$_0$ and (c) shear stress–strain for specimen at sizes 12-15-35 a$_0$ without defects and at scratching depths 0.5, 1, 2, 3 and 5a$_0$.

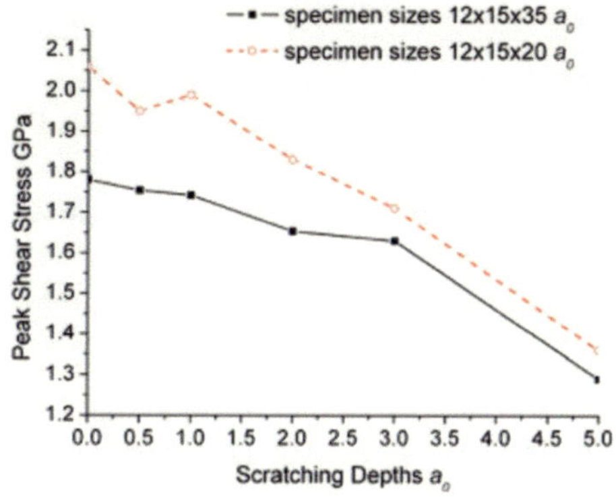

Fig. 16. Peak shear stress τ_{xz} obtained with of different sizes at different scratching depths specimens

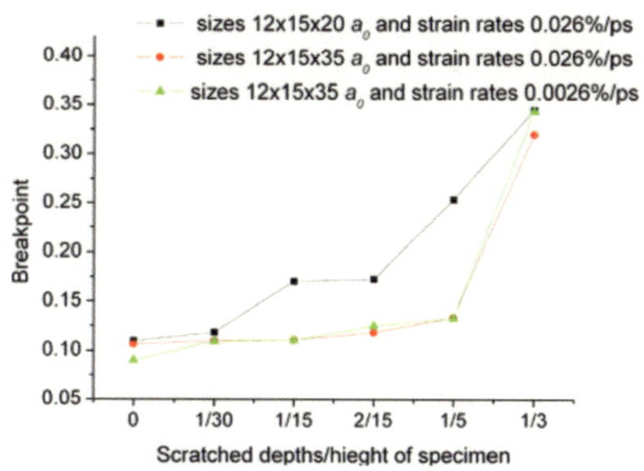

Fig. 17. The breakpoint of specimen under high-speed shear loads.

PSS of specimen for different sizes and different depths are shown in Fig. 15. From Fig. 15, it can be seen that PSS decreases with the increasing depths of scratching and an ideal specimen is the highest. This is mainly because of that dislocations are likely to nucleate in the specimen with many defects on nano scale, which results in the decreasing of shear intensity. It also can be seen from Fig. 15 that the yield stress of scratched specimen also decreases with the increasing specimen size. The yield stress is observed to decrease with the increasing specimen size, which is due to the enhanced opportunities for dislocation motion to occur at a larger size. For small specimen sizes, PSS increases at the same scratch depths, which show size effect (namely the scale dependence, that the size and quantity of defects inside the nanostructure decrease and the intensity improve when its scale becomes smaller and smaller, and thus the nanostructure will show different mechanical properties compared with body material)

For an ideal specimen or a specimen with shallow scratch depths, the stress gradients is flat near the groove, the dislocations resulting from the stress field at the corner are the main sources of plastic deformation, and the breakpoint of failure is close to the either end of the specimen. With the increasing scratch depths, the gradient of stress is steep near the groove, are likely to overtake the bafflement to cause the movement of dislocations, and thus the stress gradients near the corners and/or the groove induce the dislocations to determine the yielding point. In this way, the yield plastic deformation is likely to occur near the groove, and the breakpoint of failure is close to the groove of the specimen. Fig.17 shows the breakpoints of specimen under shear loads. It can be seen from Fig. 17 that the breakpoint is close gradually to the groove as the depth of groove increases. At the same scratched depths, the breakpoint is closer to the groove for a small specimen. It is noted that the breakpoint of specimen can be determined by groove when the ratio of the depths of groove to the height of specimen is up to one third (scratched depths $5a_0$) regardless of size and strain rates [36].

When the specimen is continuously subjected to the shear load, the specimen begins to yield and break down. The snapshots of the rupture processes of specimen are as shown in Fig. 18. From Fig.18 we can see that the formations of rupture of nanostructure are that the development of vacant clusters in the neck region which causes the rupture of specimen, or gradual thinning of theneck region of the scratched specimen into a single chain of atoms.

Fig. 18. (a) and (b) the formation of rupture of specimen with no defects for shear simulation, (c) and (d) that of specimen with scratching depths of $5a_0$.

5. Conclusion

Because of the smaller size of the nanostructure and the lack of related instrument, the traditional experimental method is not suitable to study the mechanical behavior of nanostructure. However, lots of researching work proves that the MD simulation method is a feasible method to study the mechanical and physical properties and plastic deformation of materials in nanoscale. based on the simulating results we provide above, we find that the removal mechanism of material, and states of the stress and strain of structure in nanoscale show obvious differences compared with macrostructures, such as the size effects, surface effects, and interface effects, two yield peaks et al. and the defects induced by machining play an important role during the dislocations moving and material yielding, and lead to the rupture of nanostructure loaded at last. MD simulation method is able to make the researchers to observe directly and clearly this phenomenon and find the hiding regulations in micro processing. The deep and wide study will improve the development of modern electrical industry and the whole manufacturing industry in the future.

6. Acknowledgment

We would like to thank the National Science Fund for Distinguished Young Scholars of China (50925521), the Natural Science Foundation of China (51075092), the Natural Science Foundation of Heilongjiang Province in China (E200903) and China Postdoctoral Science Foundation (20100471020), and Harbin Innovative Talents Fund (2011RFQXG017) for their financial supports.

7. References

[1] G. Binnig, C.F. Quate, and C. Gerber, Physics Review Letter 56 (1986) 930.
[2] S.G. Corcoran, R.J. Colton, E.T. Lilleodden, and W.W. Gerberich, Physical Review B 55 (1997) 16057.
[3] R. Smith, D. Christopher, S. D. Kenny, A. Richter, and B. Wolf, Physical Review B 67 (2003) 245405.
[4] T. Sumomogi, T. Endo, and K. Kuwahara, Journal of Vacuum Science and Technology B 12 (1994) 1876.
[5] Y.D. Yan, T. Sun, Y.C. Liang, and S. Dong, International Journal of Machine Tools & Manufacture 47 (2007) 1651.
A. Gannepalli, and S. K. Mallapragada, Physics Review B, 66 (2002) 104103.
[6] R Komanduri, N Chandrasekaran, and L.M Raff, Physical Review B, 61 (2000) 14007.
[7] K. Cheng, X. Luo, R. Ward, and R. Holt, Wear, 255 (2003) 1427.
[8] S. Shimada, SPIE, 2576 (1995) 396.
[9] R. Komanduri, N. Chandrasekaran, and L. M. Raff, International Journal of Mechanical Sciences, 43 (2001) 2237.
[10] H. A. Wu, European Journal of Mechanics - A/Solids, 25 (2006) 370.
[11] Y. Champion, C. Langlois, G. M. Sandrine, P. Langlois, J. L. Bonnentien, and M. J. Hytch, Science 300 (2003) 1357.
[12] S. J. A. Koh, H. P. Lee, C. Lu, and Q. H. Cheng, Physical Review B 72 (2005) 085414.
[13] S. P. Ju, J. S. Lin, and W. J. Lee, Nanotechnology 15 (2004) 1221.
[14] J. W. Kan and H. J. Hwang, Nanotechnology 13 (2002) 524.

[15] S. A. Kotrechko, A. V. Filatov, and A. V. Ovsjan-nikov, Theoretical and Applied Fracture Mechanics 45 (2006) 92.
[16] D. S. Xu, R. Yang, J. Li, J. P. Chang, H. Wang, D. Li,and S. Yip, Computational Materials Science, 36 (2006) 60.
[17] F. Golovnev, E. I. Golovneva, and V. M. Fomin, Computational Materials Science, 36 (2006) 176.
[18] H. S. Park and J. A. Zimmerman, Physical Review B 72 (2005) 054106.
[19] K. A. Ifat, R. C. Sidney, G. Konstantin, I. Viktoria, H.Thomas, S. Gotthard, W. Inna, H. D. Wagner, and T.Reshef, Proceedings of the National Academy of Sciences, 103 (2006) 523.
[20] P. Heino, H. Ha, K. Kinen, and K. Kask, Physical Review B 58 (1998) 641.
[21] Y. W. Zhang, T. C. Wang, and Q. H. Tang, Journal of Applied Physics, 77 (1995) 2393.
[22] S. Shimada, International Journal Japan Society Precision Engineering, 29 (1995) 283.
[23] S.Shimada, SPIE 2576 (1995) 396.
[24] Y. Y. Ye, R. Biswas, J. R. Morris, A. Bastawros, and A. Chandra, Nanotechnology 14 (2003) 390.
[25] P. Heino, H. Ha. K. kinen, and K. Kask, Physical Review B, 58 (1998) 641.
[26] Win. Jin. Chang, Microelectronic Engineering, 65 (2003) 239.
[27] R. Komanduri, M. Lee, and L.M. Raff, International Journal of Machine Tools & Manufacture, 44 (2004) 1115.
[28] Komanduri, R., Chandrasekaran, N. and Raff, L.M. Philosophical Magazine B, 79 (1999) 955.
[29] Komanduri, R., Chandrasekaran, N. and Raff, L.M. Physical Review B, 6 (2000) 14007.
[30] Komanduri, R., Chandrasekaran, N. and Raff, L.M. Wear, 240 (2000)113.
[31] M.S. Daw, and M.I. Baskes, Physical Review Letter, 50 (1983) 1285.
[32] M.S. Daw, and M.I. Baskes, Physical Review B, 29 (1984) 6443.
[33] R.A. Johnson, Physical Review B, 37 (1988) 3924.
[34] R.A. Johnson, Physical Review B, 39 (1989) 12554.
[35] Y.C. Liang, J.X. Chen, M.J. Chen, Y.L. Tang and Q.S. Bai, Computational Materials Science, 43 (2008) 1130.
[36] S.Nose, Journal Chemical. Physics, 81 (1984) 551.
[37] W.G.Hoover, Physical Review A, 31 (1985) 1695.
[38] J.X. Chen, Y.C. Liang, Q.S. Bai, Y.L. Tang. M.J. Chen. Journal of Computational and Theoretical Nanoscience, 5 (2008) 1485
[39] JX. Chen, Y.C. Liang, L.Q. Wang, M.J. Chen, Z. Tong, W.Q. Chen, Proceedings of SPIE, 7655 (2010) 76550J
[40] Y.C. Liang, J.X. Chen, M.J. Chen , Y.L. Tang and Q.S. Bai, Chinese Journal of Chemical Physics, 20 (2007) 649
[41] Y.C. Liang, J.X. Chen, Q.S. Bai, Y.L. Tang, M.J. Chen. Acta Metallurgica Sinica, 44 (2008) 119

10

Novel Mechanochemical Process for Aqueous - Phase Synthesis of Superparamagnetic Magnetite Nanoparticles

Tomohiro Iwasaki
Department of Chemical Engineering,
Osaka Prefecture University
Japan

1. Introduction

Nano-sized magnetite (Fe_3O_4) powder has been widely used in various industrial products: for instance, pigments, recording materials, printing and electrophotography such as copying toner and carrier powders, etc. (Buxbaum & Pfaff, 2005; Hakata, 2002; Ochiai et al., 1994) because of the excellent physicochemical properties. In particular, Fe_3O_4 nanoparticles with a size of smaller than about 20 nm exhibit a unique magnetic property, namely, superparamagnetism. In recent years, its non-toxicity property and high chemical stability have attracted much attention, and the use of superparamagnetic Fe_3O_4 nanoparticles have rapidly expanded in biomedical fields, such as magnetic resonance imaging (Hu et al., 2010; Lee et al., 2009), drug delivery (Guo et al., 2009; Zhou et al., 2009), hyperthermia (Gao et al., 2010; Muzquiz-Ramos et al., 2010), immobilization of enzymes (Wang et al., 2008) and proteins (Can et al., 2009), and magnetic separation of cells (Li et al., 2009).

For industrial production of superparamagnetic Fe_3O_4 nanoparticles, chemical coprecipitation is often employed (Buyukhatipoglu et al., 2009; Compeán-Jasso et al., 2008; Iwasaki et al., 2011a; Mizutani et al., 2010; Yang et al., 2009; Yu et al., 2010). In coprecipitation methods, ferrous and ferric ions are simultaneously precipitated as ferrous hydroxide ($Fe(OH)_2$) and goethite (α-FeOOH) in an alkaline solution, respectively, resulting in the formation of Fe_3O_4. In order to prepare superparamagnetic Fe_3O_4 nanoparticles with good dispersibility, not only the primary size of Fe_3O_4 nanoparticles must be controlled but also the aggregation of Fe_3O_4 nanoparticles must be avoided. Thus, anti-aggregation agents (e.g., surfactant) and toxic organic solvents are added to the staring solution in many cases (Cheng et al., 2005; Ge et al., 2007; Hua et al., 2008; Wan et al., 2007; Wen et al., 2008). Therefore, the Fe_3O_4 suspensions thus obtained should be washed enough before its use in the applications. In addition, for improving the crystallinity, heating treatments such as annealing and hydrothermal treatment are often performed, leading to the enhancement of the ferromagnetism (Mizutani et al., 2008; Wu et al., 2007, 2008; Zheng et al., 2006). Unfortunately, these required treatments complicate the preparation process and may increase the environmental impact and the production cost.

In order to overcome these problems, a novel synthesis method using a mechanochemical effect has been developed for the production of superparamagnetic Fe_3O_4 nanoparticles with high crystallinity without using any environmental-unfriendly additives (Iwasaki et al., 2008, 2009, 2010, 2011b). This chapter describes the outline of this method and the kinetic analysis of the mechanochemical process.

2. Mechanochemical synthesis of Fe_3O_4 nanoparticles via coprecipitation

In this method, a tumbling ball mill is used as a reactor, and a suspension of $Fe(OH)_2$ and α-FeOOH as a precursor, which is prepared via coprecipitation, is ball-milled in an organic solvent-free water system. The ball-milling treatment is performed under a cooling condition. Thus, high mechanical energy generated by collision of ball media is applied to the precursor instead of the heat energy, which promotes the solid phase reaction between $Fe(OH)_2$ and α-FeOOH forming Fe_3O_4 and the crystallization process without the crystal growth caused by the heat energy. This corresponds to the mechanochemical effect. Accordingly, this method does not need any additional heating treatment to improve the crystallinity of the product. In addition, any additives such as surfactants and oxidizing and reducing agents are not required. Consequently, this method provides successfully superparamagnetic Fe_3O_4 nanocrystals with a size of less than 15 nm. The details of this method are described below.

Typically, 1.5 mmol of ferrous sulfate heptahydrate ($FeSO_4 \cdot 7H_2O$) and 3.0 mmol of ferric chloride hexahydrate ($FeCl_3 \cdot 6H_2O$) are dissolved in 60 ml of deionized and deoxygenated water in a beaker. The molar ratio of ferrous ion to ferric ion is 0.5, corresponding to the chemical stoichiometric ratio of the Fe_3O_4 formation reaction. 30 ml of 1.0 kmol/m^3 sodium hydroxide (NaOH) solution is added into the acid solution at a constant addition rate of 3 ml/min under vigorous stirring using a magnetic stirrer in an argon atmosphere. As the pH of the solution increased, $Fe(OH)_2$ and α-FeOOH coprecipitate, according to Eqs. (1) to (3).

$$Fe^{2+} + 2OH^- \rightarrow Fe(OH)_2 \tag{1}$$

$$Fe^{3+} + 3OH^- \rightarrow Fe(OH)_3 \tag{2}$$

$$Fe(OH)_3 \rightarrow \alpha\text{-}FeOOH + H_2O \tag{3}$$

When adding the NaOH solution, the solution temperature is kept below 5°C by ice-cooling in order to avoid the solid phase reaction forming Fe_3O_4 from $Fe(OH)_2$ and α-FeOOH according to Eq. (4) (Lian et al., 2004).

$$Fe(OH)_2 + 2\alpha\text{-}FeOOH \rightarrow Fe_3O_4 + 2H_2O \tag{4}$$

However, even under cooling, $Fe(OH)_2$ and α-FeOOH partially take place the Fe_3O_4 formation reaction. This results in a dark brown suspension with a pH of higher than 12 containing $Fe(OH)_2$, α-FeOOH, and a tiny amount of Fe_3O_4. The suspension thus prepared is subjected to the following ball-milling treatment.

The starting suspension is poured into a milling pot with an inner diameter of 90 mm and a capacity of 500 ml, made of stainless steel (18%Cr-8%Ni). Stainless steel balls with a diameter of 3.2 mm are used as the milling media. The charged volume of balls containing the void formed among them is 40% of the pot capacity, as illustrated in Fig. 1.

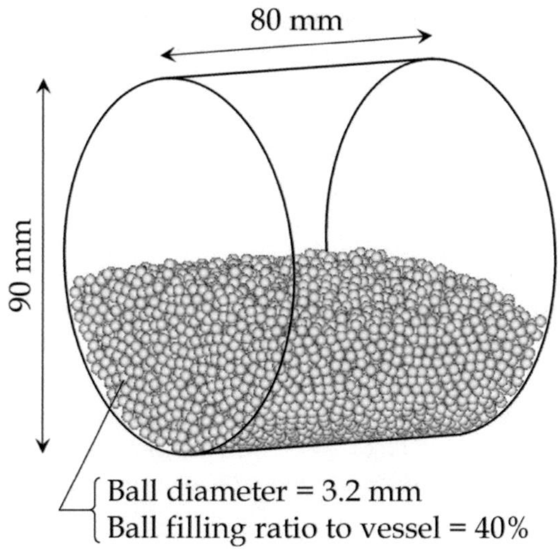

Fig. 1. Schematic illustration of tumbling ball mill used in this work.

After replacement of air in the milling pot with argon, the milling pot is sealed. In order to promote the reaction between $Fe(OH)_2$ and α-FeOOH, the ball-milling treatment is then carried out by rotating the milling pot at rotational speeds of 35 to 140 rpm (corresponding to 16% to 64% of the critical rotational speed (= 220 rpm) determined experimentally based on the behavior of balls containing the suspension) for a given time. During the ball-milling treatment, the milling pot is cooled from its outside in a water bath. Temperature of the water bath is kept at 1.0±0.1°C, and temperature of the suspension is between 1.6°C and 1.7°C within the rotational speed range; this means that the milling pot is cooled enough. After the ball-milling treatment, the obtained precipitate is washed and then dried at 30°C under vacuum overnight.

The dried samples thus obtained were characterized according to standard methods. The powder X-ray diffraction (XRD) pattern of samples was measured with CuKα radiation ranging from 2θ = 10 to 80° at a scanning rate of 1.0°/min using a Rigaku RINT-1500 powder X-ray diffractometer. Fig. 2 shows the XRD pattern of samples obtained at various rotational speeds of the milling pot. In all the XRD patterns, clear diffractions indicating $Fe(OH)_2$ phase were not observed because it tends to form amorphous phase. Before the ball-milling treatment, the sample contained amorphous α-FeOOH and Fe_3O_4 phases. As the milling time elapsed, α-FeOOH gradually disappeared and finally the single-phase of Fe_3O_4 formed. This reaction can be attributed to the application of the mechanical energy generated by collision of the balls to α-FeOOH and $Fe(OH)_2$. The reaction rate depended strongly on the rotational speed. The time required for completing the Fe_3O_4 formation reaction was reduced with increasing in the rotational speed; the reaction almost completed in 12 h at 35 rpm, in 9 h at 70 rpm, in 7.5 h at 105 rpm, and in 6 h at 140 rpm. On the other hand, when the ball-milling treatment was not conducted, i.e., the suspension was kept cooling statically at below 2°C in the water bath, the XRD pattern was almost the same as that of the staring precipitate even after 12 h. It was confirmed that the ball-milling treatment promoted the solid phase reaction by the mechanochemical effect. At early stages of the ball-milling treatment, α-FeOOH seemed to increase from the initial. Actually, suspensions at early stages had light brown colors as compared to the staring suspension. This implies that $Fe(OH)_2$ is oxidized and Fe_3O_4 is decomposed (hydrolyzed) because $Fe(OH)_2$ and Fe_3O_4 in the starting solution are relatively unstable. As the milling time elapsed, the color of the suspension became darker brown, and finally black precipitates were obtained. This result also implies the reduction of α-FeOOH. Fe_3O_4 formed by the mechanical energy is hardly decomposed during ball-milling because it is well crystallized and becomes stable.

The samples had the typical diffraction angles showing relatively high peaks, agreed well with those of Fe_3O_4 phase. The average crystallite size was calculated from the full-width at half-maximum (FWHM) of the Fe_3O_4 (311) diffraction peak at $2\theta \approx 35.5°$ using the Scherrer's formula. The lattice constant was also determined from several diffraction angles showing high intensity peaks. The lattice constant was determined to be between 8.374 and 8.395 Å and close to the standard value of Fe_3O_4 (= 8.396 Å) rather than γ-Fe_2O_3 (= 8.345 Å). These results support that the samples had a single Fe_3O_4 phase rather than γ-Fe_2O_3 regardless of the milling conditions.

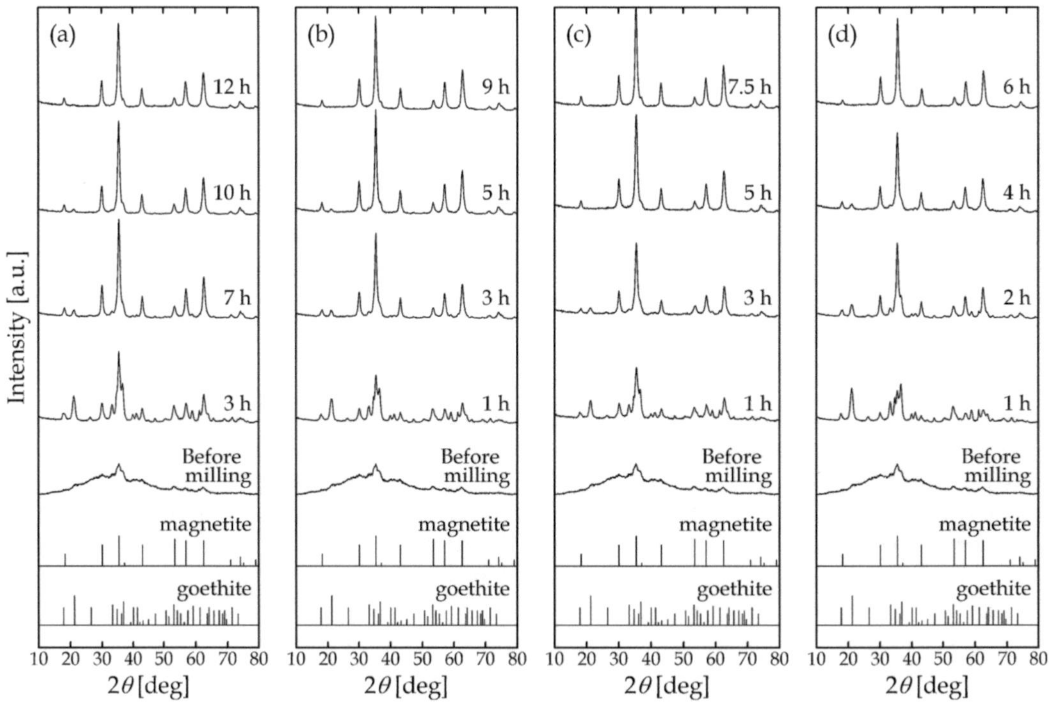

Fig. 2. XRD pattern of samples obtained at (a) 35, (b) 70, (c) 105, and (d) 140 rpm.

The morphology of samples was observed with a field emission scanning electron microscope (FE-SEM; JSM-6700F, JEOL). Fig. 3 shows the SEM image of samples. The samples were spherical nanoparticles with a size of about 10–20 nm.

The hydrodynamic particle size distribution of samples was measured by dynamic light scattering (DLS-700, Otsuka Electronics) for the sample-redispersed aqueous suspension containing a small amount of sodium dodecyl sulphate as a dispersion stabilizer. The median diameter (number basis) was determined from the obtained size distribution. Fig. 4 indicates the particle size distribution of samples. The samples had a narrow size distribution with a median diameter of about 11 nm. The median size almost agreed with the average crystallite size and the particle size observed by SEM; this means that the crystallinity of samples was high.

Fig. 3. SEM images of samples obtained (a) at 35 rpm in 12 h and (b) at 140 rpm in 6 h.

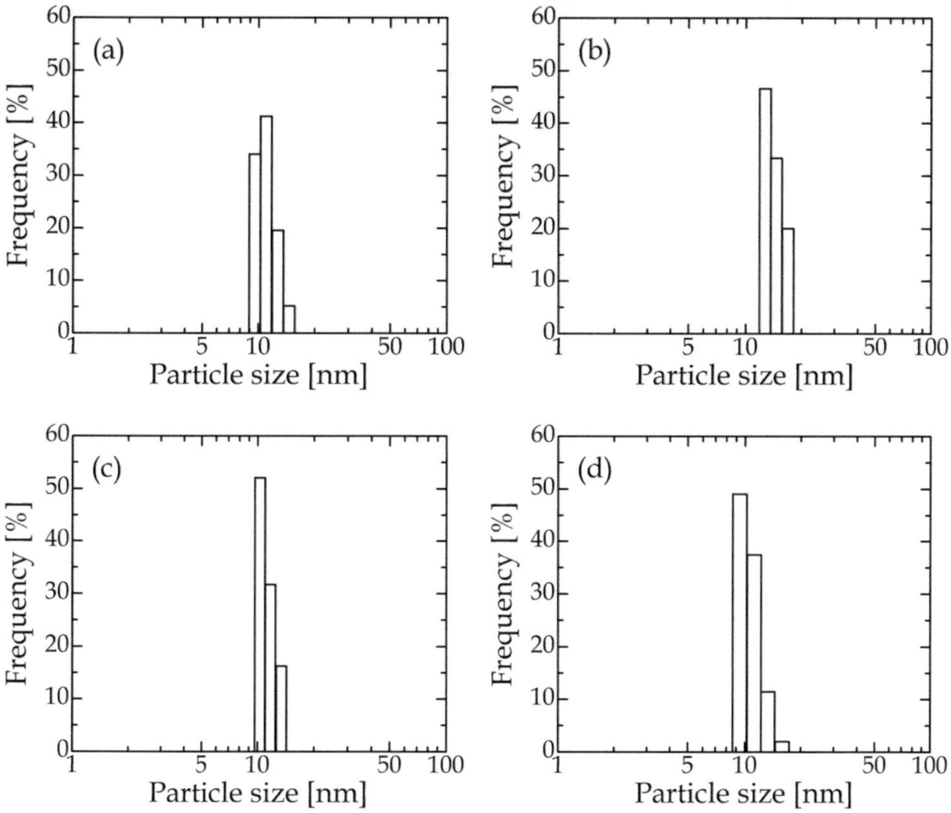

Fig. 4. Particle size distribution of samples obtained (a) at 35 rpm in 12 h, (b) at 70 rpm in 9 h, (c) at 105 rpm in 7.5 h, and (d) at 140 rpm in 6 h.

The magnetic property (magnetization-magnetic field hysteretic cycle) was analyzed using a superconducting quantum interference device (SQUID) magnetometer (Quantum Design model MPMS) at room temperature in the rage of magnetic field between –10 kOe and 10 kOe. Fig. 5 shows the magnetization-magnetic field curve of samples. The samples had a low coercivity, showing superparamagnetism. The saturation magnetization was a little lower than that of the corresponding bulk (= 92 emu/g) because of the smaller size (Lee et al., 1996).

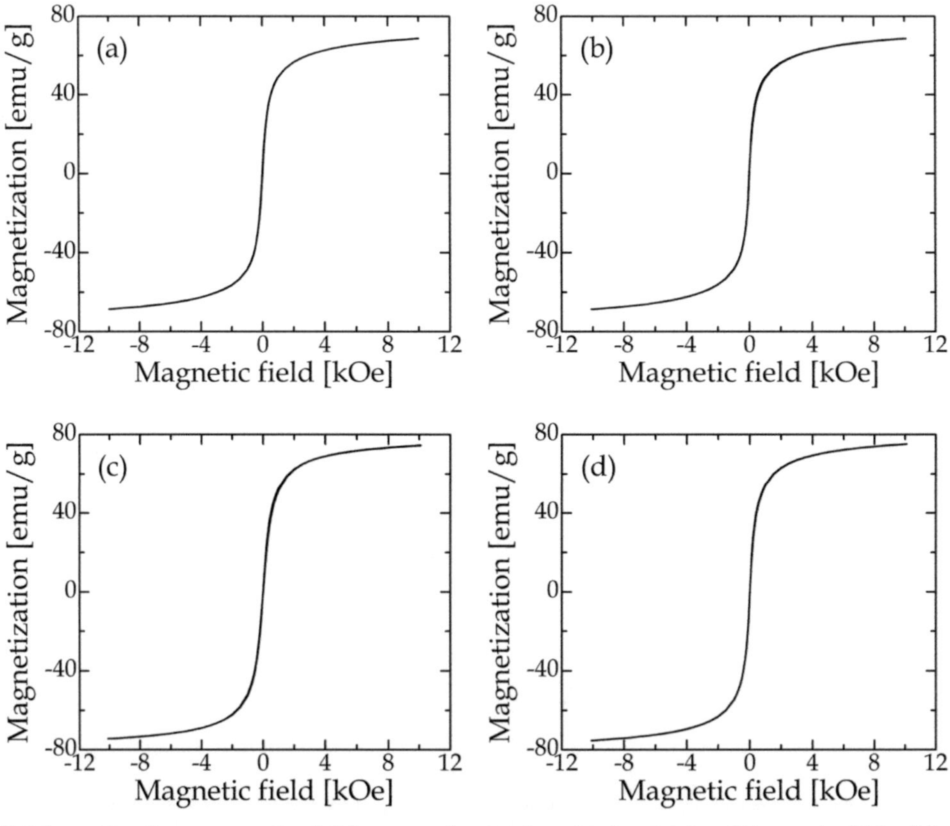

Fig. 5. Magnetization-magnetic field curve of samples obtained (a) at 35 rpm in 12 h, (b) at 70 rpm in 9 h, (c) at 105 rpm in 7.5 h, and (d) at 140 rpm in 6 h.

Table 1 summarizes the properties. The zeta potential was measured with a zeta potential analyzer (Zetasizer Nano ZS, Malvern Instruments). As can be seen in Table 1, the Fe_3O_4 nanoparticles with similar properties were obtained regardless of the rotational speed. The zeta potential of samples was relatively high with adding neither anti-aggregation agents nor organic solvents in the synthesis and almost the same even when the mechanical energy applied to the suspension per unit time was varied. This reveals that the applied mechanical energy hardly affected the dispersibility of Fe_3O_4 nanoparticles.

In this method, the sample contamination caused by the wear of milling pot and balls is concerned. Therefore, the chemical component of samples was determined by means of an energy dispersive X-ray spectrometer (EDS; JED-2300F, JEOL) equipped with the FE-SEM.

Table 2 gives the impurity content in the samples. The sample contamination had a tendency to increase at higher rotational speeds. However, the use of fluorocarbon resin-lined pot and carbon steel balls can decrease the incorporation of Cr and Ni into the product even at higher rotational speeds.

Milling conditions				
Rotational speed [rpm]	35	70	105	140
Milling time [h]	12	9	7.5	6
Average crystallite size [nm]	11.2	11.3	10.8	11.9
Lattice constant [Å]	8.395	8.385	8.392	8.374
Median size [nm]	9.5	10.9	10.3	8.8
Saturation magnetization [emu/g]	68.7	68.6	74.6	75.2
Coercivity [Oe]	9	7	2	1
Zeta potential [mV]	–16.3	–15.6	–20.6	-18.1

Table 1. Properties of samples obtained under various conditions.

Milling conditions				
Rotational speed [rpm]	35	70	105	140
Milling time [h]	12	9	7.5	6
Cr [wt.%]	0.39	0.62	1.26	1.21
Ni [wt.%]	0.10	0.37	0.61	0.69
Na [wt.%]	0.53	0.19	0.25	0.32
S [wt.%]	0.08	0.03	0.03	0.03
Cl [wt.%]	0.15	0.02	0.02	0.04

Table 2. Impurity content in samples.

Generally, preparation of nanoparticles with a size of less than 20 nm is very difficult by means of grinding techniques using ball mills and bead mills (i.e., break-down methods) even under wet conditions. In particular, tumbling mills are impossible to provide such nanoparticles by grinding because the mechanical energy generated in tumbling mills is too low to reach a particle size less than several micrometers. Accordingly, the mechanical energy is used not for the grinding of coarse (grown) particles but for the Fe_3O_4 formation reaction; the Fe_3O_4 nanoparticles are prepared by means of a build-up method. Consequently, the obtained results shown above demonstrate that this synthesis method is effective for the production of superparamagnetic Fe_3O_4 nanoparticles with good dispersibility.

3. Analysis of mechanochemical process

3.1 Reaction rate equation in mechanochemical process

For kinetically analyzing this mechanochemical process, the reaction rate equation must be derived. Therefore, the change in the concentration of the starting materials with the reaction time is required. In this investigation, the concentration of the starting materials was approximately estimated from the XRD data. As mentioned earlier, in this mechanochemical process, the formation and conversion of α-FeOOH plays an important role rather than $Fe(OH)_2$. Accordingly, the temporal change in the content of α-FeOOH in the product was focused on. In order to estimate the content of α-FeOOH from the XRD data, the relationship between the content of α-FeOOH and the diffraction intensity is needed as an analytical curve. Therefore, virtual products with various compositions of α-FeOOH and Fe_3O_4, expressed by the molar ratio γ of α-FeOOH to the α-FeOOH–Fe_3O_4 mixture, were artificially prepared by mixing appropriate amounts of α-FeOOH and Fe_3O_4. The XRD analysis for the virtual products was conducted, and the diffraction intensity I at $2\theta = 21.2°$ corresponding to the (011) plane of α-FeOOH was measured. The intensity ratio ξ was calculated, defined by

$$\xi = (I - I_m)/(I_g - I_m) \tag{5}$$

where I_g and I_m indicate the peak intensity at $2\theta = 21.2°$ for α-FeOOH and Fe_3O_4, respectively. Fig. 6 shows the intensity ratio ξ as a function of the molar ratio γ. Using this analytical curve, the concentration C_g of α-FeOOH in the suspension was estimated.

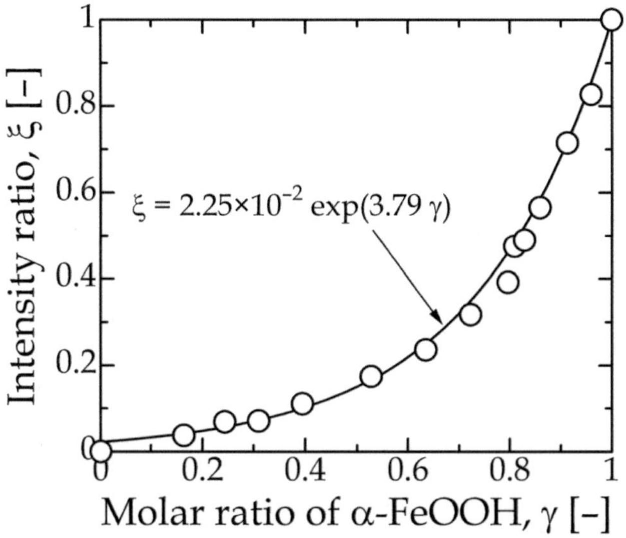

Fig. 6. Relationship between intensity ratio ξ and molar ratio γ of α-FeOOH.

Fig. 7 shows the change in C_g with the milling time t_m at the rotational speed of 140 rpm as an example. C_g increased immediately after the ball-milling treatment began and then decreased with the milling time. The similar tendency was observed at other rotational speeds. The highest C_g observed in 1800 s almost agreed with the concentration of total iron in the starting solution ($\approx$ 4.5 mmol/90 ml = 50 mol/m^3). This implies that the total iron in the starting solution may be converted to α-FeOOH at an initial stage of the ball-milling treatment, and after that, Fe_3O_4 may form from α-FeOOH by the mechanochemical effect. Thus, the formation path of Fe_3O_4 in this mechanochemical process can differ from that in conventional coprecipitation processes.

From these results shown above, a possible reaction mechanism of Fe_3O_4 formation can be constructed as follows. At an initial stage of the ball-milling treatment, amorphous phases of $Fe(OH)_2$ and Fe_3O_4 in the staring suspension are rapidly oxidized and hydrolyzed by the mechanochemical effect according to Eqs. (6) and (7), respectively, resulting to the formation of single α-FeOOH phase.

$$Fe(OH)_2 \rightarrow \alpha\text{-}FeOOH + ½H_2 \quad (6)$$

$$Fe_3O_4 + 2H_2O \rightarrow 2\alpha\text{-}FeOOH + Fe(OH)_2 \quad (7)$$

After that, α-FeOOH is partially reduced to $Fe(OH)_2$ by the generated H_2 gas, and the formed $Fe(OH)_2$ reacts immediately with α-FeOOH, resulting to the formation of Fe_3O_4.

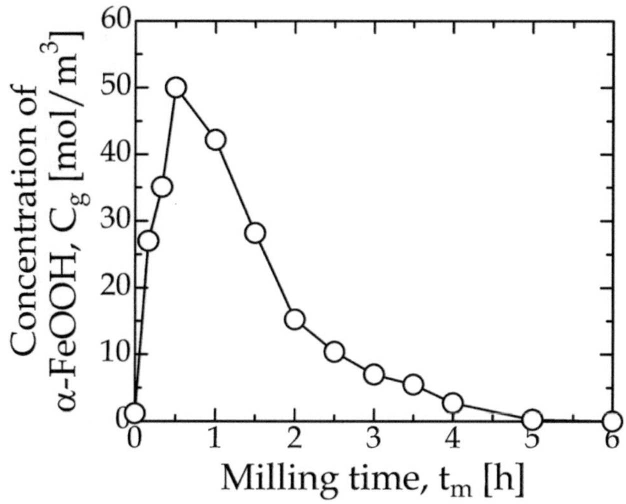

Fig. 7. Change in concentration C_g of α-FeOOH with milling time t_m at rotational speed of 140 rpm.

$$\alpha\text{-}FeOOH + ½H_2 \rightarrow Fe(OH)_2 \quad (8)$$

$$2\alpha\text{-}FeOOH + Fe(OH)_2 \rightarrow Fe_3O_4 + 2H_2O \quad (4)$$

From these reaction equations, the overall apparent equation for the formation of Fe_3O_4 can be expressed by

$$3\alpha\text{-}FeOOH + ½H_2 \rightarrow Fe_3O_4 + 2H_2O \quad (9)$$

Consequently, in this analysis, it can be assumed that the single α-FeOOH phase is gradually converted to Fe_3O_4 after the milling time of 1800 s according to Eq.(9). Here, the reaction time t was newly defined as

$$t = t_m - 1800 \quad (10)$$

Fig. 8 shows the change in the concentration ratio C_g/C_0 with the reaction time t, where C_0 indicates the initial concentration of α-FeOOH at t = 0. C_g/C_0 decreased exponentially with t, suggesting that the conversion reaction from α-FeOOH to Fe_3O_4 may be described by the n-th order rate equation. In this case, the reaction rate equation is expressed by

$$dC_g/dt = -kC_g^n \quad (11)$$

where k and n are the rate constant and the order of reaction, respectively. By solving this differential equation using the boundary condition, $C_g = C_0$ at $t = 0$, the concentration ratio C_g/C_0 is expressed by

$$C_g/C_0 = \{1+(n-1)kC_0^{n-1}t\}^{1/(1-n)} \tag{12}$$

As shown in Fig. 8, the data of the concentration of α-FeOOH against the reaction time were fitted to Eq. (12), and the values of n and k were determined. Fig. 9 shows n and k as a function of the rotational speed, respectively. n was kept almost constant, about 0.6, regardless of the rotational speed. Accordingly, the formation reaction of Fe_3O_4 expressed by Eq. (9) may be described by the 0.6th-order rate equation. On the other hand, k increased with increasing in the rotational speed. At higher rotational speeds, higher mechanical energy is generated per unit time in the ball-milling treatment. Thus, this phenomenon relating to the increase in k in this mechanochemical process is analogous to an increase in the reaction rate at higher temperatures, at which greater amounts of heat energy are given to the system.

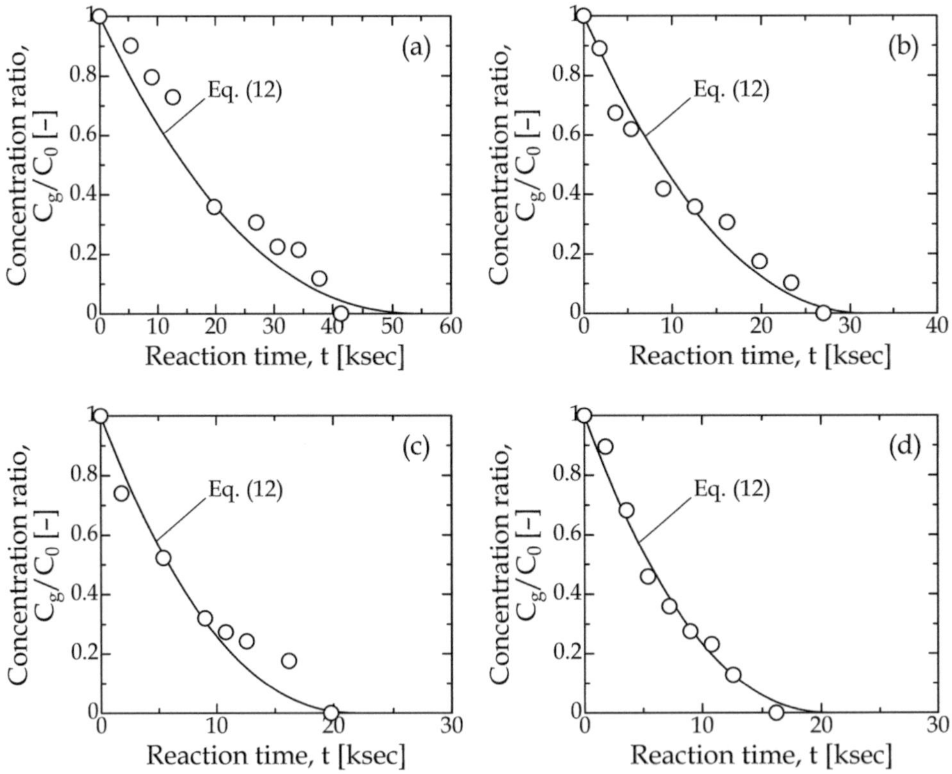

Fig. 8. Change in concentration ratio C_g/C_0 with reaction time t at rotational speeds of (a) 35 rpm, (b) 70 rpm, (c) 105 rpm, and (d) 140 rpm.

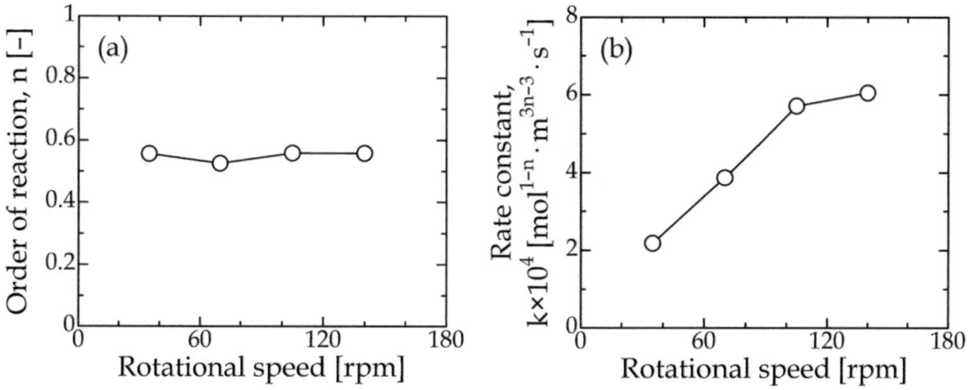

Fig. 9. Changes in (a) order of reaction n and (b) rate constant k with rotational speed.

Fig. 10 shows the effect of the initial iron concentration C_0 on the reaction rate. There was no noticeable difference among the results. Fig. 11 shows the change in n and k with C_0. n and k were almost constant against the variation in C_0. In this mechanochemical process, the order of reaction was relatively small. This means that the concentration dependence of the reaction rate is low in this system where the reaction proceeds using the mechanical energy, unlike those using the heat energy. Generally, in liquid-phase reaction processes using heating treatments, the reaction solution is heated overall and the reaction proceeds everywhere in the solution. Thus, higher concentration of staring materials tends to lead to faster reaction rate. On the other hand, in this mechanochemical process, the mechanical energy promoting the reaction is applied to the suspension on the impact points of the balls, which distribute discretely in the milling pot. Thus, even when the concentration is high, the suspension which can receive the mechanical energy is limited. This makes the effect of the initial concentration on the reaction rate small.

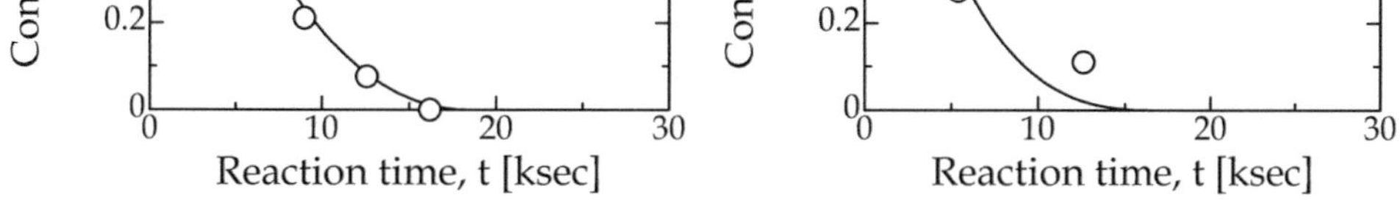

Fig. 10. Effect of initial concentration C_0 on reaction rate at 140 rpm: (a) $C_0 = 25\ mol/m^3$ and (b) $C_0 = 12.5\ mol/m^3$.

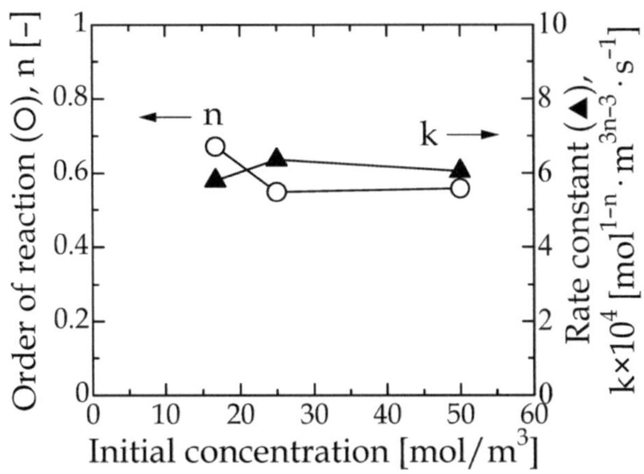

Fig. 11. Changes in n and k with initial concentration C_0.

3.2 Mechanical energy generated in mechanochemical process

In order to analyze in more details this mechanochemical process, the mechanical energy generated by collision of the balls, i.e., the impact energy of balls, was numerically analyzed by simulating the behavior of balls in the milling pot by means of the discrete element method (DEM). Based on the analysis results, the contribution of the mechanical energy to the Fe_3O_4 formation reaction was investigated, and the reaction mechanism in this system was analyzed.

For calculating the impact energy of balls, the behavior of balls in the milling pot under wet condition was simulated using the three-dimensional DEM. This simulation model describes the motion of each ball based on Newton's second law for individual ball, allowing for the external forces acting on the ball (Cundall and Strack, 1979). In this model, the interaction between ball and atmosphere gas (argon) was neglected because the contact force acting on the colliding balls is much stronger than the drag force acting on the balls in the translational motion. The fundamental equations of translational and rotational motions of a ball are expressed as follows:

$$d^2\mathbf{X}/dt^2 = (\mathbf{F}/M)+\mathbf{g} \tag{13}$$

$$d\boldsymbol{\omega}/dt = \mathbf{T}/I_b \tag{14}$$

where $\mathbf{X}$, M, $\boldsymbol{\omega}$, and I_b are mass, position, inertia moment, and angular velocity of a ball, respectively. t, $\mathbf{g}$, $\mathbf{F}$, and $\mathbf{T}$ are time, gravity acceleration, contact force, and torque caused by the tangential contact force, respectively. $\mathbf{X}$ and $\boldsymbol{\omega}$ were calculated by integrating Eqs. (13) and (14) with respect to time between t and t+Δt.

For estimating the contact force acting on a ball, the Hertz-Mindlin contact model was used. The contact forces of normal and tangential directions, $\mathbf{F}_n$ and $\mathbf{F}_t$, were estimated using the following equations:

$$\mathbf{F}_n = (-\kappa_n \boldsymbol{\delta}_n^{3/2} - \eta_n \mathbf{v}_n \cdot \mathbf{n})\mathbf{n} \tag{15}$$

$$\mathbf{F}_t = -\kappa_t \boldsymbol{\delta}_t - \eta_t \mathbf{v}_t \quad (\text{when } |\mathbf{F}_t| \le \mu |\mathbf{F}_n|) \tag{16}$$

$$\mathbf{F}_t = -\mu |\mathbf{F}_n| (\mathbf{v}_t / |\mathbf{v}_t|) \quad (\text{when } |\mathbf{F}_t| > \mu |\mathbf{F}_n|) \tag{17}$$

$$\mathbf{v}_t = \mathbf{v} - (\mathbf{v} \cdot \mathbf{n})\mathbf{n} + r(\boldsymbol{\omega}_i + \boldsymbol{\omega}_j) \times \mathbf{n} \tag{18}$$

where $\boldsymbol{\delta}$, $\mathbf{v}$, κ, η, μ, r, and $\mathbf{n}$ are overlap displacement between contacting balls, relative velocity of contacting balls, stiffness, damping coefficient, sliding friction coefficient, radius of a ball, and unit vector of normal direction at a contact point, respectively. The subscripts n and t mean the components of normal and tangential directions at a contact point, respectively. The subscripts i and j indicate the number of contacting balls. κ and η are determined from the following equations (Tsuji et al., 1992).

Stiffness for the ball-to-ball collision:

$$\kappa_n = (2r)^{1/2} Y_b / [3(1-\sigma_b^2)] \tag{19}$$

$$\kappa_t = 2(2r)^{1/2} Y_b \boldsymbol{\delta}_n^{1/2} / [2(1+\sigma_b)(2-\sigma_b)] \tag{20}$$

Stiffness for the ball-to-pot wall collision:

$$\kappa_n = (4/3) r^{1/2} / [(1-\sigma_b^2)/Y_b + (1-\sigma_w^2)/Y_w] \tag{21}$$

$$\kappa_t = 8 r^{1/2} Y_b \boldsymbol{\delta}_n^{1/2} / [2(1+\sigma_b)(2-\sigma_b)] \tag{22}$$

Damping coefficient:

$$\eta_n = \eta_t = \alpha (M\kappa_n)^{1/2} \boldsymbol{\delta}_n^{1/4} \tag{23}$$

where Y and σ are Young's modulus and Poisson's ratio, respectively. The subscripts b and w indicate ball and pot wall, respectively. α is the constant depending on the restitution coefficient and was determined to be 0.20 based on the experimental value of restitution coefficient (= 0.75) according to the method proposed by Tsuji et al. (1992). The simulation parameters are summarized in Table 3. The sliding friction coefficient used in the calculation was determined based on the critical rotational speed measured experimentally under wet condition (Gudin et al., 2007).

The impact energy of each ball in a single collision, E_i, was defined as the kinetic energy of a ball contacting with another ball or the pot wall (Kano et al., 2000).

$$E_i = (1/2) M |\mathbf{v}|^2 \tag{24}$$

Here, $\mathbf{v}$ is the relative velocity of a ball at the moment when contacting. The ball behavior was simulated using the calculation parameters presented in Table 3, and the impact energy of balls was determined. The balls collide with each other at various relative velocities during ball-milling, resulting in the generation of various amounts of the impact energy. Thus, the impact energy of balls is distributed, as shown in Fig. 12. The distribution of the impact energy shifted to a high energy range with increasing in the rotational speed because of vigorous motion of the balls at higher rotational speeds.

Next, the impact energy of balls per unit time, E, was defined as the total kinetic energy of contacting balls.

$$E = \sum^{f} (1/2)M|\mathbf{v}|^2 \quad (25)$$

Number of balls	7218
Ball diameter	3.2 mm
Pot diameter	90 mm
Pot depth	80 mm
Ball filling ratio to pot capacity	40%
Density of ball and pot	7930 kg/m^3
Young's modulus of ball and pot	197 GPa
Poisson's ratio of ball and pot	0.30
Sliding friction coefficient	0.23
Time step	1.0 μs

Table 3. Simulation parameters used in calculations.

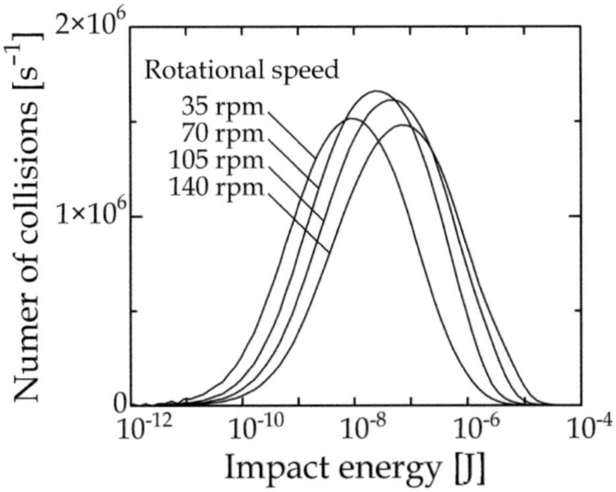

Fig. 12. Effect of rotational speed on impact energy distribution.

f is the average number of contact points per unit time. Eq.(25) means that the kinetic energy of balls just before contacting (i.e., at $|\boldsymbol{\delta}| = 0$) is summed up for all the contact points within unit time but the kinetic energy during contacting (i.e., within $|\boldsymbol{\delta}| > 0$) is not calculated. Therefore, the impact energy thus defined corresponds to the maximum kinetic energy of balls when colliding. Fig. 13 shows the calculation result of the impact energy per unit time. The impact energy was approximately proportional to the rotational speed. This implies that the impact energy depends on the number of revolutions of the pot per unit time rather than the kinetic energy of the rotational motion of the pot. The impact energy considerably varied depending on the rotational speed, resulting in variation of the milling time required for completing the Fe_3O_4 formation reaction. However, the properties of obtained Fe_3O_4 nanoparticles were almost the same even when the rotational speed was varied as shown above. This reveals that the grinding of Fe_3O_4 nanoparticles do not occur in this process and that the impact energy greatly influences the formation process of Fe_3O_4 nanoparticles (in particular, the reaction rate) rather than the properties of products.

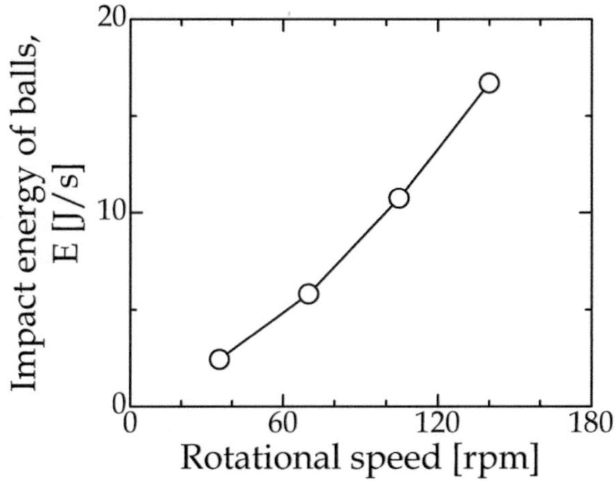

Fig. 13. Change in impact energy of balls with rotational speed.

Fig. 14 shows the relationship between the rate constant k and the impact energy E. k increased with increasing in E. This result reveals that the reaction rate increases under high mechanical energy fields. As mentioned above, this is analogous to increase of the reaction rate caused by temperature rise of the system, i.e., increase of the heat energy given to the system. As can be seen in Fig. 14, the value of k at the rotational speed of 140 rpm was not so large while relatively great amount of mechanical energy applied to the suspension. This suggests that the impact energy was not effectively used for progress of the reaction at high rotational speeds.

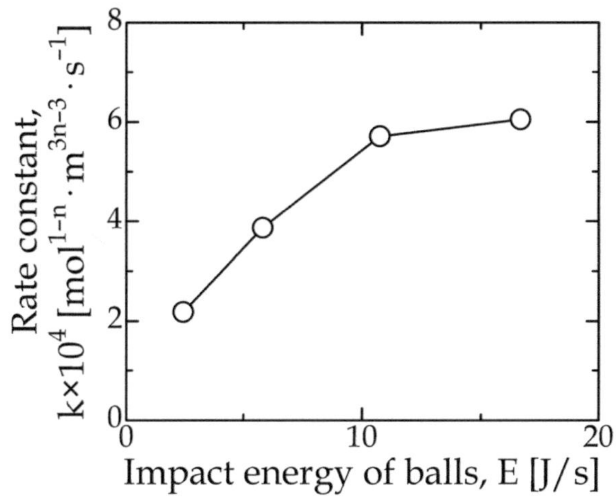

Fig. 14. Relationship between rate constant k and impact energy E of balls per unit time.

3.3 Mechanical energy required for Fe_3O_4 formation reaction

The Fe_3O_4 formation reaction can occur when the impact energy exceeding a threshold value (corresponding to the activation energy) applies to α-FeOOH at the contact points. Smaller impact energy than the threshold value cannot promote the reaction. Assuming that the region where the impact energy applies is extremely small, even though the impact energy larger than the threshold value generates at the contact point, α-FeOOH far away from there may not react using the surplus energy. Thus, it is considered that progress of the reaction at the contact points depends on whether the impact energy exceeding the threshold value is given to α-FeOOH or not, and that the surplus energy cannot promote the reaction. Accordingly, the reaction rate can be proportional to the number of collisions that the reaction occurred. The relationship between the rate constant k and the number n_t of collisions per unit time with the energy exceeding predetermined threshold values is shown in Fig. 15. As can be seen in Fig. 15, the correlation between k and n_t was expressed by

$$k = a n_t^b \tag{26}$$

where a and b are the coefficients depending on the threshold value. Fig. 16 shows the relationship between the predetermined threshold value and the coefficient b. As indicated in Eq. (26), when b = 1, k is proportional to n_t. From Fig. 16, the threshold value giving b = 1 was determined to be 93.7 nJ. The analysis result reveals that the Fe_3O_4 formation reaction occurs at the contact points where the impact energy exceeding 93.7 nJ generates, regardless of the rotational speed. Consequently, this threshold value may be closely related to the activation energy in this reaction system. Fig. 17 shows the number of collisions per unit time that the impact energy more than 93.7 nJ generates as a function of the rotational speed. The number of collisions increased with increasing in the rotational speed.

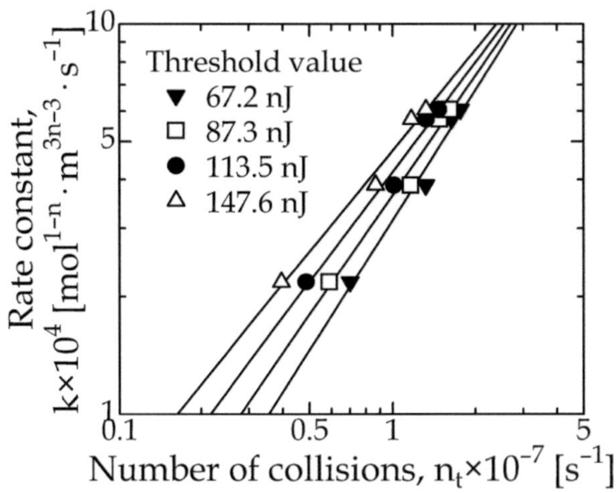

Fig. 15. Relationship between rate constant k and number n_t of collisions of balls per unit time for various threshold values of impact energy.

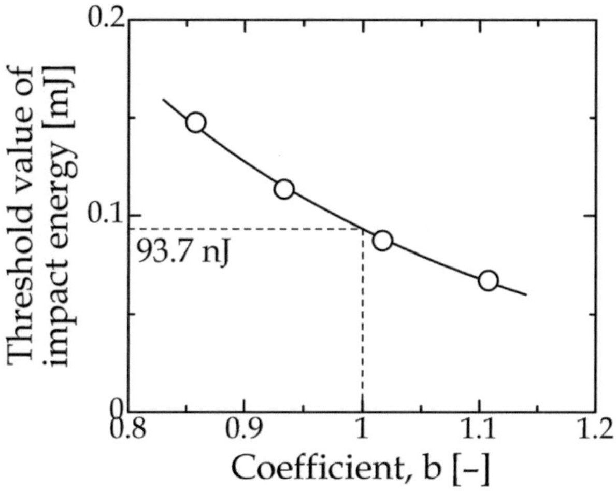

Fig. 16. Relationship between threshold value of impact energy and coefficient b.

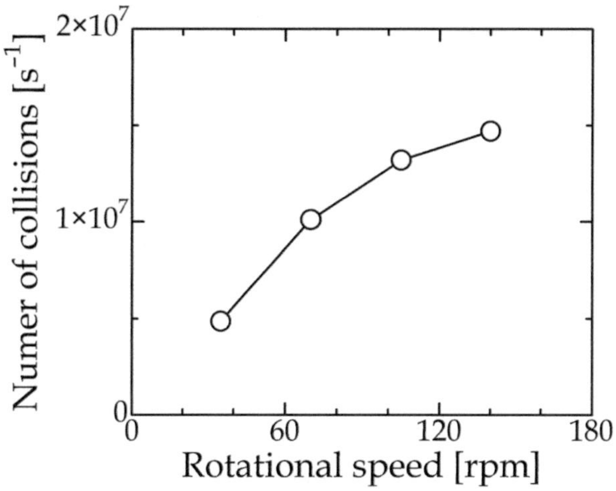

Fig. 17. Change in number of collisions generating impact energy more than 93.7 nJ with rotational speed.

This threshold value of the impact energy in a single collision, the number of collisions shown in Fig. 17, and the completion time of the Fe_3O_4 formation reaction shown in Fig. 8 were used to estimate the accumulative impact energy required for completing the Fe_3O_4 formation reaction. Fig. 18 shows the variation in the accumulative impact energy with the rotational speed. It was found that the accumulative impact energy was almost constant regardless of the rotational speed. Because the accumulative impact energy is the mechanical energy required for synthesizing 1.5 mmol of Fe_3O_4, the accumulative impact energy per 1 mol of Fe_3O_4 is defined as the apparent activation energy. Using the average of the accumulative impact energy, the apparent activation energy was determined to be 15.6 MJ/mol, which was independently of the rotational speed. This result is also analogous to the fact that generally the activation energy is independently of the reaction temperature in reaction systems using heat energy. However, the apparent activation energy thus determined was considerably larger than the activation energy in conventional liquid-phase reaction systems because the impact energy was the maximum mechanical energy which α-FeOOH is able to receive. When both the net mechanical energy transferred from the balls to α-FeOOH and the amount of α-FeOOH receiving the energy are known, the true activation energy of the mechanochemical reaction can be determined.

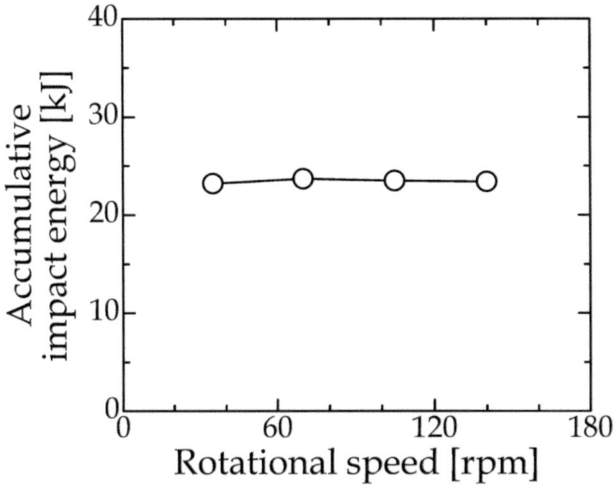

Fig. 18. Variation in accumulative impact energy required for completing Fe_3O_4 formation reaction with rotational speed.

In general mechanochemical processes, when the mechanical energy caused by shear, compression and friction actions of balls applies to particulate materials under dry condition, the surface energy of particles can increase due to the physical change such as distortion of the crystal lattice, increase of the surface area, and appearance of newly formed crystal surface. This causes the mechanochemical activation of particles. Under wet condition, however, the particles are difficult to undergo the mechanochemical activation because the increased surface energy is reduced by the solvent. Furthermore, it is very difficult to apply the mechanical energy effectively to nanoparticles. Accordingly, in this synthesis process, the mechanochemical activation of nanoparticles is difficult to occur, and the solid phase reaction from α-FeOOH to Fe_3O_4 may hardly proceed by direct contribution of the mechanical energy. However, at the rotational speed of 35 rpm, i.e. in a low mechanical energy field, the Fe_3O_4 formation reaction surely proceeded while the reaction rate was relatively low. There is no doubt that the applied mechanical energy promotes the reaction; the reaction mechanism in this synthesis process is considered that the reaction may proceed not by the mechanochemical activation of α-FeOOH but by local and rapid heating and/or through a different reaction path. In the conventional methods for synthesizing Fe_3O_4 nanoparticles in water system, for promoting the formation reaction and increasing the crystallinity, the starting suspension in the vessel is heated from the outside by conductive heat transfer, resulting in temperature rise of the whole system. The heating is continued to keep the reaction temperature. This causes aggregation of the precipitates, leading to the growth of Fe_3O_4 nanoparticles. On the contrary, in the synthesis process with the ball-milling treatment, the suspension may be heated in the contact points between balls, which are extremely small regions, and then is cooled immediately because the pot and the balls are cooled enough. Therefore, local temperature rise of the suspension occurs instantaneously, and the heat energy is hardly stored in the system. In addition, the contact points exist discretely in the pot; the discrete heating occurs everywhere. Consequently, the heating type in the synthesis process is the internal heating, which is regarded as a non-uniform heating from a microscopic viewpoint but a uniform heating from a macroscopic one. This inhibits aggregation of the precipitates effectively, and the nucleation frequently occurs rather than the particle growth, resulting in the formation of the ultrafine Fe_3O_4 nanoparticles with high crystallinity and relatively narrow size distribution. Even at low rotational speeds, the Fe_3O_4 nanoparticles with a size of about 10 nm can be formed; this means that the aggregation-inhibition effect is confirmed in low energy fields.

4. Conclusion

A novel process for preparing superparamagnetic Fe_3O_4 nanoparticles with high crystallinity in water system has been developed, in which a cooled tumbling ball mill is used as the reaction field. It has been confirmed that this method provides successfully the Fe_3O_4 nanoparticles having a size of less than 15 nm without using any conventional heating techniques. This mechanochemical process was kinetically analyzed, indicating that the Fe_3O_4 formation reaction obeys the 0.6th-order rate equation. In addition, the mechanical energy (i.e., the impact energy of balls) promoting the Fe_3O_4 formation reaction was also analyzed using the numerical simulation method. The rate constant of the reaction was investigated based on the mechanical energy. As a result, the apparent activation energy of the reaction was estimated. This mechanochemical process may contribute to the production of superparamagnetic Fe_3O_4 nanoparticles under environmentally friendly conditions and be applied to another reaction systems synthesizing functional nanoparticles.

5. References

Buxbaum, G. & Pfaff, G. (2005). *Industrial Inorganic Pigments.* Third edition, Wiley-Vch, ISBN 978-3527303632, Weinheim, Germany.

Buyukhatipoglu, K.; Miller, T.A. & Morss Clyne, A. (2009). Flame synthesis and in vitro biocompatibility assessment of superparamagnetic iron oxide nanoparticles: Cellular uptake, toxicity and proliferation studies. *J. Nanosci. Nanotechnol.*, 9, 6834–6843.

Can, K.; Ozmen, M. & Ersoz, M. (2009). Immobilization of albumin on aminosilane modified superparamagnetic magnetite nanoparticles and its characterization. *Colloids Surf. B*, 71, 154–159.

Cheng, F.-Y.; Su, C.-H.; Yang, Y.-S.; Yeh, C.-S.; Tsai, C.-Y.; Wu, C.-L.; Wu, M.-T. & Shieh, D.-B. (2005). Characterization of aqueous dispersions of Fe_3O_4 nanoparticles and their biomedical applications. *Biomater.*, 26, 729–738.

Compeán-Jasso, M.E.; Ruiz, F.; Martínez, J.R. & Herrera-Gómez, A. (2008). Magnetic properties of magnetite nanoparticles synthesized by forced hydrolysis. *Mater. Lett.*, 62, 4248–4250.

Cundall, P.A. & Strack, O.D.L. (1979). A discrete numerical model for granular assemblies. *Geotechnique*, 29, 47–65.

Gao, F.; Cai, Y.; Zhou, J.; Xie, X.; Ouyang, W.; Zhang, Y.; Wang, X.; Zhang, X.; Wang, X.; Zhao, L. & Tang, J. (2010). Pullulan acetate coated magnetite nanoparticles for hyper-thermia: Preparation, characterization and in vitro experiments. *Nano Res.*, 3, 23–31.

Ge, J.; Hu, Y.; Biasini, M.; Beyermann, W.P. & Yin, Y. (2007). Superparamagnetic magnetite colloidal nanocrystal clusters. *Angew. Chem. Int. Ed.*, 46, 4342–4345.

Gudin, D.; Kano, J. & Saito, F. (2007). Effect of the friction coefficient in the discrete element method simulation on media motion in a wet bead mill. *Adv. Powder Technol.*, 18, 555–565.

Guo, S.; Li, D.; Zhang, L.; Li, J. & Wang, E. (2009). Monodisperse mesoporous superparamagnetic single-crystal magnetite nanoparticles for drug delivery. *Biomater.*, 30, 1881–1889.

Hakata, T. (2002). A novel composite carrier for electrophotographic developers. *J. Imaging Sci. Techn.*, 46, 591–597.

Hu, F.; MacRenaris, K.W.; Waters, E.A.; Schultz-Sikma, E.A.; Eckermann, A.L. & Meade, T.J. (2010). Highly dispersible, superparamagnetic magnetite nanoflowers for magnetic resonance imaging. *Chem. Commun.*, 46, 73–75.

Hua, C.C.; Zakaria, S.; Farahiyan, R.; Khong, L.T.; Nguyen, K.L.; Abdullah, M. & Ahmad, S. (2008). Size-controlled synthesis and characterization of Fe_3O_4 nanoparticles by chemical coprecipitation method. *Sains Malays.*, 37, 389–394.

Iwasaki, T.; Kosaka, K.; Mizutani, N.; Watano, S.; Yanagida, T.; Tanaka, H. & Kawai, T. (2008). Mechanochemical preparation of magnetite nanoparticles by coprecipitation. *Mater. Lett.*, 62, 4155–4157.

Iwasaki, T.; Kosaka, K.; Yabuuchi, T.; Watano, S.; Yanagida, T. & Kawai, T. (2009). Novel mechanochemical process for synthesis of magnetite nanoparticles using coprecipitation method. *Adv. Powder Technol.*, 20, 521–528.

Iwasaki, T.; Kosaka, K.; Watano, S.; Yanagida, T. & Kawai, T. (2010). Novel environmentally friendly synthesis of superparamagnetic magnetite nanoparticles using mechanochemical effect. *Mater. Res. Bull.*, 45, 481–485.

Iwasaki, T.; Mizutani, N.; Watano, S.; Yanagida, T. & Kawai, T. (2011a). Size control of magnetite nanoparticles by organic solvent-free chemical coprecipitation at room temperature. *J. Exp. Nanosci.*, in press.

Iwasaki, T.; Sato, N.; Kosaka, K.; Watano, S.; Yanagida, T. & Kawai, T. (2011b). Direct transformation from goethite to magnetite nanoparticles by mechanochemical reduction. *J. Alloys Compd.*, 509, L34–L37.

Kano, J.; Mio, H. & Saito, F. (2000). Correlation of grinding rate of gibbsite with impact energy of balls. *AIChE J.*, 46, 1694–1697.

Lee, C.M.; Jeong, H.J.; Kim, E.M.; Kim, D.W.; Lim, S.T.; Kim, H.T.; Park, I.K.; Jeong, Y.Y.; Kim, J.W. & Sohn, M.H. (2009). Superparamagnetic iron oxide nanoparticles as a dual imaging probe for targeting hepatocytes in vivo. *Magn. Reson. Med.*, 62, 1440–1446.

Lee, J.; Isobe, T. & Senna, M. (1996). Magnetic properties of ultrafine magnetite particles and their slurries prepared via in-situ precipitation. *Colloids Surf. A*, 109, 121–127.

Li, Y.G.; Gao, H.S.; Li, W.L.; Xing, J.M. & Liu, H.Z. (2009). In situ magnetic separation and immobilization of dibenzothiophene-desulfurizing bacteria. *Bioresource Technol.*, 100, 5092–5096.

Lian, S.; Wang, E.; Kang, Z.; Bai, Y.; Gao, L.; Jiang, M.; Hu, C. & Xu, L. (2004). Synthesis of magnetite nanorods and porous hematite nanorods. *Solid State Commun.*, 129, 485–490.

Mizutani, N.; Iwasaki, T.; Watano, S.; Yanagida, T.; Tanaka, H. & Kawai, T. (2008). Effect of the ferrous/ferric ions molar ratio on the reaction mechanism for the hydrothermal synthesis of magnetite nanoparticles. *Bull. Mater. Sci.*, 31, 713–717.

Mizutani, N.; Iwasaki, T.; Watano, S.; Yanagida, T. & Kawai, T. (2010). Size control of magnetite nanoparticles in hydrothermal synthesis by coexistence of lactate and sulfate ions. *Curr. Appl. Phys.*, 10, 801–806.

Muzquiz-Ramos, E.M.; Cortes-Hernandez, D.A. & Escobedo-Bocardo, J. (2010). Biomimetic apatite coating on magnetite particles. *Mater. Lett.*, 64, 1117–1119.

Ochiai, M.; Masui, M.; Asanae, M.; Tokunaga, M. & Iimura, T. (1994). Magnetic toner prepared by the suspension polymerization method. *J. Imaging Sci. Techn.*, 38, 415–420.

Tsuji, Y.; Tanaka, T. & Ishida, T. (1992). Lagrangian numerical simulation of plug flow of cohesionless particles in a horizontal pipe. *Powder Technol.*, 71, 239–250.

Wan, J.; Tang, G. & Qian, Y. (2007). Room temperature synthesis of single-crystal Fe_3O_4 nanoparticles with superparamagnetic property. *Appl. Phys. A*, 86, 261–264.

Wang, S.; Bao, H.; Yang, P. & Chen, G. (2008). Immobilization of trypsin in polyaniline-coated nano-Fe_3O_4/carbon nanotube composite for protein digestion. *Anal. Chim. Acta*, 612, 182–189.

Wen, X.; Yang, J.; He, B. & Gu, Z. (2008). Preparation of monodisperse magnetite nanoparticles under mild conditions. *Curr. Appl. Phys.*, 8, 535–541.

Wu, J.-H.; Ko, S.P.; Liu, H.-L.; Kim, S.; Ju, J.-S. & Kim, Y.K. (2007). Sub 5 nm magnetite nanoparticles: Synthesis, microstructure, and magnetic properties. *Mater. Lett.*, 61, 3124–3129.

Wu, J.H.; Ko, S.P.; Liu, H.L.; Jung, M.-H.; Lee, J.H.; Ju, J.-S. & Kim, Y.K. (2008). Sub 5 nm Fe_3O_4 nanocrystals via coprecipitation method. *Colloids Surf. A*, 313-314, 268–272.

Yu, S.; Wan, J.; Yu, X. & Chen, K. (2010). Preparation and characterization of hydrophobic magnetite microspheres by a simple solvothermal method. *J. Phys. Chem. Solids*, 71, 412–415.

Yang, D.P.; Gao, F.; Cui, D.X. & Yang, M. (2009). Microwave rapid synthesis of nanoporous Fe_3O_4 magnetic microspheres. *Curr. Nanosci.*, 5, 485–488.

Zheng, Y.-H.; Cheng, Y.; Bao, F. & Wang, Y.-S. (2006). Synthesis and magnetic properties of Fe_3O_4 nanoparticles. *Mater. Res. Bull.*, 41, 525–529.

Zhou, L.; Yuan, J.; Yuan, W.; Zhou, M.; Wu, S.; Li, Z.; Xing, X. & Shen, D. (2009). Synthesis and characterization of multi-functional hybrid magnetite nanoparticles with biodegradability, superparamagnetism, and fluorescence. *Mater. Lett.*, 63, 1567–1570.

Part 4

Optical Materials

11

Time Resolved Investigation of Fast Phase-Change Phenomena in Rewritable Optical Recording Media

Shigeru Kimura[1], Yoshihito Tanaka[1,2], Shinji Kohara[1] and Masaki Takata[1,2,3]
[1]*Research & Utilization Division, Japan Synchrotron Radiation Research Institute*
[2]*RIKEN SPring-8 Center, RIKEN Harima Institute*
[3]*School of Frontier Science, The University of Tokyo*
Japan

1. Introduction

The worldwide hunger for secure and smarter electronic data storage grows daily to satisfy today's extremely demanding business and entertainment requirements, and this drives the need to pack data into an increasingly smaller volume at ever faster speeds. Digital versatile disks (DVDs) are one of the most convenient storage devices for large amount of information such as video and digital graphs. One of the keys to develop the DVDs was to create the fast phase-change material, which gives rise to faster bi-directional phase change between crystal phase and amorphous phase by laser heating within the recording mark. Another key was to minimize the size of the recording mark for high density data storage. The DVDs have already been a mass product as one of the indispensable home electronic appliances, even though the mechanism of fast phase change has been the remained problem to be solved. Then, to develop the next generation higher density and faster optical data storage system, the atomic and electron level study was required in the last decades. This chapter will describe the challenge of the atomic level research for the fast phase-change mechanism of the DVD materials via X-ray pinpoint structural measurement, which is the state-of-the-art synchrotron radiation (SR) materials-analysis techniques developed at SPring-8. The X-ray pinpoint structural measurement and it's data analysis technique has been further upgraded by combining with theoretical structure modelling and became a tool to investigate how laser pulses alter the atomic structure of a class of materials useful for data storage and uncover a mechanism that could help the development of even faster information storage in the future.

So far, the great efforts and developments have been paid for the progress of the DVD materials (Wutting & Yamada, 2007). The development of new phase-change materials, leading to the discovery of GeTe-Sb_2Te_3 (Yamada et al., 1987) and of $Ag_{11}In_{11}Sb_{55}Te_{23}$ (Iwasaki et al., 1992), has allowed us to produce rewritable compact discs, DVDs, and Blu-ray discs. In the commercial rewritable DVDs such as DVD-random access memory (RAM) and DVD-rewriteable (RW), information is written to the phase-change materials composed of the alloys of Ge, Sb and Te, or those of Ag, In, Sb and Te by changing their phase locally between amorphous and crystalline states using a sub-micrometer sized laser irradiation.

The recording and erasing are performed by controlling the laser intensity for irradiation, yielding a difference in the transient heating and cooling rates of the materials. A higher cooling rate from a temperature above the melting point generates the amorphous phase; a lower rate produces the crystalline phase from the amorphous phase. Because the optical reflectivity of the two states is different, this information can be easily read from the disk by measuring the reflectance with a low power laser. Recently developed DVD materials can complete the phase change with a 20 ns laser irradiation (Fig. 1).

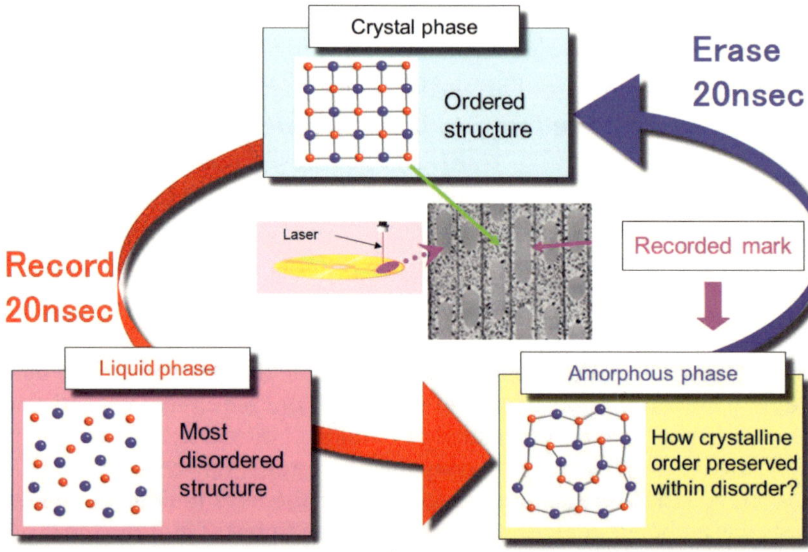

Fig. 1. Schematic illustration of recording and erasing process in rapid phase-change materials.

In contrast to the reliability and high-speed-performance of DVD media, the mechanism of rapid phase change, especially crystallization from the amorphous phase is still not fully understood, in spite of a lot of investigation using transmission electron microscopy (Park et al., 1999; Naito et al., 2004), fluctuation electron microscopy (Kwon et al., 2007), optical (Wei & Gan, 2003), electronic (Lee et al., 2005), and structural (Yamada & Matsunaga, 2000; Kolobov et al., 2004; Kohara et al., 2006) studies. We therefore developed the X-ray pinpoint structural measurement system (Kimura et al., 2006), which enables the 40 picosecond time resolved pump and probe X-ray diffraction experiment using 100 nm scale SR beam. Using the system, we achieved the real-time *in situ* structural observation of the amorphous-crystal phase change process in nanosecond time-scale (Fukuyama et al., 2008a). The obtained results using the time-resolved X-ray diffraction apparatus coupled with *in situ* optical reflectivity monitor showed the strong correlation between the crystal growth and optical reflectivity, and the difference in the crystal growth process between $Ge_2Sb_2Te_5$ (GST) and $Ag_{3.5}In_{3.8}Sb_{75.0}Te_{17.7}$ (AIST).

In the following section, we describe the detail of the X-ray pinpoint structural measurement system for investigation of optical recording process and its performance (Tanaka et al., 2009). Moreover, we describe the recent progress for fully understanding the atomic structure of AIST and compare it to GST through the research with X-ray diffraction, extended X-ray absorption fine structure (EXAFS), hard X-ray photoelectron spectroscopy (HXPS) measurements and computer simulations (Matsunaga et al., 2011).

2. X-ray pinpoint structural measurement system for investigation of optical recording process

The pulse characteristic and high coherent X-ray beam of SPring-8 allow us to investigate dynamics of chemical reactions and phase transition of materials caused by applied field. In order to realize the direct investigation, we developed the X-ray pinpoint structural measurement system at BL40XU of SPring-8, which enables the advanced X-ray measurement technique in nanometer spatial scale and/or picosecond time scale. This system was used not only for investigation of optical recording process but also for charge density level visualization of photo-induced phase transition (Fukuyama et al., 2010) and sub-micron single crystal structure analysis (Yasuda et al., 2008; Yasuda et al., 2010).

In order to investigate the amorphous-crystal phase change on recording DVD media we optimized the X-ray pinpoint structural measurement system. Figure 2 shows the schematic illustration of experimental setup for taking a snapshot of X-ray diffraction profile of optical phase-change materials. The main optimized components are laser-SR timing control system, a sample rotation system, an optical reflectivity probe system, and an X-ray diffractometer. The following sub-sections describe the components and their performance.

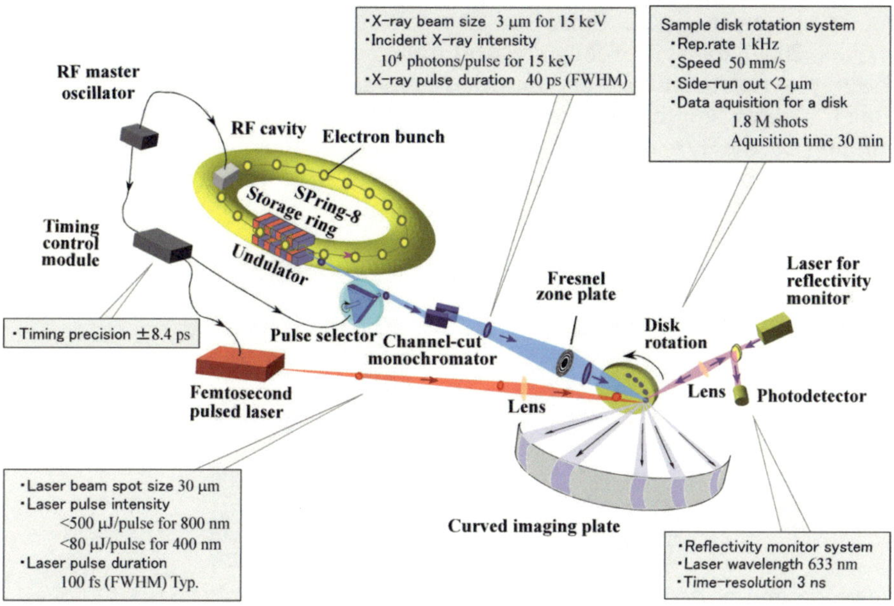

Fig. 2. Experimental setup for taking a snapshot of X-ray diffraction profile of optical phase-change materials.

2.1 Laser-pump and X-ray SR probe system

A pulsed nature of synchrotron X-ray beam due to the acceleration mechanism of charged particles has enabled us to conduct the stroboscopic X-ray measurement in picosecond time-scale. In the stroboscopic X-ray diffraction measurement, the pulsed X-rays from a SR source are diffracted at a sample after a short-pulsed laser synchronized to the SR X-ray pulses illuminates it, as schematically shown in Fig. 3. The time-dependent snapshot can be taken by changing the time delay indicated as τ in Fig. 3.

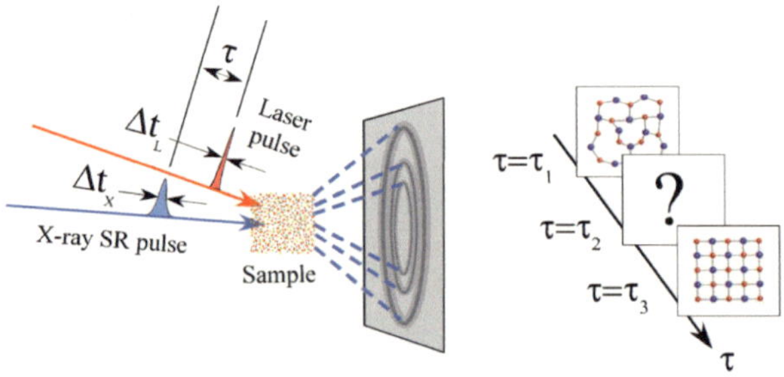

Fig. 3. Schematic illustration of time-resolved X-ray diffraction measurement using a laser-pump X-ray-probe method.

To measure an impulse response of the crystallization, a mode-locked Ti:sapphire laser with a pulse width, Δt_L, of 130 fs was used as a pump laser. A photo of the laser system is shown in Fig. 4. The laser system is equipped with a regenerative amplifier which produces pulses with a pulse energy of 1 mJ, a wavelength of 800 nm, and a repetition rate of 1 kHz. The laser was synchronized to the RF master oscillator of the SR storage ring for acceleration of electrons, as shown in Fig. 2. Since the detailed synchronization system was published (Tanaka ęt al., 2000), the synchronization and timing control system are briefly summarized here. Timing synchronization of the mode-locked laser oscillator was achieved by controlling the cavity length, with the precision of a few ps. The repetition rate is 84.76 MHz, corresponding to the 1/6 of RF of the storage ring master oscillator. Output timing of regeneratively amplified laser pulses is controlled in consideration of the revolving frequency (the round trip is ~5 μs) of the storage ring.

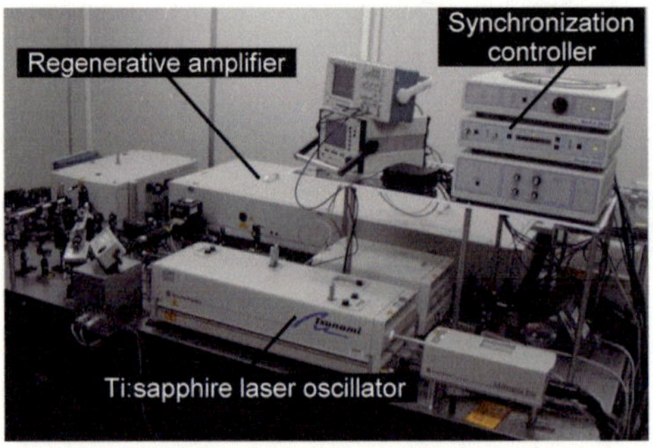

Fig. 4. Femtosecond pulsed laser system. The laser system is mainly composed of a Ti:sapphire laser oscillator and a regenerative amplifier. The regenerative amplifier produces optical pulses with a pulse energy of 1 mJ, a wavelength of 800 nm, and a repetition rate of 1 kHz.

Figure 5 shows a time chart of the laser and the SR pulses in the experiment. The time structure of SR pulses is dependent on the filling pattern of electron bunches in the storage ring. Because the repetition rate of the measurement was ~1 kHz, only one bunch in the storage ring was used for the measurement. The filling pattern was set to be hybrid mode, in which the high current electron bunch (3 mA) exists with lower current bunch train (see Fig. 5(a)). In the filling pattern, the pulse width of SR, which is determined by the electron bunch length was ~40 ps (FWHM).

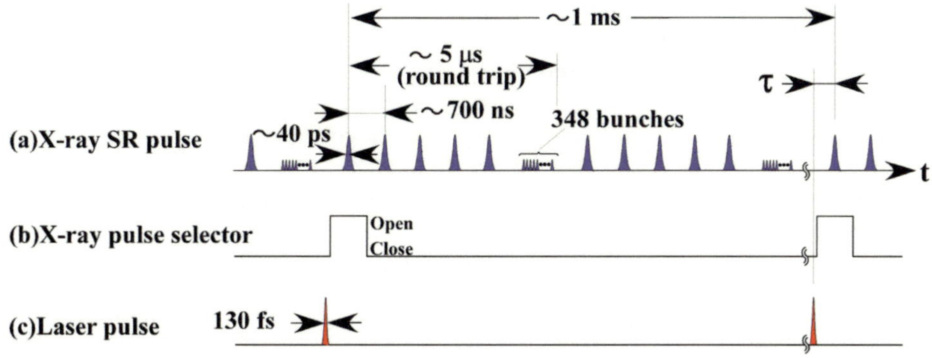

Fig. 5. Time chart of the pulsed laser and X-ray SR irradiation for a snapshot measurement with a time delay of τ.

The control technique of the timing delay, τ, is also a key to make a stroboscopic measurement. As the time-scale of completing the phase change is a microsecond, the large delay is required, which is generally difficult to obtain with an optical delay. Thus, we developed a high precision RF phase control delay circuit to scan the time delay without degrading the precision (Fukuyama et al., 2008b). Figure 6 shows the photograph of the RF phase control circuit, whose timing jitter between the output and the input RF signal is about 3 ps (in rms), which was evaluated with a 16 GHz digital oscilloscope (Tanaka et al., 2010). The timing control precision in the whole system was 8.4 ps (in rms), which was estimated by time-resolved diffraction measurement using a fast lattice response of semiconductor single crystal. Since the distribution of the crossing points is narrower than the pulse width of the SR, the pump-probe system has a time resolution of < 40 ps.

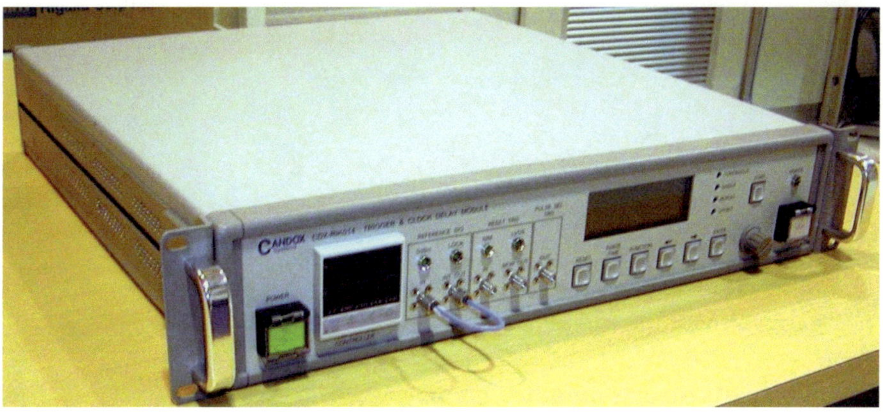

Fig. 6. High precision RF phase control circuit. The circuit provides a delayed RF signal from the output port with a precision of better than 3 ps (in rms) which is independent of the delay time.

2.2 Rotating stage for sample holder

Repetitive measurement is necessary when the diffraction intensity for one shot was too low to obtain the profiles with high time resolution. Therefore the sample was rotated to give a virgin sample for every measurement, as shown in Fig. 2. To perform 1 kHz-repetitive measurement, the rotation speed of the disk and the laser beam size on the sample are adjusted to satisfy the requirement of v (m/s) > f (Hz) • s (m), where the v, f, and s are the moving speed of the disk, the repetition rate, and a beam diameter of the laser or X-rays, respectively. In addition, an X-ray microbeam technique with a phase zone plate was applied to achieve a few micron X-ray beam size, since the smaller beam gives a larger number of data for one disk. The laser beam diameter on the sample was adjusted to be about 30 μm in the moving direction with a 40 mm-focal lens. The parameters in the experiment are summarized in Table 1, together with the parameters for 5 Hz-measurement without X-ray microbeam (Fukuyama et al., 2008a) for comparison.

Repetition rate (Hz)	Disk rotation speed (m/s)	Laser beam size (μm)	X-ray beam size (μm)	Number of repetition for one disk
1×10^3	8×10^{-2}	30	3	1.8×10^6
5	5×10^{-3}	300	50	3×10^4

Table 1. Optical parameters of the time-resolved X-ray diffraction measurement system.

Figure 7 shows the photograph of the DVD sample attached on a rotating stage. The measurement with high repetition rate using microbeams enabled us to obtain a diffraction intensity profile within an hour for one disk sample, and to get systematic data.

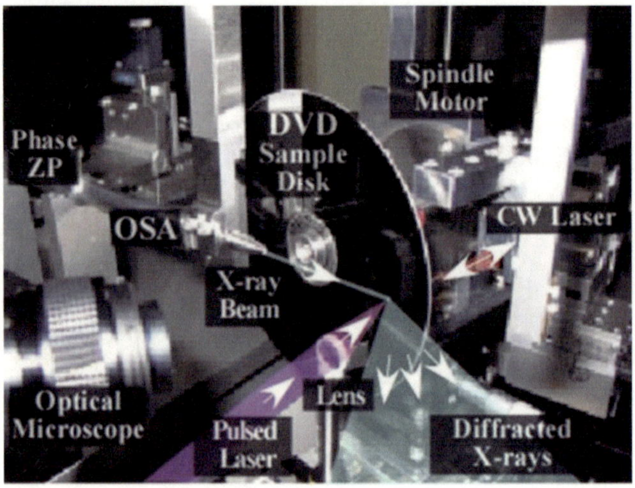

Fig. 7. Photograph around the sample disk.

2.3 Optical reflectivity measurement system

On-line monitor system of the time profile of optical reflectivity was installed to check the phase-change condition of the DVD sample. As shown in Fig. 7, a He-Ne laser beam was guided through the transparent substrate onto the backside of the sample, since the clearance is too small on the right side due to a relatively shorter working distance of the lens for the pump laser. The reflected light intensity was monitored through a beam splitter with a photodiode having a response time of 10 ns, and the signal was stored in a digital oscilloscope.

2.4 High-precision X-ray diffractometer

The intense X-ray beam produced from an undulator is guided to the sample through a double curved mirror system, a pulse selector (rotating shutter), a Si(111) channel-cut monochromator, and a high-precision diffractometer with a phase zone plate (ZP) focusing system (Yasuda et al., 2008), as shown in Fig. 8. The ZP has the diameter of 100 μm, the innermost zone radius of 5.0 μm, the outermost zone width of 250 nm, and the tantalum thickness of 2.5 μm. The photon energy, E, and energy resolution, $\Delta E/E$, tuned by the channel-cut monochromator, were 15 keV, and ~10^{-4}, respectively. The X-ray beam was focused by the ZP, and the 1st order focused beam is only selected with the order sorting aperture placed in front of the sample position. The beam size at the sample position was ~3 μm, evaluated by a knife edge scan method using Au meshes. The focused beam intensity for one pulse at the sample position was ~10^4 photons/pulse, which is ~10^{-11} of the number of photons for a second in the upstream of the pulse selector. As described in Table 1, the total number of photons incident on the sample are then estimated to be ~10^{10} photons for one sample disk.

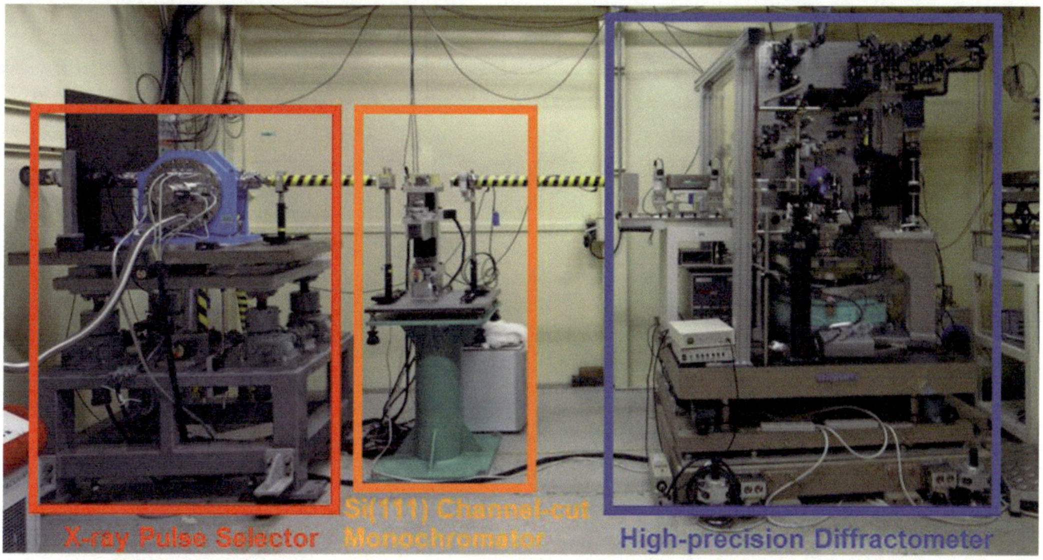

Fig. 8. Photograph of the X-ray pulse selector, the Si(111) channel-cut monochromator and the high-precision diffractometer.

A curved imaging plate (IP) was employed to obtain an X-ray diffraction profile. The use of an IP is advantageous in such longer exposure time as ~30 min in the present experiment. To obtain one-dimensional diffraction profile, the signal intensity was integrated using the full area of IP to compensate the smaller number of incident photons on the sample.

2.5 Performance of the X-ray pinpoint structural measurement system for investigation of optical recording process

Here we briefly summarize the X-ray pinpoint structural measurement system for investigation of optical recording process (see Fig. 2). The time-resolution and time precision are ~40 ps and ±8.4 ps, respectively. The spatial resolution is determined by the area of the laser beam of 30 μm, which is larger than the X-ray beam size of 3 μm. The maximum repetition rate is 1 kHz, in which the fresh sample area on the rotating disk appears at the irradiating position for every measurement. All parameters shown here are described in full width at half maximum.

3. Time-resolved investigation of nanosecond crystal growth in rapid-phase-change materials

Using the X-ray pinpoint structural measurement system described in previous section, we investigated the crystallization process of GST and AIST (Fukuyama et al., 2008a), which are thought to exhibit different crystallization behavior. To reveal the crystallization process of the amorphous phase in these rapid-phase-change materials, it is necessary to know (i) the time constants of both crystallization and optical reflectivity changes, and (ii) crystallization behavior. We therefore employed avalanche photodiode (APD)/multi-channel scaling (MCS) measurement with a time resolution of 3.2 ns coupled with optical reflectivity measurement for (i) and imaging plate (IP)/pump-probe measurement of a time resolution of 40 ps for (ii) as schematically shown in Fig. 9(a). Figure 9(b) shows the experimental method and the time chart for pump-probe measurement using 40 ps X-ray pulses and synchronous femtosecond laser pulses. The DVD samples were prepared by depositing GST and AIST on a SiO_2 glass substrate (12 cm diameter and 0.6 mm thick) with a 2-nm-thick 80 mol% ZnS-20 mol% SiO_2 cap layer. The thickness of the sample was 300 nm, which gave sufficient X-ray diffraction intensities for structure characterization and sufficient stability of the amorphous phase to avoid spontaneous crystallization.

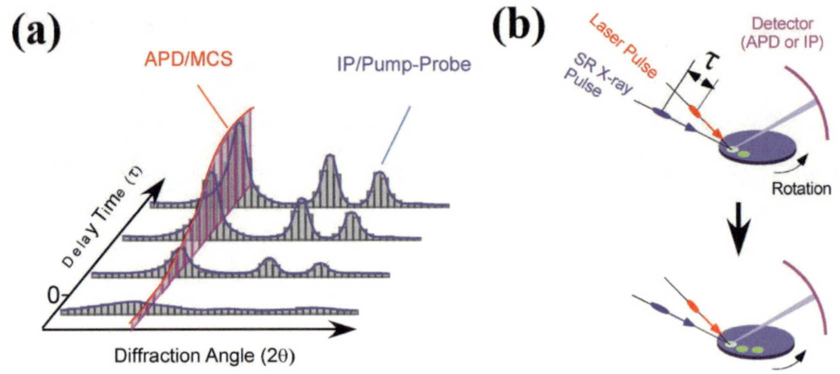

Fig. 9. (a) Schematic diagram of APD/MCS and IP/pump-probe measurement and (b) scheme of DVD rotating system and the time chart of pump-probe measurement.

Figure 10 shows the time-profiles of the X-ray diffraction intensity of Bragg peaks (black and blue lines) together with those of the optical reflectivity profiles (red lines) of (a) GST and (b) AIST, respectively, measured by APD/MCS measurement. These show good accordance, which prove the close relation between the X-ray diffraction intensity and the optical reflectivity of the phase-change materials, *i.e.*, the structure and the electronic properties.

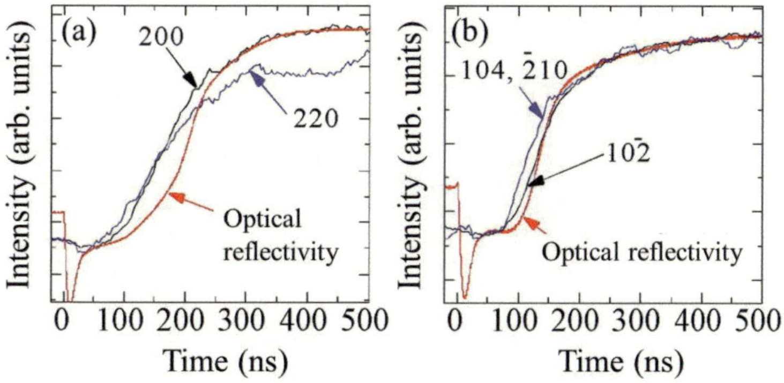

Fig. 10. Time-evolution of the optical reflectivity and X-ray diffraction intensity of (a) GST and (b) AIST.

From the intensity profile measured by the APD/MCS method, the delay times, τ, were determined for the IP/pump-probe method. Figure 11(a) shows the diffraction patterns obtained from the IP/pump-probe method for a 40 ps snapshot. Since the intensity of each diffraction peak has a uniform time-dependent increase, there is no crystal-crystal phase transition in GST and AIST during the crystal growth. We estimated the grain sizes from the line width of Bragg reflection, as shown in Fig. 11(b). In GST, the grain size is nearly constant (=70 nm), while the grain size significantly increases up to 1 μs in AIST. It is remarkable that the volume fraction of the crystal phase is almost saturated at 300 ns (see Fig. 10). Thus these observations suggest the coalescence of the crystal domains after 300 ns.

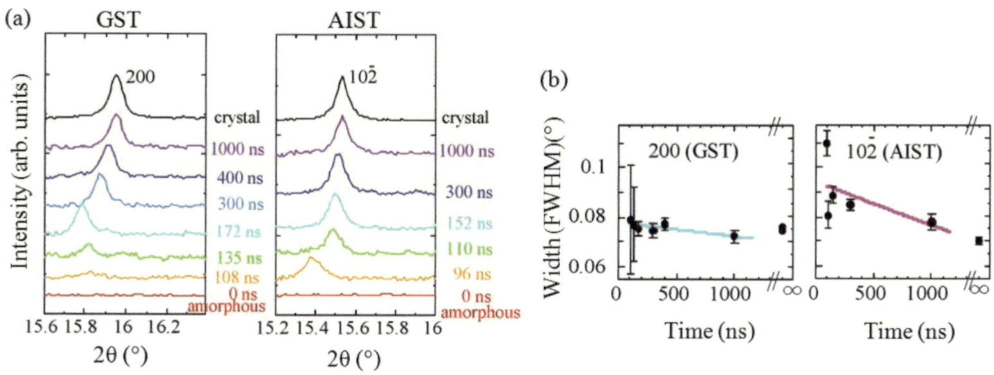

Fig. 11. (a) Snapshots of X-ray diffraction patterns and (b) the changes in peak width.

From the above experimental findings, we propose schematic models for the crystallization processes of GST and AIST as shown in Fig. 12. In the case of GST, nucleation takes place in the whole area in the amorphous phase after laser irradiation (A), and the number of newly formed crystallites of 70 nm diameter increases during the cooling process until 300 ns (B-

C). The crystal growth is then disturbed by the impingement of crystallites with each other (D). Our schematic model is consistent with the nucleation-driven crystallization process (Zaou et al., 2000).

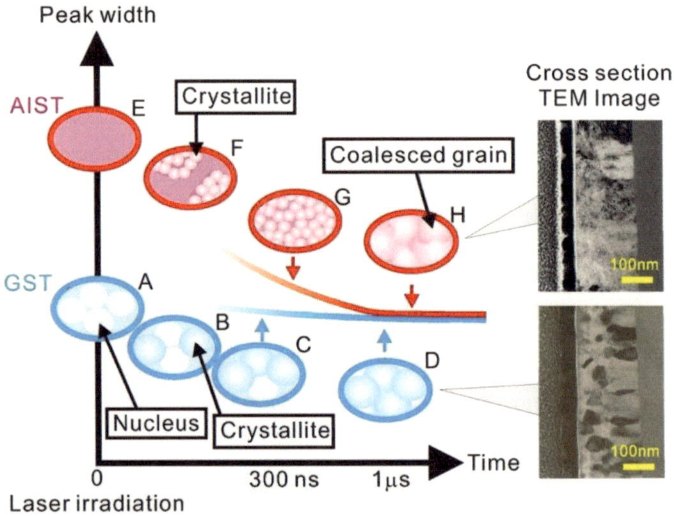

Fig. 12. (a) Schematic models with TEM image for the crystallization processes in GST and AIST.

On the other hand, the nuclei of AIST can immediately transform to smaller crystallites (<< 60 nm), which form domains (F). These domains are enlarged by edge-growth crystallization (F-G) and the crystallites coalesce at the final stage (300 ns~) of crystallization in AIST (G-H). These proposed schematic models are consistent with the TEM pictures, in which GST has a grainy texture filled with 100-nm-size grains, whereas AIST has a fine texture.

4. Nanosecond recrystallization dynamics in GST and AIST

To uncover recrystallization dynamics in GST and AIST at atomic level, we investigated the atomic structure of the amorphous phase by using a combination of advanced synchrotron radiation measurements (X-ray diffraction, EXAFS, HXPS) and reverse Monte Carlo simulation (RMC)/density function theory (DF)-molecular dynamics (MD) simulations (Matsunaga et al., 2011).

Figure 13 shows the structure factors $S(Q)$ of AIST and GST obtained using X-ray diffraction. The crystalline forms of both materials have sharp Bragg peaks (red lines), and the amorphous forms (blue lines) have typical halo patterns. However, oscillations up to the maximum Q value in a-AIST indicate a structure with well defined short-range order. The total correlation functions $T(r)$ for AIST and GST are shown in Fig. 14. The $T(r)$ for crystalline (c-) AIST and c-GST, which are very similar beyond 4 Å. Small differences between the two crystalline forms are found at shorter distances, for example the double peak in c-AIST (2.93 Å and 3.30 Å) and a single peak in c-GST (2.97 Å). The $T(r)$ for the amorphous materials, however, are significantly different: the first peak in amorphous (a-)

AIST (2.86 Å) is only slightly shorter than that found in c-AIST (2.93 Å), whereas the first peak in a-GST (2.79 Å) is shorter than that in c-GST (2.97 Å). We note also that the shoulder on the second peak in a-AIST (3.50 Å) is near that observed in the crystalline form (3.30 Å). The pronounced difference between the diffraction patterns of the two materials is strong evidence that they crystallize differently. The atomic motion and/or diffusion accompanying the phase change are larger in GST than in AIST, where the phase change is accompanied by small changes in bond lengths.

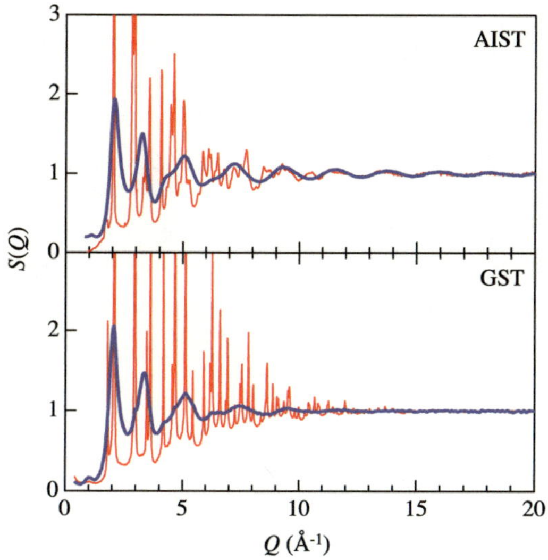

Fig. 13. Total structure factors, $S(Q)$, for AIST and GST. Blue line, amorphous form; red line, crystalline form.

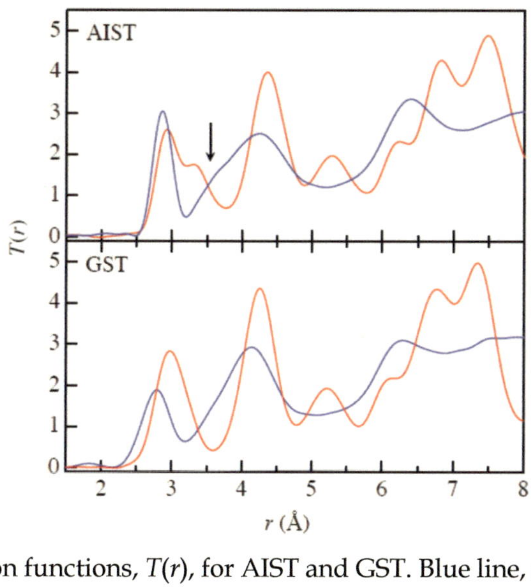

Fig. 14. Total correlation functions, $T(r)$, for AIST and GST. Blue line, amorphous form; red line, crystalline form.

Atomic configurations for a-AIST and a-GST obtained from the RMC / DF-MD simulations are shown in Fig. 15. It is found from the atomic configurations that the local environment of the predominant element Sb (75%) is very similar to that in the crystal (A7, a distorted octahedron) in AIST whereas large fraction of ABAB squares (manifested by ring) is observed in a-GST.

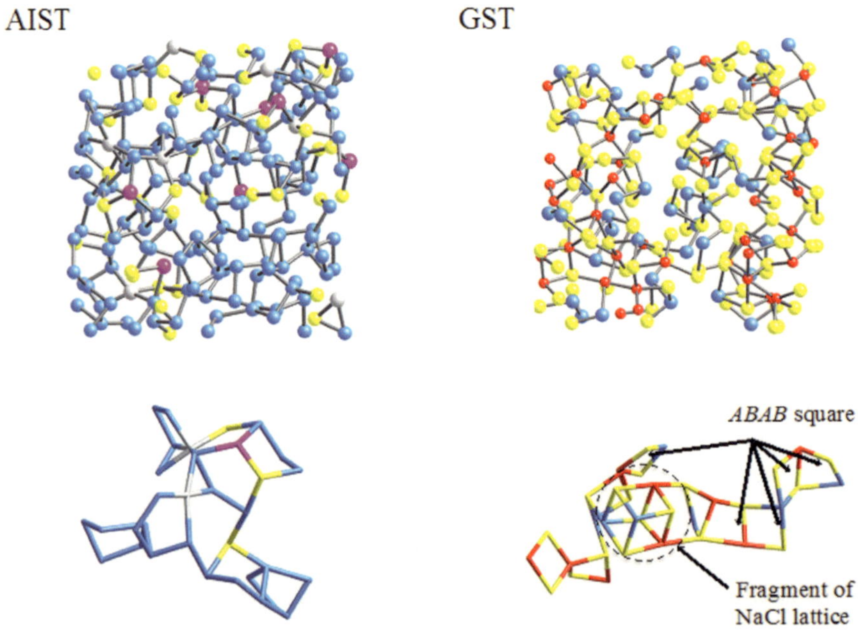

Fig. 15. Atomic configurations for a-AIST and a-GST obtained from the RMC / DF-MD simulations. Ag, silver; In, magenta; Sb, blue; Te, yellow; Ge, red.

Figure 16(a) shows ring statistics in a-AIST and a-GST derived from the RMC/DF-MD models which are consistent with the results of X-ray diffraction, EXAFS and HXPS measurements. In a-GST, 40% of the rings are fourfold or sixfold, whereas the distribution in a-AIST is much broader; the most common (fivefold) rings make up only 15% of the total. On the basis of structural features mentioned above, possible phase-change schemes in both materials are shown in Fig. 16(b). In a-AIST, it is suggested that a sequence of ring reconstructions by way of bond interchanges results in sixfold rings with short Sb bonds accompanied by small changes in the bond lengths, because RMC/DF-MD model for a-AIST has locally distorted 3+3 octahedron, which resembles c-AIST. By contrast, many four- and sixfold rings in a-GST act as nuclei for crystallization and require larger atomic displacements than in a-AIST. Crystallization starts simultaneously from many such nuclei in the amorphous mark (NaCl fragments) and lead to an aggregation of small crystal grains. These features propose a "bond-interchange" model as the origin of "growth-dominated" crystallization of a-AIST, whereas the large fraction of "crystalline nuclei" in a-GST is the origin of the "nucleation-driven" crystallization in GST. The structural finding and possible phase-change mechanism at atomic level obtained in this study is in line with the results of time resolved measurements mentioned above. A combination of advanced synchrotron measurements and computer simulation has provided insight into the atomistic differences between two phase-change materials, which is related to their different crystallization mechanisms.

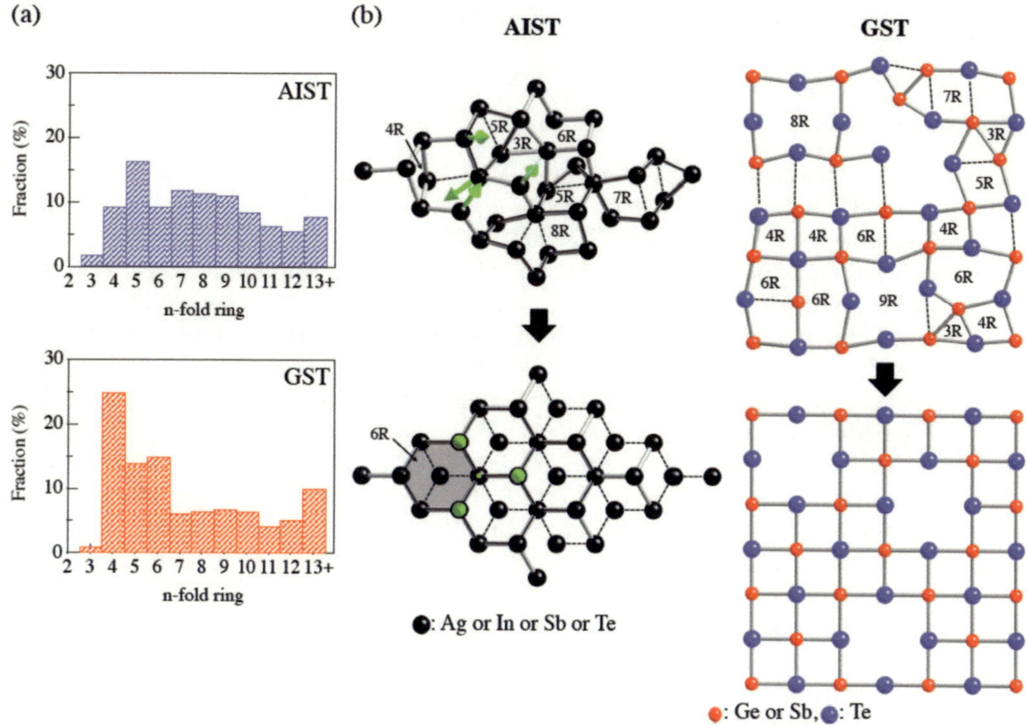

Fig. 16. Ring statistics (a) and possible schemes for fast phase change (b) in a-AIST and a-GST.

5. Conclusion

We described the detail of the X-ray pinpoint structural measurement system for investigation of optical recording process. Furthermore, we showed the recent progress for fully understanding the atomic structure of AIST and compare it to GST with the research combined X-ray diffraction, EXAFS, HXPS measurements and computer simulations. These demonstrated that the time resolved X-ray diffraction technique using SR is very powerful for the structural investigation of crystal growth phenomena.

6. Acknowledgment

This work was supported by Core Research for Evolutional Science and Technology (CREST) "X-ray pinpoint structural measurement project -Development of the spatial- and time-resolved structural study for nano-materials and devices-" and by the Academy of Finland and the Japan Science and Technology Agency through the Strategic Japanese - Finnish Cooperative Program on "Functional materials". The synchrotron radiation experiments were carried out with the approval of the Japan Synchrotron Radiation Research Institute (Proposal Nos. 2006A1746, 2006B1726, 2007A1864, 2007B1876, 2008A1856, 2008B1996, 2009A1891, 2009B1931 and 2009B1991). We would like to thank and acknowledge the help and advice of Noboru Yamada, Toshiyuki Matsunaga, Rie Kojima, Nobuhiro Yasuda, Yoshimitsu Fukuyama, Hitoshi Osawa, Jungeun Kim, Hitoshi Tanaka, Takashi Ohshima, Haruno Murayama, Kenichi Kato, Koshiro Toriumi and Yutaka Moritomo.

7. References

Iwasaki, H.; Ide, Y.; Harigaya, M.; Kageyama, Y. & Fujimura, I. (1992). Completely Erasable Phase Change Optical Disk. *Japanese Journal of Applied Physics,* Vol.31, No.2B, (February 1992), pp.461-465, ISSN 0021-4922

Fukuyama, Y.; Yasuda, N.; Kim, J.; Murayama, H.; Tanaka, Y.; Kimura, S.; Kato, K.; Kohara, S.; Moritomo, Y.; Matsunaga, T.; Kojima, R.; Yamada, N.; Tanaka, H.; Ohshima, T. & Takata, M. (2008a). Time-Resolved Investigation of Nanosecond Crystal Growth in Rapid-Phase-Change Materials: Correlation with the Recording Speed of Digital Versatile Disc Media. *Applied Physics Express,* Vol.1, No.4, (March 2008), pp. 045001-1-3, ISSN 1882-0778

Fukuyama, Y.; Yasuda, N.; Kim, J.; Murayama, H.; Ohshima, T.; Tanaka, Y.; Kimura, S.; Kamioka, H.; Moritomo, Y.; Toriumi, K.; Tanaka, H.; Kato, K.; Ishikawa, T. & Takata, M. (2008b). Ultra-high-precision Time Control System over Any Long Time Delay for Laser Pump and Synchrotron X-ray Probe Experiment. *Review of Scientific Instruments,* Vol.79, No.4, (April 2008), pp. 045107-1-4, ISSN 0034-6748

Fukuyama, Y.; Yasuda, N.; Kamioka, H.; Kim, J.; Shibata, T.; Osawa, H.; Nakagawa, T.; Murayama, H.; Kato, K.; Tanaka, Y.; Kimura, S.; Ohshima, T.; Tanaka, H.; Takata, M. & Moritomo, Y. (2010). Simultaneous Measurements of Picosecond Lattice and Charge Dynamics in Co-Fe Cyanides. *Applied Physics Express,* Vol.3, No.1, (January 2010), pp. 016601-1-3, ISSN 1882-0778

Kimura, S.; Moritomo, Y.; Tanaka, Y.; Tanaka, H.; Toriumi, K.; Kato, K.; Yasuda, N.; Fukuyama, Y.; Kim, J; Murayama, H. & Takata, M. (2006). X-ray Pinpoint Structural Measurement for Nanomaterials and Devices at BL40XU of the SPring-8. *AIP Conference Proceedings,* Vol.879, (2006), pp. 61-66, ISSN 0094-243X

Kohara, S.; Kato, K.; Kimura, S.; Tanaka, H.; Usuki, T.; Suzuya, K.; Tanaka, H.; Moritomo, Y.; Matsunaga, T.; Yamada, N.; Tanaka, Y.; Suematsu, H. & Takata, M. (2006). Structural Basis for the Fast Phase Change of $Ge_2Sb_2Te_5$: Ring Statistics Analogy between the Crystal and Amorphous States. *Applied Physics Letters,* Vol.89, No.20, (November 2006), pp. 021910-1-3, ISSN 0003-6951

Kolobov, A.V.; Fons, P.; Frenkel, A.I.; Ankudinov, A.L.; Tominaga, J. & Uruga, T. (2004). Understanding the Phase-change Mechanism of Rewritable Optical Media. *Nature Materials,* Vol.3, No.10, (October 2004), pp. 703-708, ISSN 1476-1122

Kwon, M.-H.; Lee, B.-S.; Bogle, S.N.; Nittala, L.N.; Bishop, S.G.; Abelson, J.R.; Raoux, S.; Cheong, B.-K. & Kim, K.-B. (2007). Nanometer-scale Order in Amorphous $Ge_2Sb_2Te_5$ Analyzed by Fluctuation Electron Microscopy. *Applied Physics Letters,* Vol.90, No.2, (January 2007), pp. 021923-1-3, ISSN 0003-6951

Lee, B.-S.; Abelson, J.R.; Bishop, S.G.; Kang, D.-H.; Cheong, B.-K. & Kim, K.-B. Investigation of the Optical and Electronic Properties of $Ge_2Sb_2Te_5$ Phase Change Material in Its

Amorphous, Cubic, and Hexagonal Phases. (2005). *Journal of Applied Physics,* Vol.97, No.9, (May 2005), pp. 093509-1-8, ISSN 0021-8979

Matsunaga, T.; Akola, J.; Kohara, S.; Honma, T.; Kobayashi, K.; Ikenaga, E.; Jones, R.O.; Yamada, N.; Takata, M. & Kojima, R. (2011). From local structure to nanosecond recrystallization dynamics in AgInSbTe phase-change materials. *Nature Materials,* Vol.10, No.2, (February 2011), pp. 129-134, ISSN 1476-1122

Naito, M.; Ishimaru, M.; Hirotsu, Y. & Takashima, M. (2004). Local Structure Analysis of Ge-Sb-Te Phase Change Materials using High-resolution Electron Microscopy and Nanobeam diffraction. *Journal of Applied Physics,* Vol.95, No.12, (June 2004), pp. 8130-8135, ISSN 0021-8979

Park, J.; Kim, M.R.; Choi, W.S.; Seo, H. & Yeon, C. (1999). Characterization of Amorphous Phases of $Ge_2Sb_2Te_5$ Phase-change Optical Recording Material on Their Crystallization Behavior. *Japanese Journal of Applied Physics,* Vol.38, No.8B, (August 1999), pp. 4775-4779, ISSN 0021-4922

Tanaka, Y.; Hara, T.; Kitamura, H. & Ishikawa, T. (2000). Timing Control of an Intense Picosecond Pulse Laser to the SPring-8 Synchrotron Radiation Pulses. *Review of Scientific Instruments,* Vol.71, No.3, (March 2000), pp. 1268-1274, ISSN 0034-6748

Tanaka, Y.; Fukuyama, Y.; Yasuda, N.; Kim, J.; Murayama, H.; Kohara, S.; Osawa, H.; Nakagawa, T.; Kimura, S.; Kato, K.; Yoshida, F.; Kamioka, H.; Moritomo, Y.; Matsunaga, T.; Kojima, R.; Yamada, N.; Toriumi, K.; Ohshima, T.; Tanaka, H. & Takata, M. (2009). Development of Picosecond Time-resolved Microbeam X-ray Diffraction Technique for Investigation of Optical Recording Process. *Japanese Journal of Applied Physics,* Vol.48, No.3, (March 2009), pp. 03A001-1-5, ISSN 0021-4922

Tanaka, Y.; Ohshima T.; Fukuyama, Y.; Yasuda, N.; Kim, J.; Osawa, H.; Kimura, S.; Togashi, T.; Hara, T.; Kamioka, H.; Moritomo, T.; Tanaka, H.; Takata, M.; Sengoku, H. & Nonoshita, E. (2010). High-Precision Time Delay Control with Continuous Phase Shifter for Pump-Probe Experiments Using Synchrotron Radiation Pulses. *AIP Conference Proceedings,* Vol.1234, (2010), pp. 951-954, ISSN 0094-243X

Wei, J. & Gan, F. (2003). Theoretical Explanation of Different Crystallization Processes between As-deposited and Melt-quenched Amorphous $Ge_2Sb_2Te_5$ Thin Films. *Thin Solid Films,* Vol.441, No.1-2, (September 2003), pp. 292-297, ISSN 0040-6090

Wuttig, M. & Yamada, N. (2007). Phase-change Materials for Rewriteable Data Storage. *Nature Materials,* Vol.6, No.11, (November 2007), pp. 824-832, ISSN 1476-1122

Yamada, N.; Ohno, E.; Akahira, N.; Nishiuchi, K.; Nagata, K. & Takao, M. (1987). High Speed Overwritable Phase Change Optical Disk Material. *Japanese Journal of Applied Physics,* Vol.26, Supplement 26-4, (1987), pp. 61-66, ISSN 0021-4922

Yamada, N. & Matsunaga, T. (2000). Structure of Laser-crystallized $Ge_2Sb_{2+x}Te_5$ Sputtered Thin Films for Use in Optical Memory. *Journal of Applied Physics,* Vol.88, No.12, (December 2000), pp. 7020-7028, ISSN 0021-8979

Yasuda, N.; Murayama, H.; Fukuyama, Y.; Kim, J.; Kimura, S.; Toriumi, K.; Tanaka, Y.; Moritomo, Y.; Kuroiwa, Y.; Kato, K.; Tanaka, H & Takata, M. (2009). X-ray Diffractometry for the Structure Determination of a Submicrometre Single Powder Grain. *Journal of Synchrotron Radiation,* Vol.16, No.3, (May 2009), pp. 352-357, ISSN 0909-0495

Yasuda, N.; Fukuyama, Y.; Toriumi, K.; Kimura, S. & Takata, M. (2010). Submicrometer Single Crystal Diffractometry for Highly Accurate Structure Determination. *AIP Conference Proceedings,* Vol.1234, (2010), pp. 147-150, ISSN 0094-243X

Zhou, G.F.; Borg, H.J.; Rijpers, J.C.N.; Lankhorst, M.H.R. & Horikx, J.J.L. (2000). Crystallisation Behaviour of Phase Change Materials: Comparison between Nucleation- and Growth-dominated Crystallisation, *Proceedings of SPIE - The International Society for Optical Engineering,* Vol.4090, (2000), pp. 108-115, ISSN 0277-786X

12

Synthesis and Luminescence Properties of $EuMoO_4$ Octahedron-Like Microcrystals

Jagannathan Thirumalai[1], Rathinam Chandramohan[2] and Viswanthan Saaminathan[3]

[1]*Department of Physics, B.S. Abdur Rahman University, Vandalur, Chennai, Tamil Nadu,*

[2]*Department of Physics,Sree Sevugan Annamalai College, Devakottai,Tamil Nadu*

[3]*School of Materials Science and Engineering, Nanyang Technical University, Singapore,*

[1,2]*India*

[3]*Singapore*

1. Introduction

Global claim for luminescent materials as efficient sources of energy that can supply sustained competence is growing day by day. The phosphors are facing increased challenges including concerns about global rare earth, environmental and recycling issues, and how to quickly bring to market new technological developments in the industry to meet end-user requirements. A number of applications have emerged in recent years that will change the future of the industry and new technologies like micro/nanoscale innovations and specialty phosphors are harnering increased attention. The primary drivers for growth are the expansion of key end-use applications including solid-state lighting and fluorescent lighting. Current research in nanotechnology is focused on new materials, novel phenomena, new characterization technique and fabrication of micro/nano devices.

Rare-earth doped molybdates are excellent materials of current interest owing to their interesting optical and opto-electronic properties. Scheelite-type crystalline structures, such as metal molybdates, where molybdenum atoms are coordinated tetrahedrally, have recently attracted great attention because of their promising applications in the electro-optical field (Ryu et al., 2005,a; Gong et al., 2006). Recently, molybdates have been extensively studied due to its attractive luminescence behavior and interesting structural properties (Thongtem et al., 2010), which can be used in a number of fields, such as optic fibers (Ryu et al., 2005, b) laser host materials (Wang et al., 2007), luminescence materials (Zhang et al., 2005; Sen & Pramanik, 2001), efficient catalysts (Ishii et al., 1992), scintillation detectors (Marques et al., 2006), and microwave application (Yang et al., 2009). Among all these molybdates, $EuMoO_4$ is an important opto-electrical material based on its red and white LED phosphors (Rosa et al., 2009). Furthermore, the molybdates allow the doping of different rare-earth ions, which can vary its luminescence properties. Due to the rich emission spectral lines extending from ultraviolet to infrared for Eu^{3+} ions, spectroscopic properties of this ion have been extensively studied in various hosts to investigate the potential applications as functional material, in the fiber laser amplifiers, quantum cutting luminescence, up-conversion laser, and tunable ultraviolet (UV) , vacuum UV solid–solid-state lasers and High-pressure x-ray diffraction study (Errandonea et al., 2011; Koch et al., 1990; Zhang et al., 2007). New and enhanced properties are expected due to size confinement in nanoscale dimensions. In this background, this research work forms a recent avenue in material science research. The field of nanophosphor development, with rare earth doped metal oxide sytems have become a separate wide field in these days.

A detailed survey on undoped and doped molybdate phosphors discusses various synthesis techniques adopted by different research groups as follows. Recently, Eu^{3+} doped $ZnMoO_4$ with charge compensation were synthesized by solid-state reactions for phosphor-converted light-emitting diode (LED) is an important kind of solid-state illumination (Xie et al., 2010). Octahedron-like $BaMoO_4$ microcrystals were synthesized by the co-precipitation method at room temperature and processed in microwave-hydrothermal at 413 K for different time (Cavalcante et al., 2009). Single-crystalline $BaMoO_4$ microcrystals with uniform shuttle-like morphology have been successfully prepared via a facile aqueous solution mineralization process at room temperature (Wu et al., 2007). Phosphors $Na_5La(MoO_4)_4$:$x$$Eu^{3+}$ and $NaEu(MoO_4)_2$ were prepared with a solid-state reaction technique and are fabricated onto near-ultraviolet/violet-emitting InGaN chips for bright red-light-emitting diodes device, respectively (Wang et al., 2006). In order to prepare fluorescent material for white Light Emitting Diodes (LEDs), a new Eu^{3+} activated molybdate phosphor $SrMoO_4$ was fabricated with solid-state method (Xu et. al., 2007). Yttrium molybdate phosphors with fixed Er^{3+} and various Yb^{3+} concentrations were synthesized via a co-precipitation method (Lu et al., 2010). Tetragonal $La_2(MoO_4)_3$ with various novel and complex 3D hierarchical architectures self-assembled from different building blocks were successfully synthesized by a hydrothermal method in EDTA mediated processes (Xu et al., 2010). Enhancement emission intensity of $CaMoO_4$:Eu^{3+}, Na^+ phosphor via Bi co-doping and Si substitution for application to white LEDs phosphors were synthesized by the conventional solid state reaction method (Marques et al., 2008). $ZnMoO_4$ with a rhombus sheet or flower-like structure, α-$ZnMoO_4$ and needle-like $ZnMoO_4.0.8H_2O$ were successfully synthesized by simple hydrothermal crystallization processes with citric acid for opto-electronic applications (Zhang et al., 2010). The electronic structures of perfect crystals of barium molybdate (BaMoO4) and of crystals containing F and F+ color centers are studied within the framework of the fully relativistic self-consistent Dirac-Slater theory by using the numerically discrete variational (DV-Xα) method (Guo et al., 2009). Room temperature electrochemical growth of polycrystalline $BaMoO_4$ films on a metallic molybdenum substrate in a barium hydroxide aqueous solution has been studied (Xia & Fuenzalida, 2003). $BaMoO_4$ thin films were prepared on molybdenum substrates in $Ba(OH)_2$ solutions by electrochemical method at room temperature (Bi et al., 2003). Recently, on novel rare-earth activated molybdate [$K_xSr_{1-2x}MoO_4$:$Pr^{3+}{}_x$ ($0.00 \le x \le 0.04$)] light-emitting diodes (LEDs) for the replacement of current light sources (Qiuxia et al., 2011). A novel red phosphor $Ca_{1-2x}Eu_xLi_xMoO_4$ is a promising material for Solid-State Lighting Based on a GaN LED (Wang et al., 2005). CaMoO4 was a promising yellowish green phosphor for near-ultraviolet (NUV) InGaN-based white LED (Li et al., 2007). Highly ordered $SrMoO_4$ 3D spherical superstructure assembled with nanosheets was synthesized via a facile and fast sonochemical route without any template (Mao et al., 2010). The indoor lighting applications are the main potential applications using either white-light-emitting diode (LEDs) made by exciting appropriate phosphors. These wide band gap materials can be implemented commercially replacing all conventional LEDs in future. A well-crystallized AMO_4 (A=Ba, Ca, Sr; M=W, Mo) films have been prepared at room temperature through a simple solution reaction in respective alkaline solution at higher pH ~ 12–14 (Dinesh et al., 2005). X-ray diffraction and Raman scattering studies on the scheelite structured barium molybdate ($BaMoO_4$) under high pressure (~5.8 GPa) shows that, it undergoes a first order phase transition to the fergusonite structure ($I2/a$, $Z = 4$)—as also observed in iso-structural barium tungstate (Panchal et al., 2006). A study on the growth kinetics and habit modification of barium molybdate single crystals in silica gel medium has been presented (Kurien & Ittyachen, 1979). Praseodymium barium molybdate (PBM) single crystals grown by gel method, reveal multiple and isolated octahedral bipyramidal crystals as well as spherulites of PBM at different depths from the surface inside the gel (Isac et al., 1996). A series of novel red-emitting phosphors $AgLa_{0.95}Eu_{0.05}(WO_4)_{2-x}(MoO_4)_x$ [x=0–2] have been synthesized and their luminescence properties investigated (Sivakumar & Varadaraju, 2006). The effects of Ni^{2+} doping on structural, microstructural, dielectric and electrical properties of lead molybdate $PbMoO_3$ have been investigated. Ferroelectric phase transition in Ni-doped lead molybdate ceramics was studied (Palai et al., 2001).

Herein, we report the controlled synthesis of $EuMoO_4$ microcrystals using hydrothermal method by employing PVP as surfactant along with the state of the art of this field at present. By varying the reaction time, we have reported the morphology selection and the condition to derive octahedron morphology has been discussed. The rest of the parameters like, molar ratio between initial precursor / surfactant and temperature were kept as constant. The $EuMoO_4$ microcrystals are found to be an excellent matrix for photoluminescence application. The Eu^{3+} ion serves as a good red emitting luminescent center. Therefore, well-crystallized $EuMoO_4$ microcrystals have significant improvement in the phosphor properties considerably. Hence, the PL properties were also studied and reported. Also, this processing route provides the basis for a nearly low cost, low temperature method for the preparation of homogeneous nano-sized ceramics compared to any other existing methods.

2. Synthesis, structure, formation and morphology of the $EuMoO_4$ microcrystals

2.1 Synthesis

All the chemical reagents used were of analytical grade and used without further purification. In the typical synthesis, ~ 0.02 gm of PVP was first dissolved in 15 mL of deionized water under with vigorous stirring. Subsequently 0.021 mM of $(NH_4)_6Mo_7O_{24}\cdot 4H_2O$ was added drop wise. This is followed by addition of 1 mM $EuCl_2\cdot 7H_2O$ solutions drop wise to the above solution under with vigorous stirring by maintain the solution pH was adjusted up to 8. The mixture was stirred for another 10 min. Finally, the resultant solution was transferred to a Teflon-lined vessel kept in an autoclave at 135 °C for 0 - 7 hrs. The resultant solid products were centrifugally separated from the suspension, washed with deionized water and absolute ethanol several times, and dried at 60 °C in air for 6 hrs.

2.2 Structural analysis

The $BaMoO_4$ exhibit a scheelite-type tetragonal structure (Yang et al., 2008, Luo et al., 2009, Wu et al., 2007), which belongs to the tetragonal space group of $I4_1/a$. The unit-cell presents two formula units per primitive cell which includes two formula units where the Mo atom is surrounded by four equivalents O site in tetrahedral symmetry and divalent metal shares corners with eight adjacent MoO_4 tetrahedral [Luo et al., 2009, Ryu et al., 2005a]. Structure and phase purity of the $EuMoO_4$ microcrystals were examined using XRD analysis. From Fig. 1, it is seen that all the diffraction peaks matched well with the standard data for tetragonal scheelite-type $EuMoO_4$ (JCPDS, No. 22-1097; Space Group - $I4_1/a$). No traces of extra peaks from other phases were observed. Therefore, all as-prepared samples were of single phase $EuMoO_4$. The diffraction peaks (Fig. 1) are slightly broadened, which is characteristic of their fine nature. The average crystallite sizes of the primary particles that are calculated from peak broadening of the (112) line using the Scherrer formula was approximately 35 nm for $EuMoO_4$ microcrystals.

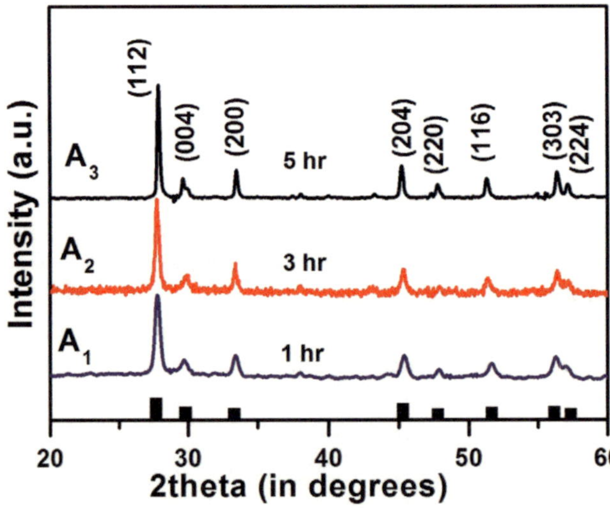

Fig. 1. X-ray diffraction (XRD) patterns of microstructures (A_1 - A_3) of $EuMoO_4$ prepared by modulating the reaction time (a) 1, (b) 3, (c) 5.

2.3 Morphological studies

The SEM and TEM images of the sample prepared with the typical procedure are shown in Fig 2. To obtain a better understanding of the formation mechanism of the $EuMoO_4$ octahedral-like microstructures, products formed at different growth stages were collected for SEM measurements. Fig 2 (a-e) implies that the concentration of PVP aqueous solutions played a crucial role in the formation of $EuMoO_4$ crystal prepared under various growth conditions. We found that the Ostwald ripening process dominated the crystal growth process at the initial stage. Fig. 2a shows some tiny equiaxial nanoparticles with polydisperse nature, at room temperature. As is well-known, at the initial stage tiny crystalline nuclei form in a supersaturated solution that acts as the nuclei for crystallization. Then the crystal growth follows, and bigger particles grow at the expense of smaller crystals, as described by the Gibbs-Thompson equation (Peng, 2003).

$$S_r = S_b \exp(2\sigma V_m / rRT) \qquad (1)$$

where, r: is the radius of the crystal; σ: the specific surface energy; V_m: molar volume of the material; S_b & S_r: solubility of the bulk crystals and crystals with a radius r, respectively; R, the gas constant; T, the absolute temperature. Fig. 2b shows that the processing performed at 1h favoured the formation of bipyramidal-like microparticles to some extent agglomerated nature. By increasing the reaction time to 1 hr, the resulting product formation is due to the interactions between the tails of surface-adsorbed surfactant molecules, influencing the growth process of the micro-octahedrons. At ~ 3 hrs, it could be seen that $EuMoO_4$ octahedral microparticles of size ~ 1 μm side length were formed (Fig 2c), which is nearly similar to the structure of micro-octahedrons (Fig 2c). The growth rate in $EuMoO_4$ microcrystal is oriented preferentially along the [001] direction than in the [100] direction. In principle, we believe that the anisotropic growth in the [001] direction for the $EuMoO_4$ micro-octahedron is caused by differences in the surface energies on each crystal face. Probably, this preferential growth direction has stronger interatomic bonds between the $[Eu^{3+}]$ and $[MoO_4]^{2-}$ clusters. The microparticles synthesized with reaction time ~ 5 hrs are highly uniform, regular with smooth surface and can be dispersed well on a large scale. The PVP may form a complex the rare earth ions Pr^{3+} at first. Subsequently it might bind to the surface of the growing particles after being dissociated from the rare earth ions slowly in the process of reaction. Hence, the main factor to influence the formation of the regular structures is estimated as time. Formation of the octahedral-like $EuMoO_4$ microstructures after hydrothermal treatment are confirmed by TEM investigation. The corresponding high-resolution TEM image in Fig. 2d displays resolved lattice fringes of (004) planes with a spacing of 0.385 nm. Insets of Fig. 2(c, d) show the selected area electron diffraction (SAD). The Fast Fourier Transform (FFT) pattern indicates that the microstructures are single crystalline in nature. It is inferred that at the higher reaction time (Fig 2e), growth of the microcrystals lead to preferential growth orientation All together, from the SEM images, we can clearly see that microstructural changes occurred at the very early stages and even at lower reaction time. As reaction time was prolonged, only similar and much denser microcrystals could be grown.

2.4 Formation mechanism

The relevant formation mechanism of $EuMoO_4$ microstructures has been interpreted in two paths: dissolution–recrystallization and as well as the effective collision rates between the small microcrystals, respectively (Cavalcante et al., 2009, Luo et al., 2009). The hydrothermal process is able to promote the effective fender-bender between the small microparticles, contributing to the growth of the microcrystals and inducing the formation of new crystallographic faces on them. Schematic (Fig. 3) shows the morphological evolution of $EuMoO_4$ microcrystals as a function of reaction time. It is proposed that the process for the formation of the octahedron $EuMoO_4$ microcrystals. The first one is the formation of the perfect small microparticles, which belongs to a typical Ostwald ripening process. Probably, the preferential growth direction has stronger interatomic bonds between the $[Eu^{3+}]$ and $[MoO_4]^{2-}$ ions were mixed, a highly supersaturated solution was formed, leading to the coexistence of the asymmetrical and the bipyramidal-like $EuMoO_4$ microparticles. It is clear that the crystalline phase of the nuclei is critical for directing the inherent shapes of the crystals owing to its distinctive symmetry and structure. As the reaction time proceeded for 1 hrs, the bipyramidal-like particles are agglomerated with a high concentration of surface

Fig. 2. Typical SEM image taken at different magnifications of $EuMoO_4$ microstructures at (a) Room temperature, (b) 1, (c) 3, (d) 5 hrs, (e) TEM image and the corresponding SAED pattern (inset) and (f) a typical HRTEM image and Inset: the corresponding FFT image.

defects and it is possible to note the initial formation of micro-octahedrons with well-defined structures. From 5 hrs of reaction time, the aggregation process between the small microparticles was induced by the coalescence process, resulting in the growth of micro-octahedrons. This proposed crystal growth mechanism is in good agreement with the literature (Cavalcante et al., 2009, Wu et al., 2007).

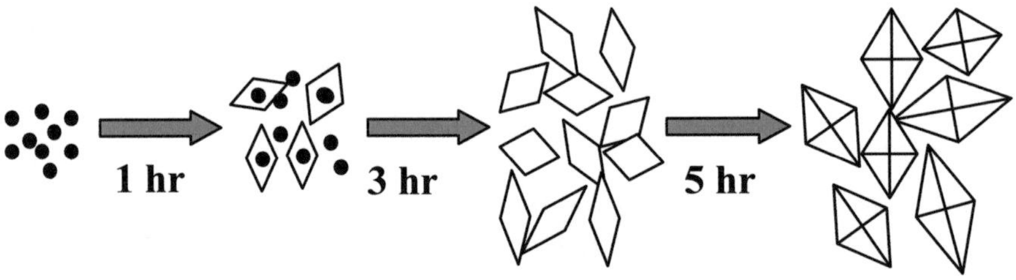

Fig. 3. Schematic representation of the formation process of the octahedron-like $EuMoO_4$ microcrystals.

2.5 Compositional analysis

It is well known that, the XPS is known to study the composition and binding energy. Therefore, in order to further prove that the as-synthesized $EuMoO_4$ microcrystals are composed of relevant elements. X-ray photoelectron spectroscopy (XPS) analysis was performed to probe the surface of the particles. The structures were probed using XPS, to investigate the elementary states and binding energy values. Fig. 4a shows the XPS survey spectra of the tetragonal $EuMoO_4$ microcrystals. The core levels of Eu, Mo and O can be seen in Fig. 2a, in which, the XPS spectra of the tetragonal $EuMoO_4$ microcrystals, obtained is the range of 100 - 600 eV (Eu 4d at 130 eV, Mo $3d_{5/2}$ at 228.5 eV, Mo $3d_{3/2}$ at 331 eV, Mo $3p_{3/2}$ at 394 eV, Mo $3d_{1/2}$ at 412 eV, Mo 3s at 506 eV, O 1s 531.5 eV) and 1100 - 1200 eV (Eu $3d_{5/2}$ at 1126 eV, Eu $3d_{3/2}$ 1156.5 eV), respectively. It can be seen that the binding energy data (calibrated using C (1s, 284.7 eV) as the reference) from $EuMoO_4$ microparticles. All these peaks are in good agreement with the literature (Moulder, et al., 1992). The composition estimated by XPS using the relative sensitivity factors of Eu, Mo and O also revealed excess oxygen in the samples. The results also reveal that the sodium ions can be eliminated from the final product after washing and calcinations. The composition estimated using XPS results are in good agreement.

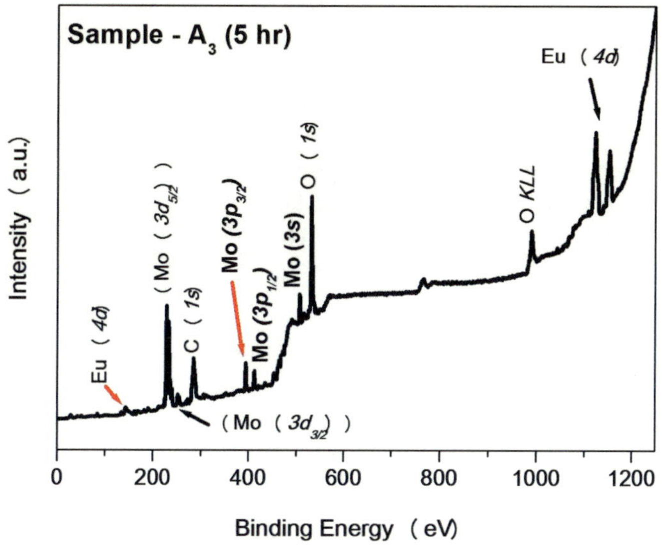

Fig. 4. XPS survey spectra of $EuMoO_4$ octahedron at 5 hr.

2.6 Optical properties

Fluorescence spectroscopy can provide valuable information about the intermolecular interactions of molecules in molecular crystals (Lifshitz, et al., 2003). The tetragonal phased $EuMoO_4$ has similar structure to that of other rare-earth doped molybdate ions with scheelite-type and is expected to be a promising phosphor candidate. The presence of strong clear peaks indicates the high crystallinity of the as-prepared products, which is very beneficial for obtaining brighter luminescence bands. The luminescence properties of the $EuMoO_4$ phosphors were then examined. Fig. 5 presents the excitation and emission spectra of $EuMoO_4$ microstructures.

The excitation spectra (Fig. 5, left) were obtained by monitoring the emission of the Eu^{3+} $^5D_0 \rightarrow {}^7F_2$ transition at 616 nm. It can be observed clearly that the excitation spectra both consist of a broadband from 200 to 350 nm, which is ascribed to the O—Mo C-T transition. The C-T band of O—Eu^{3+}, which usually appears in the range of 250–300nm in the excitation spectrum (Kiselev, et al., 2008), might have overlapped with the C-T band of molybdate group and, hence, as not observed clearly. In the longer wavelength region (360–710 nm), the sharp lines are intraconfigurational *4f–4f* transitions of Eu^{3+} in the host lattices, and the strong excitation band at 395 and 465nm is attributed to the $^7F_0 \rightarrow ^5L_6$ and $^7F_0 \rightarrow ^5D_2$ transitions of Eu^{3+}, respectively. The emission spectra of Eu^{3+} (Fig. 5, right) excited under 395nm near-UV light are mainly dominated by the hypersensitive red emission, showing a strong transition $^5D_0 \rightarrow ^7F_2$ at 616nm and a weak $^5D_0 \rightarrow ^7F_1$ transition. The two peaks are assigned to the electric dipole transition and magnetic dipole transition, respectively, and the presence of electric dipole transition confirms that Eu^{3+} ions are located at sites without inversion symmetry. From the PL emission spectra it is clearly attrributed that the transition from its major lines is greatly enhanced, and thus a bright red emission is observed. Since the absorption / emission process depends strongly on the nature of the lattice and its constituents, the luminescence behaviour of Eu^{3+} compounds is expected to vary strongly with the preparatory conditions. This is what has been observed. The strong luminescence intensity indicated the perfection of the microstructure and good crystallization. The optical property of the $EuMoO_4$ are in good agreement with the literature (Kiselev, et al., 2008), that were obtained by solid-state reaction method. The trivalent europium ion (Eu^{3+}) in solids has an intricate energy level scheme with energy gaps of various magnitude and show rich emission spectral lines (Blasse & Grabmaier, 1994). Also a systematic study related to PL and the lifetime measurements are to be undertaken. Such photoluminescent materials have potential application and an efficient red phosphor candidate in the luminescence field. The result envisages an effective route to synthesize $EuMoO_4$ microstructures have potential application in biomedical and display systems.

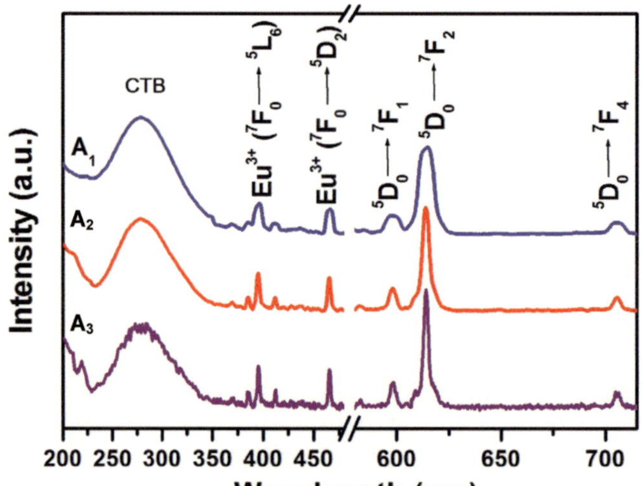

Fig. 5. A part of room temperature PL of $EuMoO_4$ microstructures ($A_1 - A_3$): excitation spectra (monitored at 643 nm) and emission spectra (under 443 nm excitation) prepared by modulating the reaction time.

In summary, this novel $EuMoO_4$ microstructures have been successfully synthesized via a facile simple and mild hydrothermal route employing PVP as surfactant. It is interesting to observe that the microparticles are highly crystalline and with uniform octahedral-like morphology in size and shape. The use of PVP as a surfactant introduced into the reaction system is essential to obtain highly crystalline and smooth surface microparticles. A strong red emission centered at 643 nm corresponding to the $^3P_0 \rightarrow ^3F_2$ transition of Eu^{3+} is observed under 440 - 500 nm excitation. The grown $EuMoO_4$ microstructures showed strong red emission upon UV illumination, finds potential application in biomedical and display devices. Altogether, this preparation methodology is found to be highly reproducible, convenient, easy, and could be extended to prepare other rare-earth doped molybdate microstructures for various applications like electroluminescent devices, optical integrated circuits, or biomarkers.

3. References

Bi, J.; Xiao, D. Q.; Gao, D. J.; Yu, P.; Yu, G. L.; Zhang, W. & Zhu, J. G. (2003). $BaMoO_4$ thin films prepared by electrochemical method at room temperature Cryst. Res. & Tech., 38., 11., (October & 2003) 935–940, ISSN: 0232-1300.

Blasse, G. & Grabmaier B.C. (1994). Luminescent Materials, Springer-Verlag Telos, 1-232, ISBN: 0387580190, New York.

Cavalcante, L. S.; Sczancoski, J. C.; Tranquilin, R. L.; Varela, J. A.; Longo, E. & Orlandi, M. O. (2009). Growth mechanism of octahedron-like BaMoO4 microcrystals processed in microwave-hydrothermal: Experimental observations and computational modeling. *Particuology*, 7., 5., (May & 2009) 353 - 362, ISSN: 1674-2001.

Dinesh, R.; Fujiwara, T.; Watanabe, T.; Byrappa, K. & Yoshimura, M. (2005). Solution synthesis of crystallized AMO_4 (A=Ba, Sr, Ca; M=W, Mo) film at room temperature. *J. Mater. Sci.* 41., 5., (March & 2005) 1541-1546, ISSN: 0022-2461.

Errandonea, D.; Santamaria-Perez, D.; Achary, S. N.; Tyagi, A. K.; Gall, P. & Gougeon, P. (2011). High-pressure x-ray diffraction study of CdMoO4 and EuMoO4, *J. Appl. Phys.* 109, February 2011) 043510–5, ISSN: 0021- 8979.

Gong, Q.; Qian, X.; Cao, H.; Du, W.; Ma, X. & Mo, M. (2006). Novel Shape Evolution of $BaMoO_4$ Microcrystals. *J. Phys. Chem. B.* 110., 39., (September & 2006) 19295–19299, ISSN: 1520-6106.

Guo, X.; Zhang, Q., Liu, T.; Song, M.; Yin, J.; Zhang, H. & Wang, X. (2009). First-principles study on electronic structures of $BaMoO_4$ crystals containing F and F^+ color centers. *Nuc. Inst. Methods. Phys. Res.: Sec. B.* 267., 7., (April & 2009) 1093–1096, ISSN: 0168-583X.

Isac, J.; Ittyachen, M. A. & Raju, K. S. (1996). Praseodymium barium molybdate-its growth and structural characterization *Bull. Mater. Sci.* 19., 3., (October & 1995) 495-504, ISSN: 0250-4707.

Ishii, M. & Kobayashi, M. (1992). Single crystals for radiation detectors. *Prog. Cryst. Growth Charact. Mater.* 23., 7-8., (August& 1991) 245–311, ISSN: 0960-8974.

Kiselev, A. P.; Shmurak, S. Z.; Sinitsyn, V. V.; Khasanov, S. S.; Red'kin, B. S.; Alekseev, A. V. & Ponyatovskii, E. G. (2008). Spectroscopy and X-Ray Diffraction Analysis of Europium Molybdate Single Crystals Subjected to Different Thermobaric Treatments, *Bull. Russian Academy Sci.: Physics,* 72., 9., (December & 2008) 1297-1302, ISSN 1062-8738.

Koch, M.E.; Kuen, A.W. & Case, W.E. (1990). Photon avalanche upconversion laser at 644 nm. Appl. Phys. Lett. 56., (January & 1990) 1083–3, ISSN: 0003-6951.

Kurien, K. V. & Ittyachen, M. A. (1979). Growth kinetics and habit modification of barium molybdate single crystals in silica gel. *J. Cryst. Growth.* 47., 5-6., (November-December & 1979) 743-745, , ISSN: 0022-0248.

Li, G.; Zhiping, Y.; Xu, L.; Chong, L. & Panlai, L. (2007). Luminescent Properties of Tb^{3+} doped $CaMoO_4$ Phosphor for UV White Light Emitting Diode. *J. Rare Earths*, 25., 7., (April & 2007) 7294 - 296, ISSN: 1002-0721.

Li, Q.; Huang, J. & Chen, D. (2011). A novel red-emitting phosphors $K_2Ba\ (MoO_4)_2$: Eu^{3+}, Sm^{3+} and improvement of luminescent properties for light emitting diodes. *J. Alloys Compd.* 509., 3., (January & 2011) 1007-1010, ISSN: 0925-838.

Lifshitz, E.; Bashouti, M.; Kloper, V.; Kigel, A.; Eisen, M.S. & Berger, S. (2003). Synthesis and Characterization of PbSe Quantum Wires, Multipods, Quantum Rods, and Cubes *Nano Lett.* 3., 6., (May & 2003) 857-862, ISSN: 1530-6984.

Lu, W.; Cheng, L.; Zhong, H.; Sun, J.; Wan, J.; Tian, Y. & Chen, B. (2010). Dependence of upconversion emission intensity on Yb^{3+} concentration in Er^{3+}/Yb^{3+} co-doped flake shaped $Y_2(MoO_4)_3$ phosphors *J. Phys. D: Appl. Phys.* 43., 8., (February & 2010) 085404, ISSN: 0022-3727.

Luo, Z.; Li, H.; Shu, H.; Wang, K.; Xia, J. & Yan, Y. (2008). Synthesis of $BaMoO_4$ Nestlike Nanostructures Under a New Growth Mechanism. *Cryst. Growth. Des.* 8., 7., (July & 2008) 2275–2281, ISSN: 1528-7483.

Mao, C-J.; Geng, J.; Wu, X-C. & Zhu, J-J.; (2010). Selective Synthesis and Luminescence Properties of Self-Assembled $SrMoO_4$ Superstructures via a Facile Sonochemical Route. J. Phys. Chem. C. 114., 5., (January & 2010) 1982 - 1988, ISSN: 1932-7447.

Marques, A. P. A.: Longo, V. M.; Melo, D. M.A.; Pizani, P. S.; Leite, E. R.; Varela, J. A. & Longo, E. (2008). Shape controlled synthesis of $CaMoO_4$ thin films and their photoluminescence property. *J. Solid State Chem.* 181., 5., (May & 2008) 1249– 1257, ISSN: 0022-4596.

Marques, A.P.A.; Melo, D.M.A.; Paskocimas, C.A.; Pizani, P.S.; Joya, M.R.; Leite, E.R. & Longo, E. (2006). Photoluminescent $BaMoO_4$ nanopowders prepared by complex polymerization method (CPM). *J. Solid State Chem.* 179., 3., (March & 2006) 671–678, ISSN: 0022-4596.

Moulder, J. F.; Stickle, W. F.; Sobol P. E. & Bomben, K. D. (1992). Handbook of X-Ray Photoelectron Spectroscopy, J. Chastain, (Ed.), 1-254, Perkin-Elmer Corporation, ISBN: 0-9627026-2-5, Eden Prairie, Minnesota, United States of America.

Palai, R.; Choudhary, RN.R. & Sharma, S. (2001). Ferroelectric phase transition in Ni-doped lead molybdate ceramics.*Ferroelectrics*, 256., 1., (April & 2006) 33–45, ISSN: 0015-0193.

Panchal, V.; Garg, N. & Sharma S. M. (2006). Raman and x-ray diffraction investigations on $BaMoO_4$ under high pressures *J. Phys.: Condens. Matter* 18., 16., (April & 2006) 3917, ISSN: 0953-8984.

Peng, X. (2003). Mechanisms for the Shape-Control and Shape-Evolution of Colloidal Semiconductor Nanocrystals. *Adv. Mater.* 15., 459, (March & 2003) ISSN: 1521-4095.

Rosa, I. L. V.; Marques, A. P. A.; Tanaka, M. T. S.; Motta, F. V.; Varela, J. A.; Leite, E. R. & Longo, E. (2009). Europium(III) Concentration Effect on the Spectroscopic and Photoluminescent Properties of BaMoO4:Eu. *J. Fluoresc.* 19., 3., (November & 2008) 495 - 500, ISSN: 1053-0509.

Ryu, J. H.; Yoon, J. W.; Lim, C. S.; Oh, W. C. & Shim, K. B. (2005a). Microwave-assisted synthesis of $CaMoO_4$ nano-powders by a citrate complex method and its photoluminescence property. *J. Alloys Compd.* 390., 1-2., (March & 2005) 245–249, ISSN: 0925-838.

Ryu, J.H.; Yoon, J.W.; Lim, C.S. & Shim, K.B. (2005b). Microwave-assisted synthesis of barium molybdate by a citrate complex method and oriented aggregation. *Mater. Res. Bull.* 40., 9., (September & 2005) 1468 - 1476.

Sen, A. & Pramanik, P. (2001). Low-temperature synthesis of nano-sized metal molybdate powders. *Mater. Lett.* 50.,5-6., (September & 2001) 287-294, ISSN: 0167-577X.

Sivakumar, V. & Varadaraju, U. V. (2006). Intense Red Phosphor for White LEDs Based on Blue GaN LEDs. *J. Electrochem. Soc.*, 153., 3., (February & 2006) H54-H57, ISSN: 0013-4651.

Thongtem, T.; Kungwankunakorn, S.; Kuntalue, B.; Phuruangrat, A. & Thongtern, S. (2010). Luminescence and absorbance of highly crystalline $CaMoO_4$, $SrMoO_4$, $CaWO_4$ and $SrWO_4$ nanoparticles synthesized by co-precipitation method at room temperature. *J. Alloys Compd.* 506., 1., (September & 2010) 475 - 481, ISSN: 0925-838.

Wang, J. G.; Jing, X. P.; Yan, C. H. & Lin, J. H. (2005). $Ca_{1-2x}Eu_xLi_x$ MoO_4: A novel red phosphor for solid-state lighting based on a GaN LED. *J. Electrochem. Soc.*, 152., 3., (January & 2005) G186, ISSN: 0013-4651.

Wang, Y.; Ma, J.; Tao, J.; Zhu, X.; Zhou, J.; Zhao, Z.; Xie, L.; Tian, H. (2007). Low temperature synthesis of $CaMoO_4$ nanoparticles. *Ceram. Int.* 33., 4., (May & 2007) 693 - 695.

Wang, Z.; Liang, H.; Wang, J.; Gong, M. & Su, Q. (2006). Red-light-emitting diodes fabricated by near-ultraviolet InGaN chips with molybdate phosphors. *Appl. Phys. Lett.* 89., 7., (August & 2006) 071921, ISSN: 0003-6951.

Wu, X.; Du, J.; Li, H.; Zhang, M.; Xi, B.; Fan, H.; Zhu, Y. & Qian, Y. (2007). Aqueous mineralization process to synthesize uniform shuttle-like $BaMoO_4$ microcrystals at room temperature. *J. Solid State Chem.* 180., 11., (November & 2007) 3288 - 3295, ISSN: 0022-4596.

Xia, C-T. & Fuenzalida, V.M. (2003). Room temperature electrochemical growth of polycrystalline $BaMoO_4$ films. *J. Euro. Ceram. Soc.* 23., 3., (March & 2003) 519 - 525, ISSN: 0955-2219.

Xie, A.; Yuan, X.; Wang, F.; Shi, Y. & Mu, Z. (2010). Enhanced red emission in $ZnMoO_4$: Eu^{3+} by charge compensation. *J. Phys. D: Appl. Phys.* 43., 5., (January & 2010) 055101, ISSN: 0022-3727.

Xu, L.; Lu, C.; Zhang, Z.; Yang, X. & Hou, W. (2010). Various self-assembled three-dimensional hierarchical architectures of $La_2(MoO_4)_3$: controlled synthesis, growth mechanisms, luminescence properties and adsorption activities. *Nanoscale*, 2., (May & 2010) 995–1005, ISSN: 2040-3364.

Xu, L.; Zhiping, Y.; Li, G.; Qinglin, G.; Sufang, H. & Panlai, L. (2007). Synthesis and Properties of Eu^{3+} Activated Strontium Molybdate Phosphor. *J. Rare Earths.* 25, 6., (December & 2007) 706-709, ISSN: 1002-0721.

Yang, X.Y. ; Liu, J.; Yang, H.; Yu, X.B.; Guo, Y.Z.; Zhou, Y.Q. & Liu, J. (2009). Synthesis and characterization of new red phosphors for white LED applications. *J. Mater. Chem.* 19., 22., (April & 2009) 3771–3774, ISSN: 0959-9428.

Yang, X.Y.; Liu, J.; Yang, H.; Yu, X.B.; Guo, Y.Z.; Zhou, Y.Q. & Liu, J. (2009). Synthesis and characterization of new red phosphors for white LED applications. *J. Mater. Chem.* 19., 22., (April & 2009)3771-3774, ISSN: 0959-9428.

Zhang, G.; Yu, S.; Yang, Y.; Jiang, W.; Zhang, S. & Huang, B. (2010) . Synthesis, morphology and phase transition of the zinc molybdates $ZnMoO_4 \cdot 0.8H_2O/\alpha\text{-}ZnMoO_4/ZnMoO_4$

by hydrothermal method. *J. Crystal Growth.* 312., 11., (May 2010) 1866–1874, ISSN: 0022-0248.

Zhang, N.; Bu, W.; Xu, Y.; Jiang, D. & Shi, J. (2007). Self-Assembled Flowerlike Europium-Doped Lanthanide Molybdate Microarchitectures and Their Photoluminescence Properties. *J. Phys. Chem. C.* 111, 5014 – 5019, ISSN: 1932-7447.

Zhang, Y.; Yang, F.; Yang, J.; Tang, Y. & Yuan, P. (2005) Synthesis of crystalline $SrMoO_4$ nanowires from polyoxometalates. *Solid State Commun.* 133., 12., (March & 2005) 759 – 763.

13

Plasma Switching by Laser Ablation

Ryosuke Hasegawa, Kazunori Fukaya and Akiyoshi Matsuzaki[1]
Mie University
Japan

1. Introduction

Plasma offers great potential for materials science [Lieberman & Lichtenberg, 2005]. Many interesting materials have been synthesized through plasma such as chemical vapour deposition (CVD). Besides, plasma etching is also known as one of the significant techniques utilizing plasma. In general, plasma is generated by direct electrical discharges of gases. On the other hand, we recently found a new method to form *pulse* plasma in an electric field *lower* than that necessary for the plasma formation by the direct discharges [Nakano et al., 2010; Tanaka et al, 2003a, 2003b, 2004]; the laser ablation of metals switches plasma formation and extinction alternately. We call this phenomenon PLASLA (plasma switching by laser ablation) and the plasma formed by this method PLASLA (plasma switched by laser ablation). The facts that PLASLA is the pulse plasma completely synchronized with laser ablation and can be formed in a lower electric field suggest PLASLA as a promising metal-catalysis reaction for materials science. For instance, the former fact suggests that we will be able to set the most suitable conditions for materials synthesis by appropriately integrating the ablation laser and the laser for the photochemical reactions and controlling their timing. As for the latter fact, the lower discharge potential of PLASLA can exclude the discharge of impurity gases. Furthermore, our previous experiments indicate that the products of the PLASLA reactions are different in the presence and absence of a magnetic field [Nakano et al., 2010]. In the present manuscript, as well as the overview of our previous study on PLASLA, we discuss PLASLA, focusing on its spatial discharge patterns and temporal waves. These ordered structures are of great interest, since plasma is a nonlinear process.

2. Generation and properties of PLASLA

Since the details of the systems and procedures of PLASLA experiments are described in the previous papers [Nakano et al., 2010; Tanaka et al, 2003a, 2003b, 2004], here we describe how to generate PLASLA with its main properties.

2.1 Generation of PLASLA

As schematically shown in Fig. 1(a), in a reaction chamber filled with CF_4 at a pressure of 0.2 Torr, with the laser ablation of Cu by the irradiation of a spinning Cu target with the fundamental beam of a Nd^{3+}:YAG laser (1.064 μm in wavelength, 45 mJ/pulse in power, 10

---

[1] Corresponding Author

Hz in repetition rate), after increasing a direct-current electric field applied between the parallel Cu plates of electrodes, stable PLASLA is formed at an electric field of 400 V, which is inadequate to form DC-plasma (direct current plasma), which is formed by the direct discharge at a potential of 500 V without laser ablation. The oscilloscope trace of PLASLA luminescence signal in Fig. 1(b) indicates that plasma is formed by the first laser ablation, quenched by the next ablation, formed again by the third ablation, quenched again by the fourth ablation, and so on, indicating that the PLASLA formation and quenching completely synchronize with laser ablation.

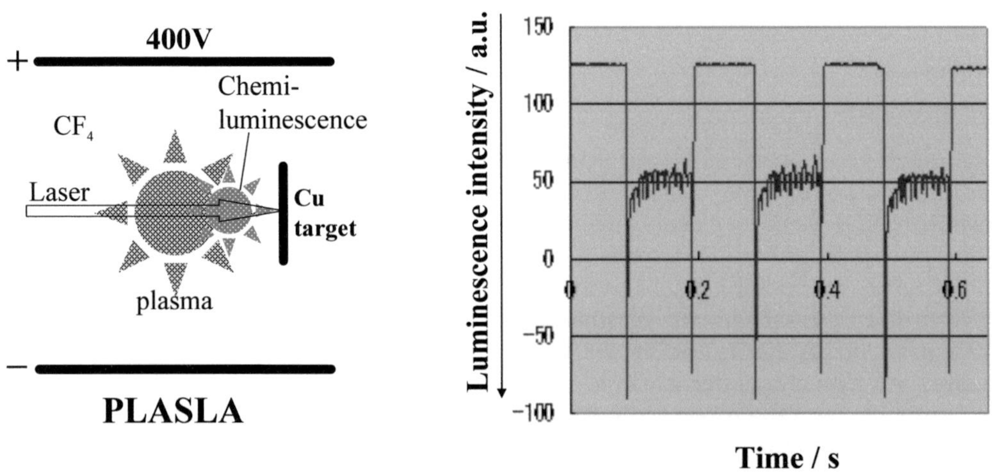

(a) (b)

Fig. 1. (a) The schematic illustration of the PLASLA experimental system. (b) The oscilloscope trace of PLASLA luminescence signal.

2.2 Properties of PLASLA

In PLASLA, the plasma can be obtained under milder conditions, since the laser ablation of metals induces the plasma in an electric field lower than that for the direct discharge in DC-plasma, suggesting the availability of the new products different from the ones of the DC-plasma. This also makes us possible to avoid the discharge of impurity gases with their low ionization potentials. For instance, in case that a minor constituent N_2 gas as the impurity discharges instead of the main constituent CF_4 gas in DC-plasma, it can be avoided in PLASLA, in which a CF_4 gas discharges as a predominant process in the same experimental system [Nakano et al., 2010]. Since the PLASLA plasma is the pulse completely synchronizing with laser ablation, the collaboration between the ablation laser and the laser for photochemical reactions will offer great potential for new materials syntheses.

The products of PLASLA in the gaseous Cu-CF_4 system are found to be polymeric carbon materials by time-resolved luminescence spectroscopy [Tanaka et al., 2003a, 2003b, 2004] and TOF (time-of-flight) mass spectrometry [Takenaka et al. unpublished data, Nakano et al., 2010]. Since various polymeric carbon materials have been extensively used for manufactures, PLASLA will be very promising in industry, as well as materials science.

PLASLA has been confirmed to form in a CF_4 gas through the laser ablation of various metals, i.e., Cu, Al, Ag, Zn, Co, Ni, Ti, Mo, and W [Nakano et al., 2010]. In view of the electronic configurations, these metals can be classified into three groups: Al, Cu, Ag, and Zn in Group-1, Co and Ni in Group-2, and Ti, Mo, and W in Group-3. The configurations in Group-1 are s^2p^1 in Al, $d^{10}s^1$ in Cu, $d^{10}s^2$ in Zn, and $d^{10}s^1$ in Ag, and have the closed highest inner sub-shells. The configurations in Group-2 are d^7s^2 in Co and d^8s^2 in Ni, and have the greater-than-half-filled highest inner sub-shells. The configurations in Group-3 are d^2s^2 in Ti, d^5s^1 in Mo, and d^4s^2 in W, and have the less-than-or-equal-to-half-filled highest inner sub-shells. On the other hand, we have experimentally found that PLASLA is most stable in Group-1, middle in Group-2, and least in Group-3 [Nakano et al., 2010], suggesting that neutral chemical reactions compete with the PLASLA plasma reactions in Groups-2 and 3 metals, since they have open highest inner sub-shells. For instance, we have confirmed that the PLASLA formation rate is fastest in Group-1, middle in Group-2, and slowest in Group-3; that the PLASLA extinction time is longest in Group-1, middle in Group-2, and shortest in Group-3.

We have found the various interesting facts on the interaction of PLASLA with a magnetic field [Nakano et al., 2010]. The magnetic field effects on PLASLA are due to the magneto-hydro-dynamics (MHD) processes and the magnetic effect on the chemical reactions. The main MHD processes in PLASLA are the ***E***x***B*** drift and the cyclotron circulation of ions and electrons. These processes increase the PLASLA potential, at which PLASLA is formed, and the activation energy of PLASLA plasma reactions, since they interrupt the movements of ions and electrons for the plasma chemical processes. As a magnetic field increases the PLASLA potential, the shape of PLASLA pulse varies considerably. By the analysis of the magnetic field effect on the activation energy of PLASLA, it is found that a magnetic field increases the activation energy of the charge-transfer processes and decreases that of the ionic chemical reactions accompanying chemical-bond cleavages. The TOF-mass spectrometry shows that the PLASLA product materials formed in a magnetic field are different from those of zero-field PLASLA. That is, for PLASLA in a magnetic field, the detected carbon clusters are rather small in size and are free from impurities such as metals and fluorine.

Thus, in view of materials science, PLASLA is a promising metal-catalysis reaction.

3. PLASLA discharge pattern

In this section, we study the mechanism of PLASLA by quantitative analysis of PLASLA discharge patterns with various metal targets.

3.1 Discharge pattern imaging

The upper panel in Fig. 2 shows the present imaging system. The lower panel in this figure shows the typical images of DC-plasma, laser ablation (CuF chemiluminescent reaction), PLASLA with a Cu target. In DC-plasma, bright plasma sheaths are formed around the electrodes. In the laser-ablation image, the bright sphere of CuF chemiluminescence is observed near the Cu target. In PLASLA, the bright plasma sheaths around the electrodes and the bright sphere of CuF chemiluminescence near the Cu target are observed.

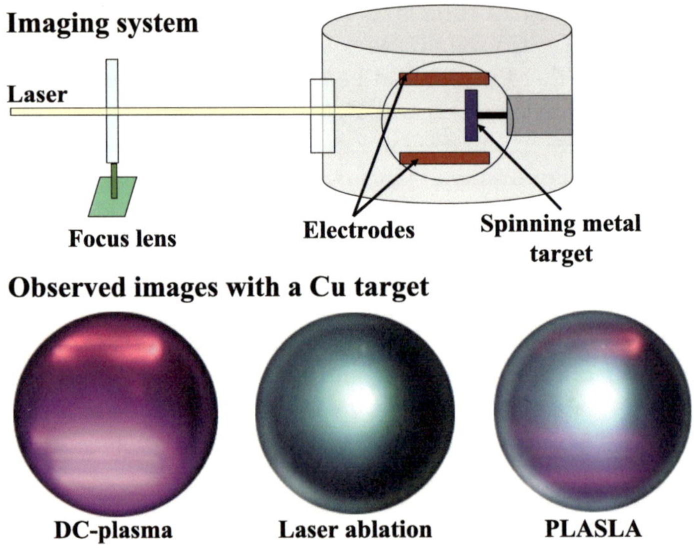

Fig. 2. Upper panel shows the imaging system. Lower panel shows the observed images of DC-plasma, laser ablation (CuF chemiluminescent reaction), and PLASLA.

3.1.1 PLASLA discharge patterns with various metals

Figure 3 shows the PLASLA discharge patterns with various target metals. PLASLA can be formed with all these metals. The deformation of the sheaths is evident in the discharge patterns of the Ti, W, and Mo targets. These target metals have been classified into Group-3, having electronic configurations of d^ns^2 ($n \leq 5$), where the electronic configurations are $d^{10}s^1$ ($d^{10}s^2$ for Zn) in Group-1 metals and d^ns^2 ($n>5$) in Group-2 metals. Such deformation cannot be observed in other metals. We suppose that the deformation is closely related to the fact that PLASLA is rather unstable with Group-3 metals. As a mechanism for the unstable PLASLA with Group-3 metals, the competition of plasma ionic chemical processes with neutral radical chemical processes is suggested from the electronic configurations [Nakano et al., 2010]. Furthermore, the figure indicates that the PLASLA potentials are rather low, middle, high in Group-1 metals (Ag, Al, Zn, Cu), Group-2 metals (Co, Ni), and Group-3 metals (Ti, W, Mo), respectively. The same mechanism as that for the PLASLA stability has been proposed for the fact [Nakano et al., 2010].

Figure 4 shows the distributions of PLASLA luminescence intensity near electrodes for Cu and Ti targets. As shown in Fig. 4(a), Regions 1-4 are set on the PLASLA discharge image. After the colour of the images is converted to grey-scale ones, the luminescence intensity is expressed in terms of a lightness scale from 1 to 25, as shown in Fig. 5. Thus, the PLASLA luminescence intensities in Regions 1-4 outside and inside the reaction site near electrodes are shown in Figs. 4(b) and (c) for Cu and Ti targets, respectively. While the distributions in Regions 1-4 are very similar between Cu-PLASLA and Ti-PLASLA inside the reaction sites near the anodes, they are evidently different outside the reaction site. In Cu-PLASLA, the distributions are rather flat over Regions 1-4 outside the reaction site, while the luminescence intensity is low in Regions 1 and 2 and steeply increases from Region 3 to Region 4 outside the reaction site in Ti-PLASLA. These results are evident from the images

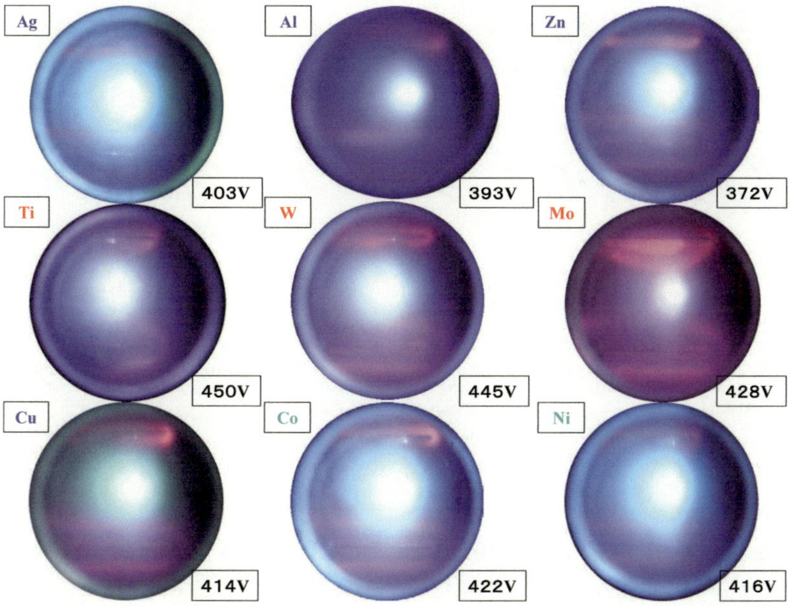

Fig. 3. PLASLA discharge patterns with various target metals. Metals and PLASLA potentials are indicated on the upper-left sides and lower-right sides of discharge images, respectively. The colours indicating metals are classifying them into groups: Group-1 is in blue, Group-2 in green, Group-3 in red.

of Cu-PLASLA and Ti-PLASLA discharge patterns in Fig. 3. This figure indicates that the stable sheaths are formed around electrodes in Cu-PLASLA, while the sheaths around the electrodes are completely quenched in Regions 1 and 2 in Ti-PLASLA. Furthermore, the sheaths peel off the electrodes in Ti-PLASLA. These facts indicate that there is a diference in the chemical processes between Cu-PLASLA and Ti-PLASLA, suggesting the additional chemical processes in Ti-PLASLA. We suppose that this additional chemical processes are the neutral radical reactions competing with the plasma ionic reactions.

The discharge image of Cu-PLASLA in Fig. 3 and its luminescence intensity distribution in Fig. 4(b) indicate that the stable, undeformed sheaths are formed arounds the anode and the cathode in Cu-PLASLA, yet the plasma luminescence disappears only in Regions 1-3 on the reaction site of the anode, suggesting that significant chemical processes occur in this area. The similar quenching of the plasma luminescence in Regions 1-3 on the reaction site of the anode is observed in the discharge patterns of Al-, Co-, Ni-PLASLA, as shown in Fig. 3.

3.1.2 PLASLA discharge patterns between Cu electrodes

In this section, we discuss the Cu-PLASLA discharge pattern between the electrodes. For the analysis, the 25 x 4 grid was set on the discharge pattern images, as shown in Fig. 5. Then the images were transformed from colour to grey-scale ones. Then, the lightness of the areas in the grid on the grey-scale images was transferred to the 1-25 quantities with use of the scale, as shown in Fig. 5. Four columns of the grid are named Regions 1-4, as shown in this figure.

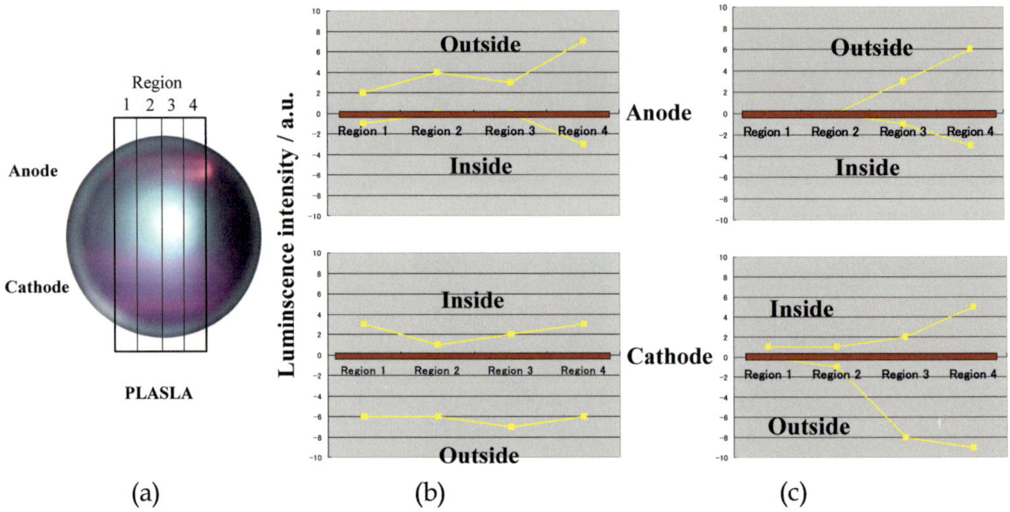

Fig. 4. Luminescence intensity distributions near electrodes in Regions 1-4, evaluated from PLASLA discharge patterns. (a) Regions 1-4. (b) Cu target. (c) Ti target.

The results of the analysis are shown in Fig. 6. In this figure, the luminescence intensity is plotted as a function of distance from the cathode in Regions 1-4 for DC-plasma, Laser ablation (CuF-chemiluminescent reaction), and PLASLA images. In the abscissa, the positions of the cathode and the anode are at 0 and 3.5 cm, respectively. In DC-plasma, the figure indicates that the sheath around the cathode is very stable and the luminescence intensity constantly decreases from the cathode to the anode. As for the resemblance among the curves in Regions 1-4, they are completely same in the area from the cathode to the anode, while there are clear differences only near the anode, indicating the differences in the sheaths on the anode. The fact that the sheaths around the anode greatly vary in Regions 1-4 is probably due to the difference in the stability in the spatial distribution of ions and electrons.

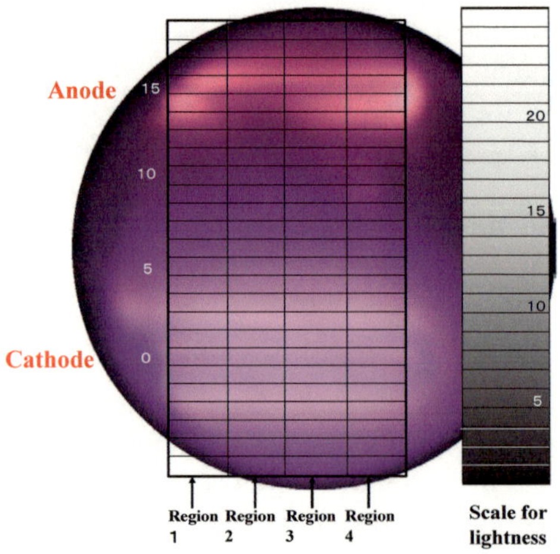

Fig. 5. Grid on the discharge pattern with the lightness scale for the analyses in Fig. 6.

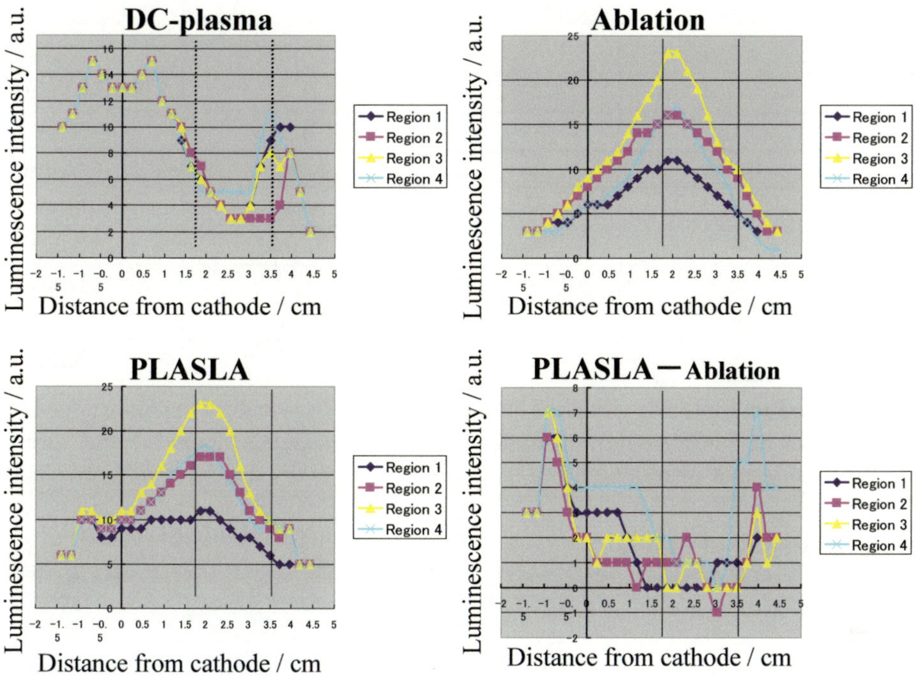

Fig. 6. Luminescence intensities between electrodes in DC-plasma, Ablation (CuF chemi-luminescence), PLASLA, and the difference in the luminescence intensity between the PLASLA and laser-ablation patterns.

In Ablation (laser ablation), the image is essentially spherical, but it is evidently extended in the flight direction of laser-ablated Cu atoms. Since the image of PLASLA includes the laser-ablation luminescence as well as the plasma luminescence of PLASLA, the latter image is extracted by subtracting the former image from the PLASLA image, as shown in the right-bottom panel of the figure. As compared with the image of DC-plasma, the luminescence intensity between the electrodes is largely reduced in this subtraction image, i.e., the plasma luminescence image of PLASLA, suggesting the more enhanced chemical reactions. The quenching is greatest and spatially widest in Regions 2, 3 and 1. The quenching in Region 4 is rather small in the inside site on the cathode and is rather large in the inside region on the anode. Furthermore, while the sheath completely surrounds the anode in DC-plasma, it is quenched in the inside site on the anode in the plasma luminescence image of PLASLA, indicating the additional chemical reactions in PLASLA. Thus the present image analysis has found that the plasma reaction occurs with laser-ablated metal atoms widely between the cathode and the anode in Regions 2, 3, and 1. The study on the product materials deposited on the electrodes will help us for the further discussion.

4. Product materials attached to the electrodes

Product materials deposited on the electrodes were stored in four bottles as samples by separately collecting the materials from the both sides of the cathode and the anode after careful observation of their colours one by one in each PLASLA experiment. The results are shown in Table 1 and in Fig. 7.

Metals	Anode/ inside	Anode/ outside	Cathode/ inside	Cathode/ outside
Ag	B, 4	B, 1	None	None
Al	B, 4	B, 2	B, 4	None
Zn	1	None	B, 1	None
Ti	B, W, 2	W, 4	None	None
W	B, 2	None	None	None
Mo	B, 2	None	None	None
Cu	B, 4	B, 4	B, 1	None
Co	B, 3	W, 4	B, 1	None
Ni	B, 3	B, 2	B, 2	None

Table 1. The product materials attached to the electrodes for various metal targets. B and W indicate the black and white coloured materials, respectively. Numbers 1-4 and "none" indicate the amounts of product materials: 1 (a little) to 4 (very much) and "none" is 0.

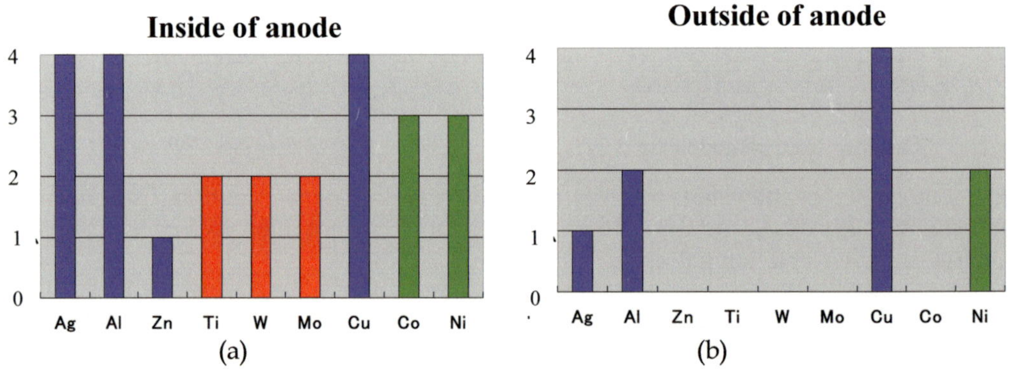

Fig. 7. The yields of carbon-like (black-coloured) product materials attached to the anode for various metal targets. In the ordinates, scales indicate the amounts of the product materials: 0 (none) to 4 (very much).

The results in Table 1 indicate that the yields of materials are evidently greater in the anode than in the cathode. Furthermore they are evidently greater in the reaction side of the both electrodes, i.e., between electrodes, than in another side. The main products on the electrodes are carbon-like (black-coloured) materials. Yet, besides the materials, white-coloured materials are yielded on the both sides of the anode in the Ti target and on the non-reaction side of the anode in the Co target. The white-coloured materials are yielded more in the non-reaction side of the anode than in its reaction side, while the black-coloured materials are more yielded in the reaction side of it. Figure 7, which shows the yields of carbon-like (black-coloured) product materials deposited on the anode for various metal targets, indicates that the carbon-like materials are yielded most in the targets of Group-1 metals, middle in the targets of Group-2 metals, and least in the targets of Group-3 metals. As described in the previous papers [Nakano et al., 2010], PLASLA is most stable in the target of Group-1 metals, middle in the targets of Group-2 metals, and least in the targets of Group-3 metals. This fact is due to the electronic configurations of the target metals [Nakano et al., 2010], which suggest that neutral chemical processes compete with the plasma reactions of PLASLA in the targets of Groups-2 and 3 metals. Hence the white-coloured product materials on the anode with the Ti and Co targets are suggested to be the products of the neutral chemical reactions [Nakano et al, 2010]. This agrees with the fact that the white-coloured materials are more yielded on the non-reaction side of the anode in the Ti and Co targets, where the plasma reaction of PLASLA is inactive.

The PLASLA products materials will be discussed more in detail with the results of the TOF mass spectrometry and Time-resolved luminescence spectroscopy measurements [Takenaka et al., unpublished data].

5. Transient signal analysis of PLASLA formation and extinction

As shown in Fig. 1(b), the PLASLA luminescence is pulse, synchronized with laser ablation. The shape of the PLASLA luminescence pulse is studied. As shown in Fig. 8, the pulse shape in the PLASLA formation is found to be classified into two types: Type-A in Fig. 8(a) and Type-BC in Fig. 8(b). In Type-A, the PLASLA ignition pulse simply decays to the constant stable luminescence, while the overshoot labelled by B and C in Fig. 8(b) is found in Type-BC. Furthermore, it is found that Type-A is observed in PLASLA with Group-1 metal targets and Type-BC is done in PLASLA with Groups-2 and 3 metal targets. However, it should be noted that Type-A was once observed in PLASLA with the Ti target, while Type-BC is usual with the target. This fact suggests that the both types are due to the common mechanism and only depend on a parameter

Thus, as shown in Fig. 9, we made the simulation study for the time features of the PLASLA ignition by using a simple model, which is the combination of the signals of the ablation luminescence and the plasma luminescence as a function of delay of the latter signal from the former signal. The results indicate that their overall signal varies from Type-A to Type-BC with the delay between these signals. Hence it is found that the plasma luminescence signal appears rather immediately after the laser ablation for Group-1 metal target, while it is delayed rather more for Groups-2 and 3 metals target. That is, it takes more time for the plasma reaction in PLASLA with Groups-2 and 3 metals target. Thus it is found that PLASLA is formed more easily with Group-1 metal target than with Groups-2 and 3 metals target, as previously described [Nakano et al., 2010].

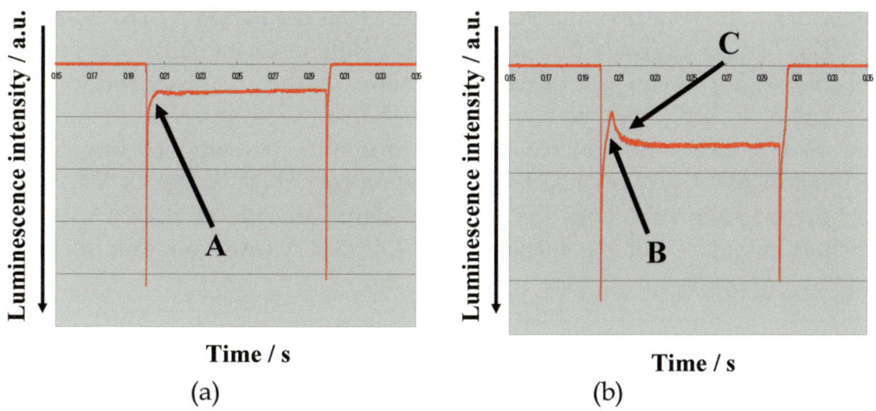

Fig. 8. The transient signals of PLASLA luminescence. (a) Type-A, (b) Type-BC, where targets: (a) Ag, (b) Ti.

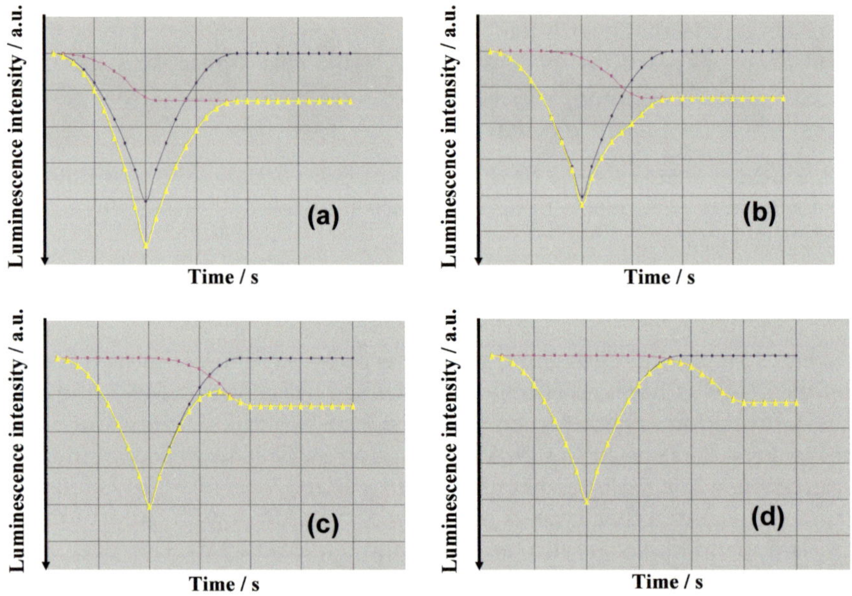

Fig. 9. The schematic illustration of the transient signals of PLASLA formation with ablation luminescence signals (blue), plasma luminescence signals (pink), and their overall signals (yellow) for various delays of the plasma signals, where the delay increases from (a) to (d).

As shown in Fig. 10(a), the decay rate of process C is found to correlate with the decay rate constant of process B for various target metals such as Al, Ni, Co, Ti, W, and Mo; the decay rate of process C exponentially increases with the decay rate constant process B. Hence it is evident that the Type-C process is the recovery process of the Type-B process. Furthermore, as described in the previous paper [Nakano et al, 2010], we have confirmed that the PLASLA formation rate in the Type-A process exponentially decreases with ionization potential of target metals and concluded that PLASLA starts from the sequent initial processes of ($M + e \rightarrow M^+ + e$) and ($M^+ + CF_4 \rightarrow MF + CF_3^+$), where M is the target metals.

On the other hand, the feature of the PLASLA pulse near the PLASLA extinction is shown in Fig. 10(b). The feature suggests that the PLASLA luminescence signal decays in the two steps probably with the sequence of two exponential decays. As described in the previous paper [Nakano et al, 2010], we also have confirmed that the PLASLA extinction time, which is defined as t in Fig. 10(b), increases with ionization potential of target metals and concluded that PLASLA is quenched through the process of ($M + CF_4^+ \rightarrow M^+ + CF_4$), where M denotes the target metals. Thus the transient signal analysis of the PLASLA formation and extinction indicates that the features of the PLASLA pulse are due to the common mechanism for various target metals.

6. Chemical waves in PLASLA

Plasma is one of the ordered structures through the nonlinear processes. In fact, the spatial waves are observed in the PLASLA photo-images, as discussed in section 3. Besides these spatial waves, temporal waves will also be formed in PLASLA. In this section, we

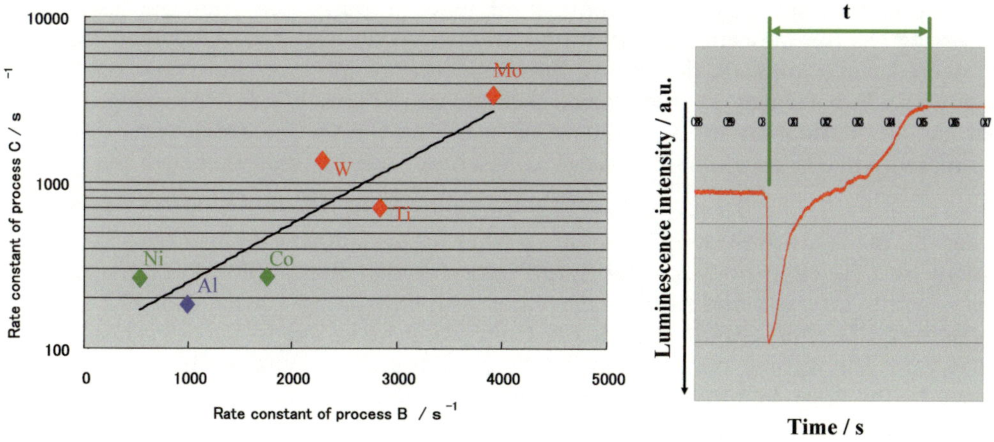

(a) (b)

Fig. 10. (a) The plot of logarithmic rate constant of process C as a function of rate constant of process B in the Type-BC PLASLA formation. (b) Transient signal of PLASLA luminescence near the extinction, indicating the PLASLA extinction time t. (Target: Ni).

investigate the temporal waves in PLASLA for the gaseous Cu-CF_4 system by analysing the waves superimposed onto the pulsing time profile of PLASLA luminescence intensity.

6.1 Systematic waves in PLASLA

As shown in Fig. 1(b), the transient signal of PLASLA luminescence pulses. It is found that the systematic waves are superimposed onto the pulses, as shown in Fig. 11.

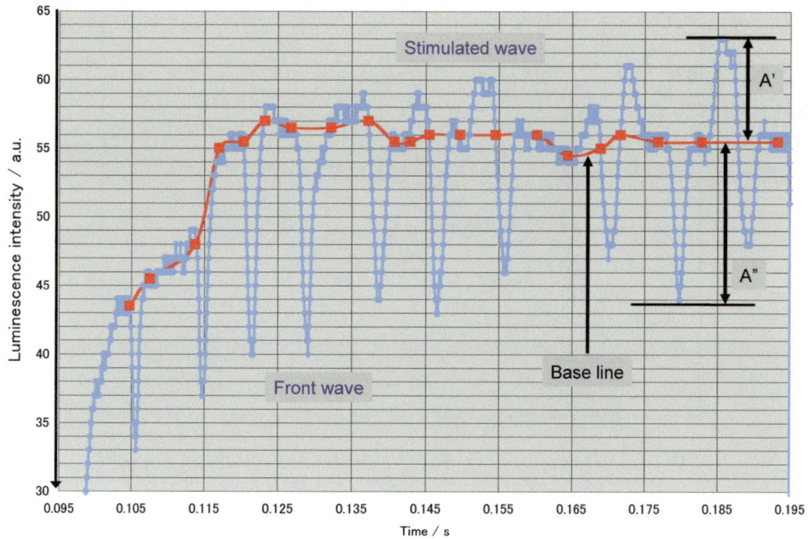

Fig. 11. Enlarged transient signal of PLASLA luminescence (Blue solid line) with the baseline (red solid line). A′ and A″ are the amplitudes of stimulated wave and front wave, respectively.

The figure indicates that the periodic front wave appears just after the laser ablation, increasing the luminescence intensity measured from the baseline. After that, the periodic stimulated wave appears, decreasing the luminescence intensity from the baseline in the front wave. In particular, its appearance is clear since 125 ms. The feature in Fig. 11 suggests that the laser ablation induces the front wave and then the front wave induces the stimulated wave to be a united wave. The baseline, which is indicated by a red line in this figure, largely varies before 125 ms and is constant after that.

In Fig. 12, the logarithmic amplitude of the front wave, which is defined by A″ in Fig. 11, is plotted as a function of time. The figure indicates that the amplitude of the front wave exponentially increases and then the decreases with time at rate constants of 23.2 and 10.1 s^{-1}, respectively, i.e., the growth rate of the front wave turns out to be twice faster than the decay rate. The turning-point region for the growth and decay of the front wave, which is shown by the green belt in Fig. 12, completely coincides with the region, where the baseline becomes constant.

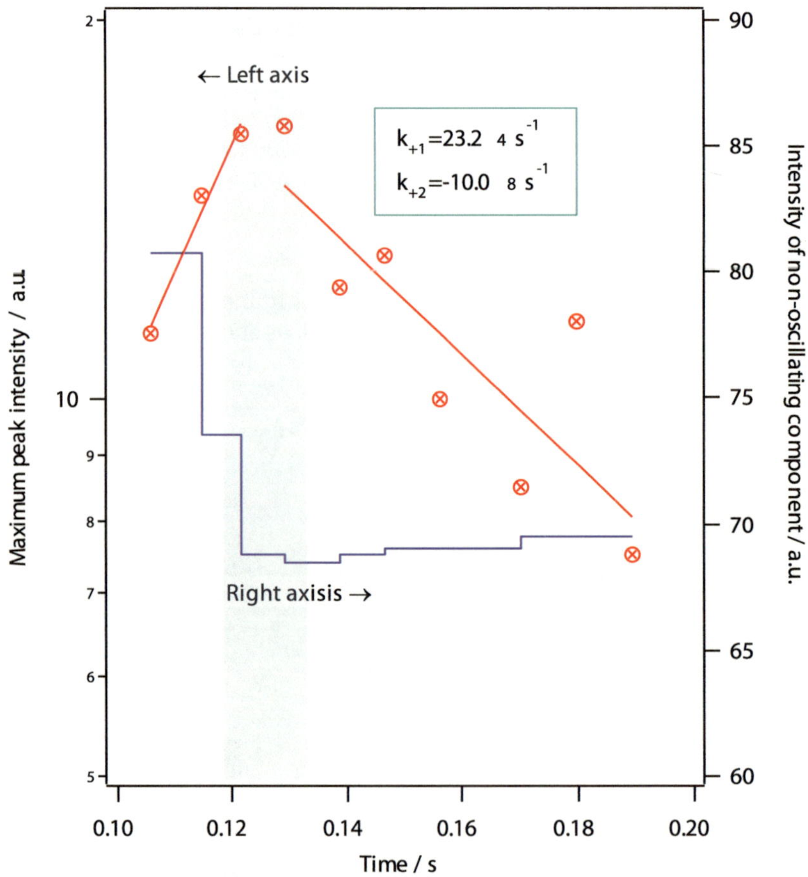

Fig. 12. Plot of the amplitude of the front wave, which is defined by A″ in Fig.11, as a function of time for the left ordinate. The blue solid line is the baseline, which is shown by a red line in Fig. 11, for the right ordinate. The green belt near 0.12-0.13 s indicates the turning point for the growth and decay of the front wave.

Figure 13 indicates the plot of the logarithmic amplitude of the stimulated wave, which is defined by A′ in Fig. 11, as a function of time. The amplitude of the stimulated wave also exponentially increases with time. The growth rate constant of the stimulated wave turns out to be 34.9 s^{-1}, which is faster by 3.5 than the decay rate constant of the front wave. Hence it is found that the change in the system is rather rapid just after the laser ablation to form the front wave, which approaches to a steady oscillation rather slowly after that, inducing the stimulated wave, which grows rather faster to be combined to the united wave with the front wave. Thus, when the front wave and the stimulated wave converge on the united wave, the rates of the changes in their amplitudes are different, while they have essentially the common period.

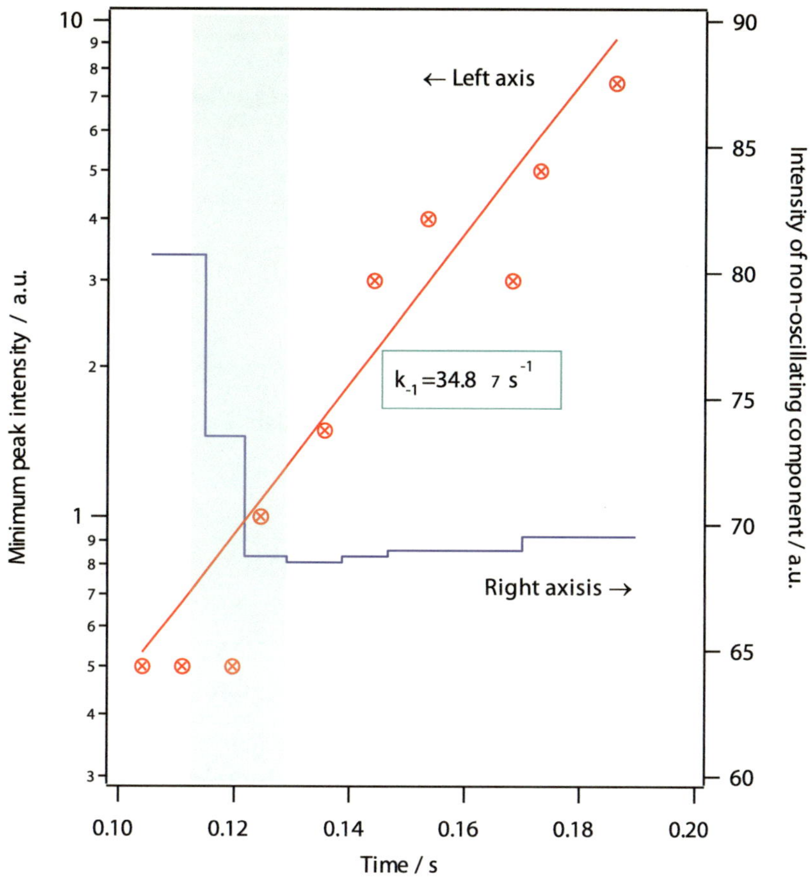

Fig. 13. Plot of the amplitude of the stimulated wave, which is defined by A′ in Fig.11, as a function of time for the left ordinate. The blue solid line is the baseline, which is shown by a red line in Fig. 11, for the right ordinate. The green belt near 0.12 s indicates the turning point for the baseline to become constant.

In Fig. 14, the total intensity, which is defied as the sum of the amplitudes of the front and stimulated waves, is plotted as a function of time. The figure indicates that the intensity linearly increases with time, expressed in terms of (amplitude = 13.9 + 8.4t), where t is time in s of its unit. The feature of the total peak intensity as a function of time suggests that the front wave and the stimulated wave are united to be a single wave, amplitude of which slightly increases with time.

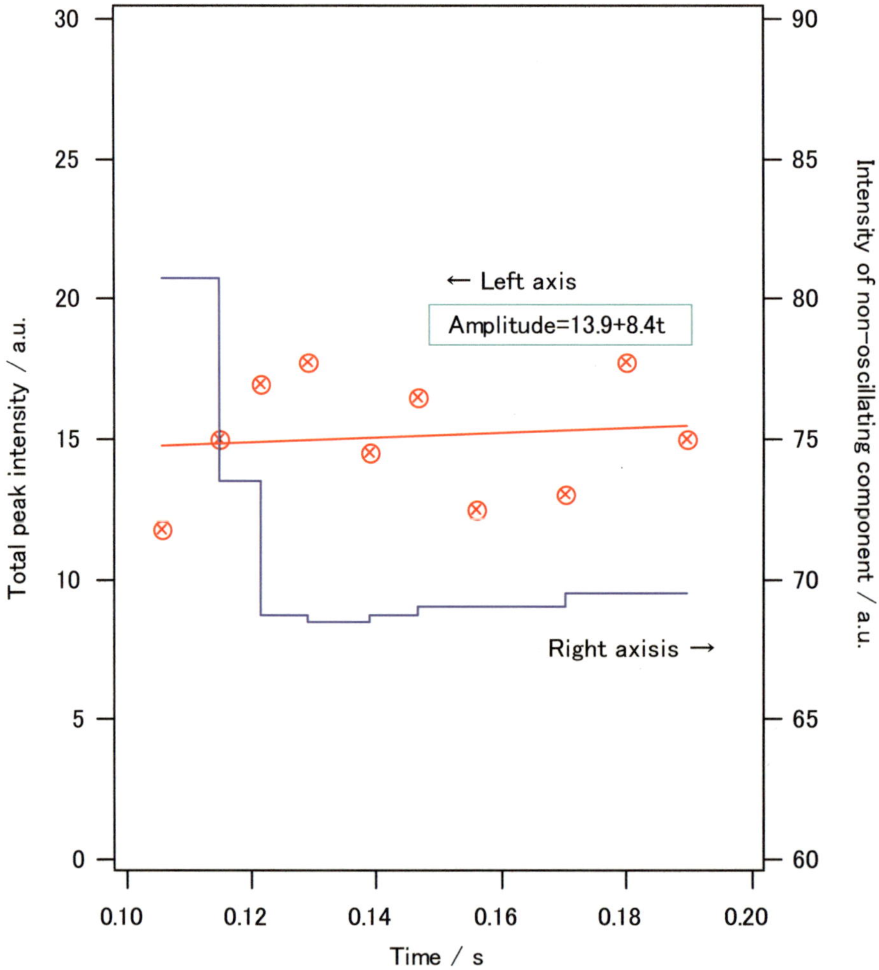

Fig. 14. Plot of the sum of the amplitudes of the front and stimulated waves, which are defined by A″ and A′ in Fig.11, as a function of time for the left ordinate. The blue solid line is the baseline, which is shown by a red line in Fig. 11, for the right ordinate.

The fact that the front and stimulated waves are united is evident also in Fig. 15; as the mid-point of the amplitudes of these waves converges to zero, the amplitudes of these waves become equal. The figure indicates that the front and stimulated waves appear alternatively. The periods of these waves fluctuate and the fluctuation is rather large in the stimulated wave.

This change in the PLASLA system is schematically illustrated in Fig. 16. At the initial point, the front wave is generated by laser ablation, grows after that, and is united with the stimulated wave to be transformed into a steady-state oscillation. Thus, by the study in this section, it is found that the waves generated in PLASLA have the systematic characteristics.

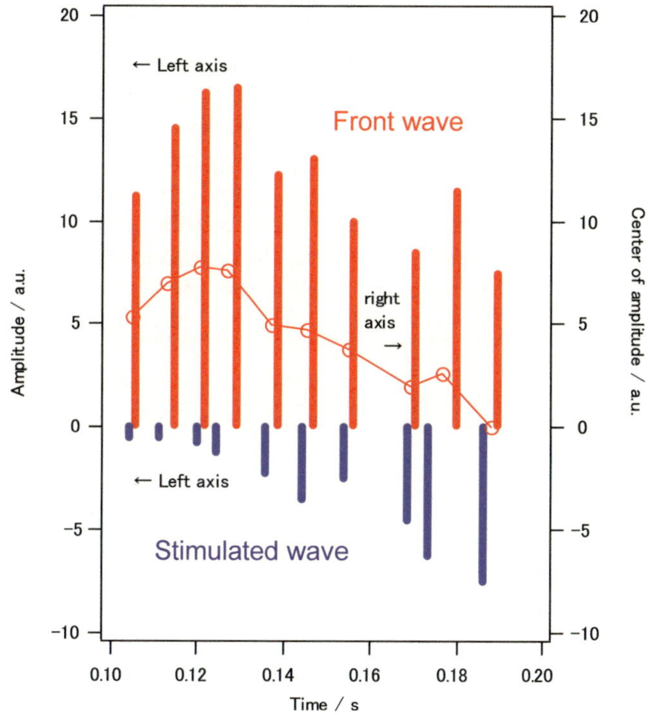

Fig. 15. Plot of the amplitudes of the front and stimulated waves, which are defined by A″ and A′ in Fig.11, as a function of time for the left ordinate. The red solid line is the mid-point of the sum of the amplitudes of the front and stimulated waves, plotted for the right ordinate.

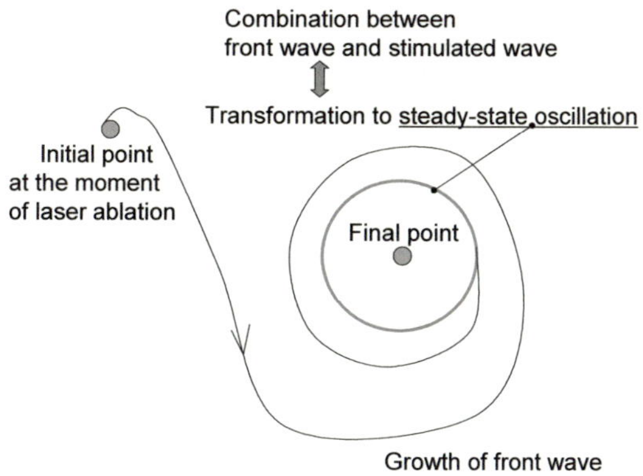

Fig. 16. Schematic illustration of the change in the PLASLA system, where, after the front wave generation by the laser ablation, the front wave is united with the stimulated wave.

6.2 Stability of systematic waves in PLASLA

The aim in this section is to make quantitative study on the stability of the systematic characteristics of the waves generated in PLASLA. In Fig. 17, we show five waves sequentially measured under the same condition in the same experiment. The waves can be classified into the waves with ten peaks and nine peaks; waves 1, 2, and 5 have ten peaks and waves 3 and 4 have nine peaks. The correlation coefficients among these waves are calculated and shown in Table 2. The results in the table indicate that the coefficients are close to 1, i.e., 0.83-0.97 among the ten-peak waves and 0.84 among the nine-peak waves. On the other hand, the coefficients between the ten-peak waves and the nine-peak waves are found to be very small, i.e., -0.044 to 0.13, where the negative value indicates the anti-phase correlation, while it is very small in the present case. These facts indicate that the waves generated in PLASLA are very systematic and classified into the two groups, i.e., the ten-peak waves and the nine-peak-waves. Therefore, they essentially support the model in Fig. 16, suggesting the additional bifurcation during the transformation to the steady-state oscillation.

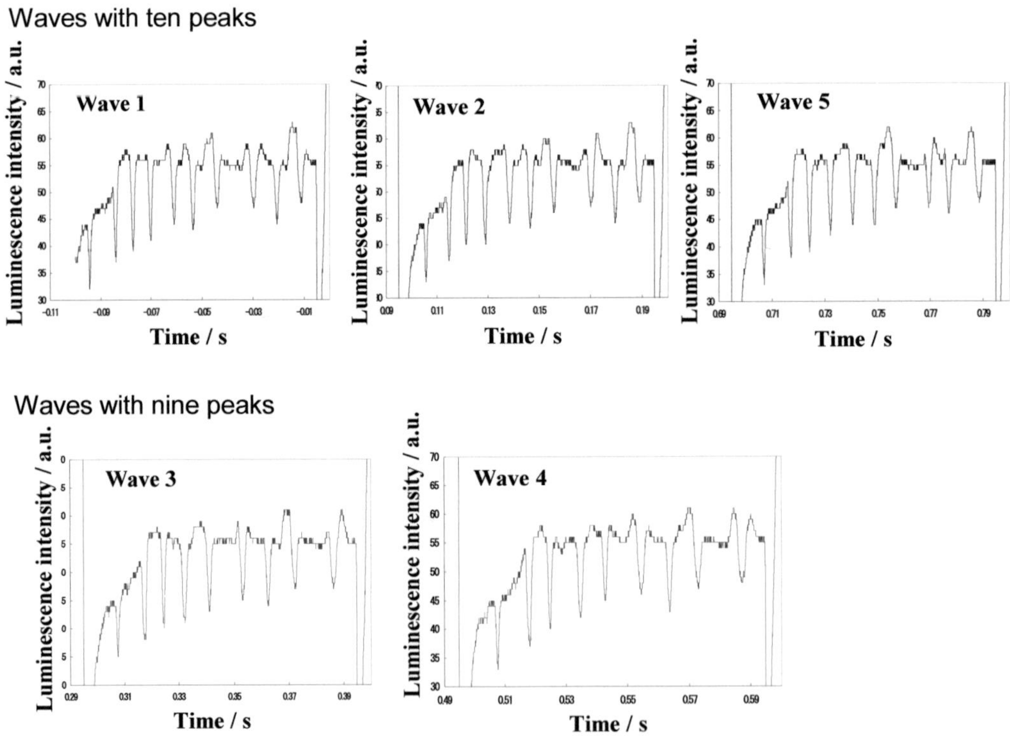

Fig. 17. PLASLA waves measured under the same conditions in the same experiment.

	Wave 1 (10 peaks)	Wave 2 (10 peaks)	Wave 3 (9 peaks)	Wave 4 (9 peaks)	Wave 5 (10 peaks)
Wave 1 (10 peaks)		**0.971**	0.127	0.074	**0.861**
Wave 2 (10 peaks)			0.182	0.149	**0.827**
Wave 3 (9 peaks)				**0.840**	0.051
Wave 4 (9 peaks)					-0.044
Wave 5 (10 peaks)					

Table 2. Correlation coefficients between the periods of waves for variations.

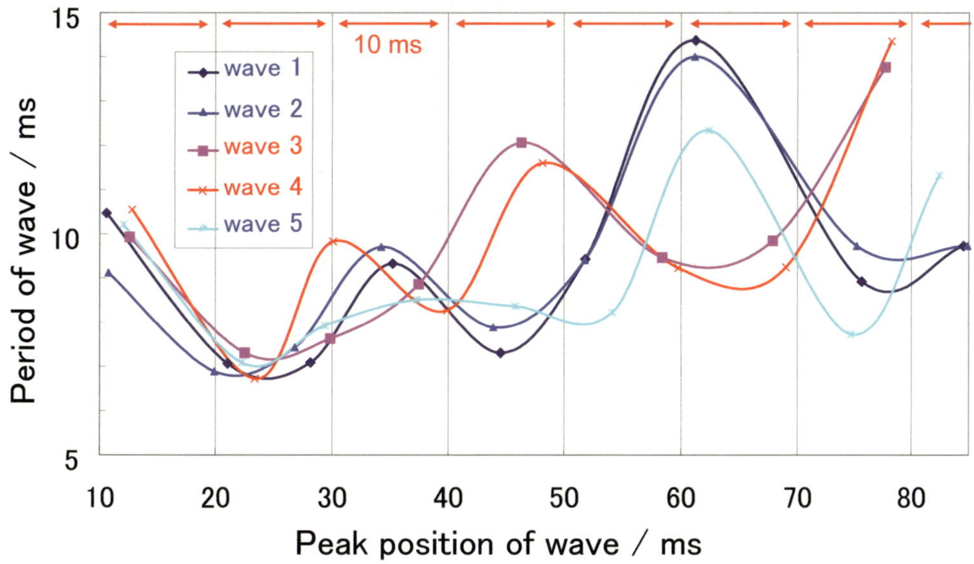

Fig. 18. Periods of waves 1-5 as a function of the peak positions of the waves.

For the study on the bifurcation into the ten-peak wave or the nine-peak wave, we investigate the temporal variations of the correlation coefficients of these five waves, i.e., the time-resolved correlation coefficients of these waves. For the time-resolved correlation coefficients, the periods of these five waves are plotted as a function of time, i.e., the peak positions of the waves, as shown in Fig. 18. In this figure, the abscissa is divided into eight sections by 10 ms. The time-resolved correlation coefficients are obtained by calculating the correlation coefficients of these curves in Fig, 18 in the individual sections of the abscissa and shown in Fig. 19. In Fig. 19, the time resolved correlation coefficients of waves 2-5 with wave 1 are shown, where the coefficients of the waves are normalized to be 1 at 10 ms, i.e., in the section between 10 and 20 ms.

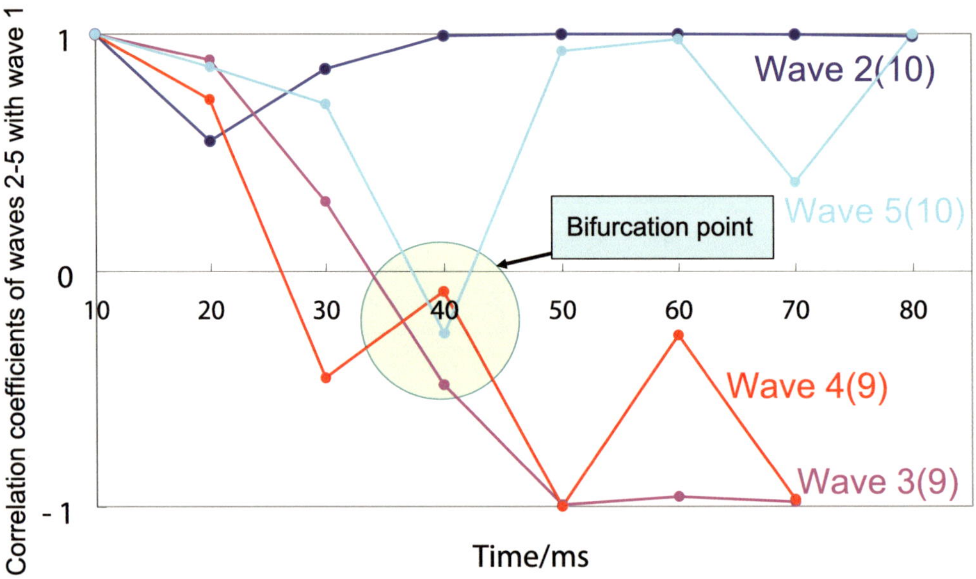

Fig. 19. Time-resolved correlation coefficients of waves 2-5 with wave 1. The numbers in the parentheses are the numbers of the peaks of the waves.

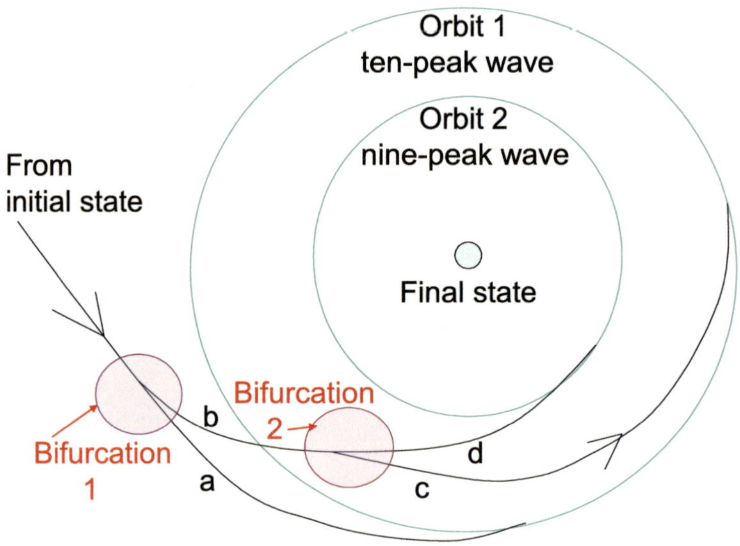

Fig. 20. Schematic illustration of the time evolution of the PLASLA system by the revision of Fig. 16 near the final state.

As shown in Fig. 19, the correlation coefficient of wave 2 decreases a little at 20 ms and after that recovers, approaching to 1. Therefore wave 2 bifurcates to the ten-peak orbit. In Fig. 20, the revised version of the schematic illustration of the time evolution of the PLASLA system in Fig. 16 is shown. In wave 2, the system goes from the initial state to orbit-1 of the ten-peak wave through bifurcation-1 and path-a.

Another ten-peak wave, wave 5, shows the very different behavior from wave 2, as shown in Fig. 19. The correlation coefficient of wave 5 decreases to zero at 40 ms and recovers to one after that. This behavior of wave 5 will be described as follows with Fig. 20. In wave 5, the system goes from the initial state to orbit-1 of the ten-peak wave through bifurcation-1, path-b, bifurcation-2, and path-c.

Let us study the behavior of the nine-peak waves, waves 3 and 4. As shown in Fig. 19, their correlation coefficients decrease to zero at 40 ms and decease to -1 after that. The behaviors of waves 3 and 4 will be described as follows with Fig. 20. In these waves, the system goes from the initial state to orbit-2 of the nine-peak wave through bifurcation-1, path-b, bifurcation-2, and path-d.

Based on the study in this section, we conclude that systematic temporal waves exist in PLASLA and their behavior is expressed in terms of bifurcation in Figs. 16 and 20. Hence the present mechanism for the temporal waves in PLASLA is found to be closely related to the general concepts for self-organization [Haken, 2006]. We suppose that the plasma chemical processes significantly contribute to the temopral waves in PLASLA, which will be discussed in Section 6.4. Self-organization through chemical processes is known in various systems. For instance, we found that the periodic pulse waves are formed in laser-induced aerosol formation in gaseous CS_2, calling it aerosol burst, and studied it experimentally and theoretically [Matsuzaki, 1994].

6.3 Effect of electrodes potential on PLASLA

It is of great interest to study the factors, which control the systematic temporal waves in PLASLA. The potential between the electrodes will be one of the important factors. In this section, we study the effect of the potential between electrodes on the systematic temporal waves in PLASLA.

In Fig. 21, the effect of the potential between the electrodes on the growth and decay rates of the amplitude of the front wave is shown. Both rates are constant under the threshold potential and steeply increase over it.

In Fig. 22, we study the effect of the potential between the electrodes on the growth rate of the amplitude of the stimulated wave. The growth rate of the amplitude of the stimulated wave decreases with an increase in the potential between the electrodes. As shown in the figure, the best fit function is the linear function with the negative slope, giving the good correlation coefficient, i.e., R^2=0.97.

Thus it is found that the potential between the electrodes affects the front wave and the stimulated wave. Hence the potential is regarded as the significant factor, which controls the waves of PLASLA. This fact will be approved, since the chemical reactions in PLASLA will be affected by the potential through the change in the kinetic energy of electrons. Yet, the mechanism for the potential effect seems to be not simple, since the dependence on the potential is quite different in these waves.

The potential dependence of the front wave suggests that the change in the front wave occurs as the plasma phase transition. The increase in the kinetic energy of electrons does not linearly enhance the front wave but cause the plasma phase transition. That is, while the increase in the electron energy will increase the significant chemical species in the plasma, the plasma phase does not change under the threshold and the phase transition drastically takes place over the threshold. The potential dependence of the front wave will be closely related to this mechanism.

On the other hand, this kind of phase transition seems to be not related to the potential dependence of the stimulated wave. While some assumptions for the mechanism of the anti-correlation between the growth rate of the amplitude of the stimulated wave and the potential between the electrodes can be proposed, here we will discuss the following one, which we consider most probable. That is, the increase in the plasma species interrupts the chemical processes enhancing the stimulated wave. For example, when the chemical processes for the stimulated wave are composed of the chemical reactions by neutral species, they must compete with the ionic chemical processes by the plasma species. As a result, the growth of the stimulated-wave amplitude will be suppressed by the increase in the potential between electrodes.

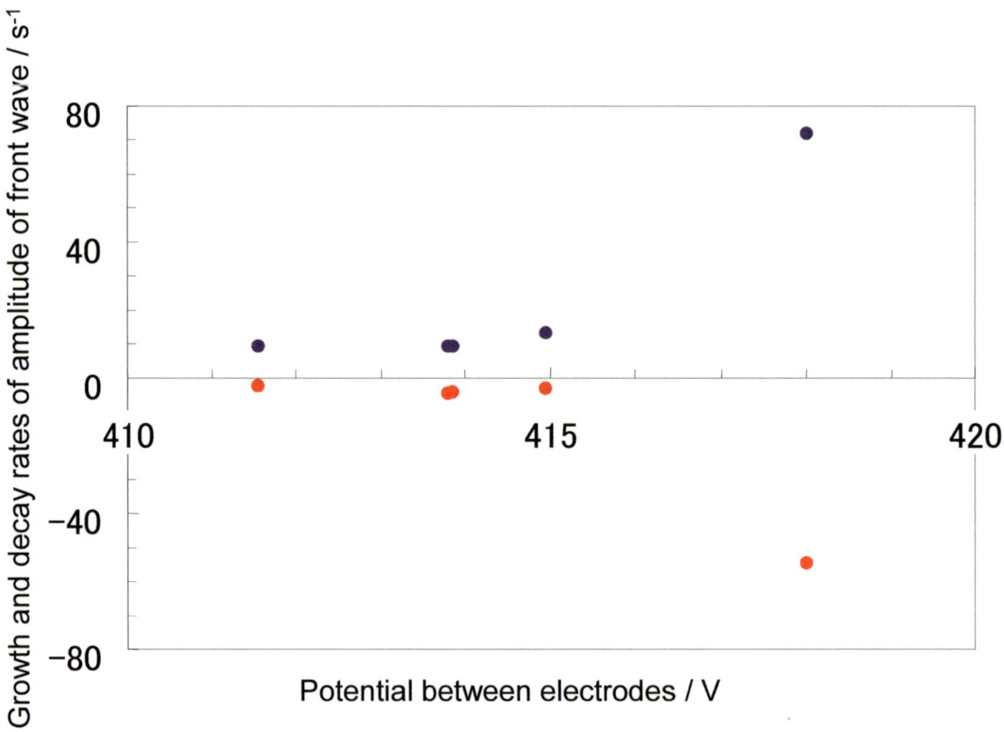

Fig. 21. Growth and decay rates of the amplitude of the front wave as a function of potential between electrodes. Blue: growth rate, red: decay rate.

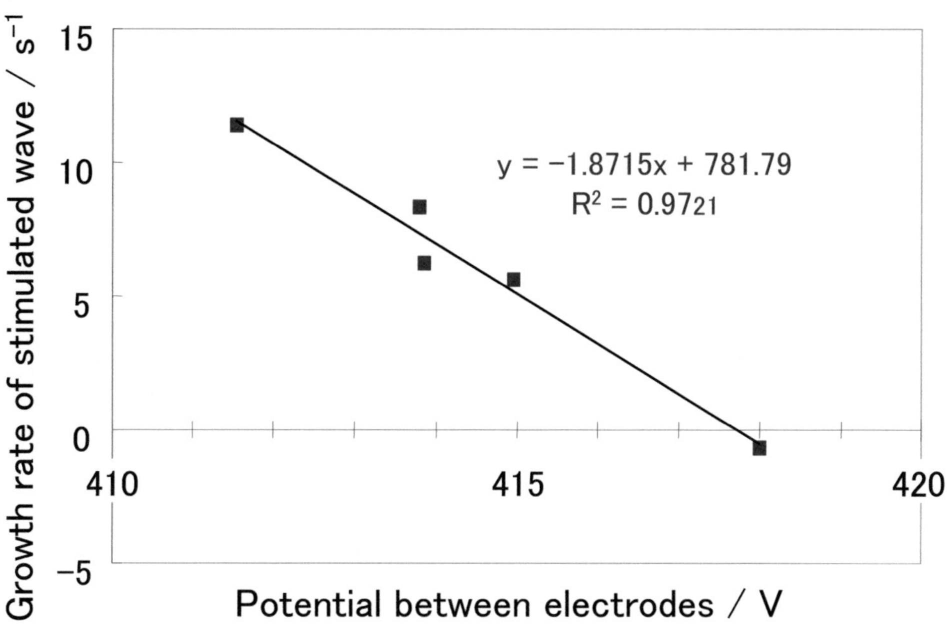

Fig. 22. Growth rate of the amplitude of the stimulated wave as a function of potential between electrodes. In the attached equation, which is the best-fit function, y and x are the growth rate of the amplitude of the stimulated wave and the potential between electrodes, respectively. R^2 is the correlation coefficient of the best-fit function.

6.4 Theoretical discussion on the temporal waves in PLASLA

On the basis of the experimental study described in sections 6.1-6.3, we have been able to confirm that the systematic temporal waves are formed in PLASLA and deduce the model for the waves. The model is that, as schematically shown in Figs.16 and 20, the front wave is generated just after the laser ablation, inducing the stimulated wave and combining with it to form the stable united wave after that. It has been confirmed that these wave transformations are nonlinear processes. In this section, we will investigate this experimentally deduced model by a theoretical approach. For the approach, we assume that these waves are closely related to the chemical processes in PLASLA, i.e., they are a kind of chemical waves.

In the present theoretical study, we assume that the PLASLA luminescence is mainly due to the luminescence from C_2^+, since the analysis of the time-resolved spectra of the PLASLA luminescence in the previous study suggests that C_2^+ ion is one of the most significant species in PLASLA. Hence, for the investigation on the temporal dependence of the PLASLA luminescence, we study the time dependence of the C_2^+ concentration in the present theoretical approach:

As shown in Eqs. (1)-(5), essentially, the following five chemical processes will be relevant to the present approach:

$$Cu + e \rightarrow Cu^+ + e \tag{1}$$

$$Cu^+ + CF_4 \rightarrow Cu + CF_4^+ \tag{2}$$

$$CF_4^+ + CF_4 \rightarrow C_2^+ + 4F_2 \tag{3}$$

$$C_2^+ + CF_4 \rightarrow C_3^+ + 2F_2 \tag{4}$$

$$C_2^+ + e \rightarrow C_2 \tag{5}$$

Here, k_1 and $-k_5$ denote the rate constants of the processes in Eqs. (1)-(5), respectively. We have the following differential equations for the chemical processes in Eqs. (1)-(5), as shown in Eqs. (6)-(8):

$$d[Cu^+]/dt = k_1[Cu] - k_2[Cu^+][CF_4] \tag{6}$$

$$d[CF_4^+]/dt = k_2[Cu^+][CF_4] - k_3[CF_4^+][CF_4] \tag{7}$$

$$d[C_2^+]/dt = k_3[CF_4^+][CF_4] - k_4[C_2^+][CF_4] - k_5[C_2^+][e] \tag{8}$$

In these equations, parentheses [ ] denote the concentrations of the species; e.g., $[C_2^+]$ indicates the concentration of C_2^+. We have Eq.(9) from Eqs.(6)-(8):

$$d^2[C_2^+]/dt^2 = k_3[CF_4]\{ k_2[Cu^+][CF_4] - ((k_3+k_4)[CF_4] + k_5[e]) [CF_4^+]\} + (k_4[CF_4] + k_5[e])^2[C_2^+] \tag{9}$$

For $[Cu^+]/[CF_4^+]=\{(k_3+k_4)[CF_4]+ k_5[e]\}/(k_2 [CF_4])$, Eq. (9) is transformed into Eq. (10):

$$d^2[C_2^+]/dt^2 = \omega^2[C_2^+] \tag{10}$$

Here ω is expressed in terms of Eq. (11):

$$\omega = k_4[CF_4] + k_5[e] \tag{11}$$

Then we have the solution of Eq. (10) as follows:

$$[C_2^+] = [C_2^+]_0 \exp(\pm\omega t) \tag{12}$$

Here $[C_2^+]_0$ is the C_2^+ concentration just after the laser ablation. Since $\omega \geq 0$ in Eqs. (11) and (12), it is found that the concentration of C_2^+, which is one of the most significant species in the PLASLA luminescence, decays and/or grows exponentially and never stably oscillates with time at a frequency of ω in Eq. (11) under the condition of $\{[Cu^+]/[CF_4^+]=\{(k_3+k_4)[CF_4]+ k_5[e]\}/(k_2 [CF_4])\}$. For instance, the formula of $\{[C_2^+]\propto[\exp(\omega t)-\exp(-\omega t)]\}$, which is one of the possible solutions, indicates that $[C_2^+]$ simply increases with time.

The result in Eq. (12) suggests that the oscillation is unstable and gradually quenching to a none-oscillating final state, which is not a limit-cycle. Therefore, the present theoretical results suggest that the oscillations in the models in Figs. 16 and 20, which have been deduced from the present experimental results, are not stable and will finally disappear. As for this discrepancy, since the present theoretical model is very simplified, its result might be inadequate for discussing the experimental results. For instance, the present theoretical model includes no spatial transference term, which is often the significant factor in self-organization process. In addition to the improvement of the theory by taking account of the spatial terms, the study on the evident correlation between the PLASLA discharge patterns and the temporal waves might be of great interest. Furthermore, we might need to use more complicated chemical processes for the theoretical study. Besides, we might need to study the temporal changes in other chemical species besides C_2^+. For instance, the significance of CF_3 and anions is suggested [Kiss & Sawin, 1992; Stoffels et al., 1998], as well as C and C_2 [Jung et al., 2006].

While Figs. 16 and 20 simply describe the transformation of the system from the initial state to the final oscillating state, more details of this process have been able to be expressed in terms of the front wave and the stimulated wave, as a result of the present experimental study. In particular, it is significant to note that these waves are due to the nonlinear processes with the exponential growth and decay and that their behaviours are not necessarily same for the common parameters such as the potential between electrodes. For example, the different behaviours of the waves for the electrode potential in Figs. 21 and 22 suggest that the processes controlling these waves are not necessarily same, as discussed previously. In spite of the fact, these waves are finally combined to be a united wave.

7. Conclusion

In the present article, first, we have reviewed PLASLA (plasma switching by laser ablation) on the basis of our previous papers. In the review, we have introduced the PLASLA phenomenon, the method for the PLASLA formation, the product materials, the properties of PLASLA, and the interaction with a magnetic field. In particular, the significance and promise of PLASLA in materials science have been pointed out. Besides the review, the new original study on PLASLA has been described. In the study, we have investigated the spatial patterns and the temporal waves of PLASLA. Since plasma is a nonlinear process and many interesting ordered structures are known in the spatial and temporal patterns of plasma, the present study will be of great interest.

First we have studied the PLASLA discharge patterns with various target metals. Based on the discharge patterns, the target metals are classified into three groups having the different electronic configurations: Groups-1, -2, -3. In particular, peeling off of the sheaths is clearly observed in the metals of Group-3. Furthermore, for Cu and Ti targets, the luminescence intensity distributions on the electrodes are found to be different outside the reaction site. These facts are due to the competition of the plasma ionic reactions with the neutral chemical reactions in Group-3 metals.

Next, we have analyzed the PLASLA discharge pattern with a Cu target more in detail, as well as those of the DC-plasma and CuF chemiluminescent reaction. In DC-plasma, the luminescence intensity constantly decreases from the cathode to the anode. Besides, it is found that the unquenched, stable sheath surrounds the cathode. As for the anode, various cases are observed; the sheath appears only in the inside, only in the outside, and in the both sides of the anode, depending on its parts. In PLASLA, the plasma luminescence intensity is much more quenched between the electrodes, where the inside is the side between the electrodes. In particular, the quenching is very large near the inside on the cathode. The sheath appears only in the outsides on the cathode and the anode and is largely, completely in many cases, quenched in the insides of the both electrodes. The characteristic discharge patterns in PLASLA are due to the chemical reactions occurring in the whole region between the electrodes.

Then we have studied the product materials attached to the electrodes for various target metals. As the common results for these metals, it has been found that the yields of black-coloured materials are evidently greater in the anode than in the cathode. Furthermore they are evidently greater in the reaction side of the both electrodes, i.e., the site including the laser-ablation target and surrounded by the electrodes, than in another side, i.e., the sites outside the reaction site. As for target metals, the yield of the black-coloured materials is highest in the Group-1 metals, middle in the Group-2 metals, and lowest in the Group-3 metals. Furthermore the white-coloured materials, which are supposed to be the products of the chemical processes competing with PLASLA chemical processes, are yielded in the metals of Groups-2 and 3, while no white materials are yielded in the Group-1 metals.

After that, we have analyzed the transient signals of PLASLA with various target metals. It is found that the rise of the PLASLA pulse is classified into two types: Type-A and Type-BC. The simulation with the simple model has demonstrated that the rise of the PLASLA pulse moves from Type-A to Type-BC with the delay of the plasma luminescence from the laser ablation. The delay depends on the electronic configurations of the target metals. While the decay rate of decay B and the rise rate of decay C depend on the target metals, the plot of the logarithmic value of the former as a function of the latter shows the good linear correlation with the positive slope. The feature of the PLASLA pulse extinction suggests that the PLASLA luminescence signal decays in the two steps probably with the sequence of two exponential decays.

Finally, we have investigated the temporal waves in PLASLA. The study has revealed how the PLASLA system moves on after the laser ablation. After the laser ablation, the front wave is generated, grows and then decays exponentially. With the decay of the front wave, the stimulated wave grows exponentially. Finally, the front wave and the stimulated wave are combined to be a united wave. The amplitude of the united wave is almost constant with a slight increase. The present theoretical study with a simple model has suggested that the oscillation of the luminescence intensity of C_2^+ is not stable and will finally disappear. Since the present theoretical model might be too simple, we need to improve the present theory. We suppose that the temporal waves are due to the chemical processes. While the potential between the electrodes are regarded as a significant parameter to control the chemical processes in PLASLA, the dependence on the potential is found to be different in the front and stimulated waves; the amplitude of the stimulated wave is constant under the threshold potential and steeply increases over it, while the one of the stimulated wave decreases with the potential. For the fact, we have supposed that the front wave is due to the ionic plasma reactions enhanced by the potential, while the chemical reactions promoting the stimulated wave compete with the ionic chemical reactions enhanced by the potential. Furthermore, we have supposed that the threshold observed in the front wave is due to the nonlinearity of the ionic plasma reactions.

8. Acknowledgment

We gratefully acknowledge the collaboration on the PLASLA experiments from Yuko Nakano, Hideyuki Kawai, Tomoko Takenaka, Naoki Mizuno, and Shuichi Komatsu of Mie University.

9. References

Lieberman, M. A. & Lichtenberg, A. J. (2005).*Frontmatter, in Principles of Plasma Discharges and Materials Processing, second Edition,* John Wiley & Sons, Inc., ISBN 9780471720010, Hoboken, NJ, USA

Nakano, Y.; Kawai, H.; Takenaka, T.; Hon, S.; Mizuguchi, T. & Matsuzaki, A. (2010).CuF-Chemiluminescent Reaction and Plasma Switching by Laser Ablation in the Gaseous Cu-CF_4 System and Their Interactions with Magnetic Field. *Journal of Physical Chemistry A,* Vol.114, No.18, (May 2010), pp. 5601-5617, ISSN 1089-5639

Tanaka, H.; Hirano, T. & Matsuzaki, A. (2003).Gas-phase Chemical Reaction of Laser Ablated Copper Atom with Carbon Tetrafluoride in Electric and Magnetic Fields. *Transaction of the Materials Research Society of Japan,* Vol.28, No.2, (June 2003), pp. 275-278, ISSN 1382-3469

Tanaka, H.; Hirano, T. & Matsuzaki, A. (2004).Gas-phase Chemical Reaction of Laser Ablated Copper Atom with Carbon Tetrafluoride in Electric Field: Plasma Switching by Laser Ablation (PLASLA). *Transaction of the Materials Research Society of Japan,* Vol.29, No.8, (December 2004), pp. 3399-3402, ISSN 1382-3469

Tanaka, H.; Hirano, T. & Matsuzaki, A. (2003).New Metal Catalytic Process of Gas-Phase Ionic Reaction : PLASLA (Plasma Switching by Laser Ablation), *Proceedings of 7th Meeting of Symposium on New Magneto-Science,* pp. 35-42, Tsukuba Magnet Laboratory, National Institute of Materials Science, Tsukuba, Japan, November 5-7, 2003

Haken, H. (2006). *Information and Self-Organization: a Macroscopic Approach to Complex Systems,* Springer, ISSN 0172-8379, Berlin Heidelberg, Germany

Matsuzaki, A. (1994).Nonlinear Dynamics in Laser-Induced Aerosol Formation from Vaporized CS_2: Period Fluctuation in Aerosol Burst. *Journal of Molecular Structure (Theochem),* Vol.310, (July 1994), pp. 83-93, ISSN 0166-1280

Kiss, L. D. B. & Sawin, H. H. (1992).Evaluation of CF_4 Plasma Chemistry by Power Modulation. *Plasma Chemistry and Plasma Processing,* Vol.12, No.4, (December 1992), pp. 523-549, ISSN 0272-4324

Stoffels, W. W.; Stoffels, E. & Tachibana, K. (1998).Polymerization of Fluorocarbons in Reactive Ion Etching Plasmas. *Journal of Vacuum Science Technology A,* Vol.16, No.1, (January/February 1998), pp. 87-95, ISSN 0734-2101

Jung, T. Y.; Kim, D. H. & Lim, H. B. (2006).Molecular Emission of CF_4 Gas in Low-pressure Inductively Coupled Plasmas. *Bulletin of Korean Chemical Society,* Vol.27, No.3, (March 2006), pp. 373-376, ISSN 0253-2964

14

Rheological Method for Determining Molecular Weight and Molecular Weight Distribution

Huiru Zhang
*Research Center for Composite Materials,
School of Materials Science and Engineering, Shanghai University,
People's Republic of China*

1. Introduction

Gel permeation chromatography (*GPC*) method is a widely used and accepted method for measuring the *MW* and *MWD* for polymers. However, the method has its limitations. The key of this method is to find the suitable solvents to dissolve the polymer well. But cellulose can not be dissolved in most of the organic solvents because of the inter- and intro-hydrogen bonding of the cellulose chains. Lithium chloride/N, N-dimethlylacetamide (LiCl/DMAc) can dissolve cellulose, but the dissolution process is very complicated, which includes pre-activation, solvent exchange, swelling and dissolution. Furthermore, more attention must be paid to each step, and cellulose has to be dissolved for 5-10 days according to the type of cellulose pulp [1-4]. Therefore, it is necessary to develop a simple and fast method which can get the *MW* and *MWD* of cellulose.

Wu [5] got the *MWD* from storage modulus G' and stress relaxation modulus $G(t)$ using approximations derived from the Doi-Edwards description of chain dynamics. Wu's method accurately predicted the *MWD* of polymers with narrow distribution. However, often, it led to a distorted shape of the *MWD* for the sample with bimodal distributions. Therefore, Tuminello [6-8] developed a theory based on a diluted assumption in 1986. His method rigorously applies only to linear polymers. Especially, it works better for linear polymers with PI < 3.5. According to his theory, the relative differential *MWD* of polymer can be determined well from dynamic modulus master curve.

Gu [9] applied the diluted assumption theory to the concentrated cellulose in N-methlymorpholine-N-oxide monohydrate ($NMMO{\cdot}H_2O$) solution. He got relative differential *MWD* curves of three kinds of cellulose pulps from the dynamic data of cellulose/$NMMO{\cdot}H_2O$ solution in 2000. But the results were not compared with the results reported by *GPC*. In 2004, the relative differential *MWD* curves of four kinds of cellulose pulps were calculated on the basis of that method and the calculated results were compared with the non-calibrated *GPC* results by Zhang [10]. In their rheology experiments, the cellulose concentration in $NMMO{\cdot}H_2O$ solution was fixed (9%, wt), and the polydispersity index (*PDI*) of cellulose was not calculated.

In the present work, the effect of cellulose concentration in $NMMO{\cdot}H_2O$ solution on prediction of the *MW* and *MWD* of cellulose using the rheology-based method was investigated. Furthermore, the calculation of the *PDI* of cellulose was developed. In addition, it also realized the conversion of the reciprocal of the frequency to the actual *MW* scale, obtaining *MWD* scale curve of cellulose [11,12].

2. Experimental

2.1 Materials

Pulp 1 was purchased from Sappi Saiccor (Durban, South Africa). Pulp 2 and 3 were supplied by Weyerhaeuser (Covington, Washington, USA). N-methlymorpholine-N-oxide (NMMO) aqueous solution from BASF (Ludwigshafen, Germany) had an initial water content of 50% (*wt*). N, N-dimethlylacetamide (DMAc) was obtained from Fluka (Lausanne, Switzerland). LiCl was obtained from Yili Finer Chemical Ltd. (Shanghai, China). Polystyrene standard was purchased from Waters (Milford, Massachusetts, USA), with nominal *MWs* of 1.3×10^4, 3.03×10^4, 6.55×10^4, 18.5×10^4, 66.8×10^4, and 101×10^4.

2.1.1 Rheological measurement method

Preparation of cellulose/NMMO·H_2O solution. A mixture of cellulose pulp in 87% NMMO·H_2O (wt) was placed in a dissolving tank maintained at 100°C and stirred continually. The mixture gradually turned into a brown and clear homogeneous liquid. The solution with 12%, 11%, and 9% cellulose in NMMO·H_2O solution (wt) was obtained, respectively.

Rheological measurements. Rheological measurements were recorded on the RS1 rheometer [13] (Thermo Haake, Karlsruhe, Germany) in a frequency scanning mode of 0.2 to 620 ($rad\cdot s^{-1}$) with a cone plate (*Ti*, 35/1^0). Dynamic rheological properties (storage modulus G' and loss modulus G'') were obtained in the linear viscoelastic region at the temperature of 75°C, 90°C, and 105°C, respectively. The data were analyzed with RheoSoft software (Thermo Haake).

Relative differential *MWD* curve and method of calculating *PDI*. Tuminello [6-8, 14] has a method of computing the relative differential *MWD* of linear polymer melts which is based on an analogy with polymer solutions. For the latter, it is well known that

$$G^0_{N,solution} \approx \varphi^2\, G^0_{N,melt} \tag{1}$$

Where φ is the volume fraction of the polymer. Tuminello assumes that the unrelaxed chains of a melt at any frequency (ω_i), are "diluted" by, but not entangled with, the relaxed (lower M_i) chains. He also assumes that each monodisperse fraction (M_i), has a single relaxation frequency (ω_i), below which it makes no contribution to the modulus - thus behaving as a step function. Hence, in Eq. (1), if one associates $G^0_{N,solution}$ with the storage modulus $G'(\omega_i)$, $G^0_{N,melt}$ with the rubbery plateau modulus G^0_N, and φ with the mass fraction of unrelaxed chains of $M \geq M_i$, then

$$C(M) = 1 - [G'(\omega_i)/\, G^0_N\,]^{0.5} \tag{2}$$

where $C(M)$ is the mass fraction of relaxed chains, and

$$MWD = \frac{d(C(M))}{d(\log M)} \tag{3}$$

where M is the molecular weight.

It is well know for the relations

$$1/\omega \propto \eta_0 \propto M^{3.4} \tag{4}$$

where ω is the frequency of dynamic rheology data, η_0 is the zero-shear viscosity of dynamic rheology data, therefore, log $(1/\omega)$ is proportional to log M, then

$$MWD = \frac{d(C(M))}{d(\log(1/\omega))} \tag{5}$$

The above equations were employed in the calculation of the relative differential *MWD* curve of the cellulose pulps.

The relative differential *MWD* curve is a Wesslan function which is the logarithm of the normal distribution function and is especially applied to measure the *MWD* of polymers. The Wessllan function is given by [15, 16]:

$$Y = \frac{1}{\sigma\sqrt{2\pi}} \exp\left(-\frac{1}{2\sigma^2}(M - M_P)^2\right) \tag{6}$$

where Y is the relative content percent of *MW*, M is the molecular weight, σ is the standard deviation, and Mp is the peak *MW* on the *MWD* curve.

There is a correlation between the peak value of the ordinate (Y_{extre}) and the σ with respect to the logarithm normal distribution, as given by *Eq*.(6) [15, 16]:

$$Y_{extre} = \frac{1}{\sqrt{2\pi\sigma}} \tag{7}$$

Then the PDI value is calculated by *Eq*.(7) [15, 16]:

$$PDI = \exp(\sigma^2) \tag{8}$$

2.1.2 GPC measurements

Dissolution of cellulose in LiCl/DMAc. A 10mg sample of each of the three pulps was placed in a 10mL centrifuge tube, respectively. Then 5mL of distilled water was added to each tube. The mixtures were stirred for 5min and left overnight to pre-activate the cellulose. The samples were centrifuged at 4000rpm for 15min. The supernatant fluid was decanted and 5mL of DMAc was added, respectively. After stirring for 15min, the centrifugation and the decantation steps were repeated. The whole solvent exchange procedure was repeated five times. Finally, 1.25mL of 8% LiCl/DMAc (wt/vol) was added, stirred for 60s, and left for approximately one week to dissolve completely, with occasional gentle stirring. The dissolved cellulose solutions were diluted to 20ml with DMAc to give a final cellulose concentration of 0.5mg/mL in 0.5% LiCl/DMAc (wt/vol). Then the solution was filtered through a 0.45μm membrane filter.

GPC analysis. The *MWDs* for the three cellulose pulps were determined by *GPC* in a liquid chromatography (Waters1525) with a refractive index detector (Waters 2410).The mobile phase of 0.5% LiCl/DMAc (wt/vol) was pumped into the system at a flow rate of 1ml/min. Columns were Waters styragel HR 3, 4, and 5 (300mm × 7.8mm) preceded by a guard column. The system was operated at 50°C controlled by a column heater (Waters column temperature system). Injection volume was 200μL. Run time was 45min. A linear calibration curve was constructed with polystyrene standards directly dissolved in 0.5 % LiCl/DMAc (wt/vol). Data acquisition and *MWD* calculations were performed using Breeze software (Waters, Milford, MA, USA). Furthermore, the *GPC* data were calibrated using the Eawkins and Maddock calibration equation, as follows [1, 2]:

$$\log M_1 = \log M_2 + \frac{2}{3}\log\frac{(\kappa_\theta)_2}{(\kappa_\theta)_1} \tag{9}$$

where subscript 1 and 2 represent the unknown sample and the standard sample, respectively. Here, they are the cellulose and the polystyrene. K_θ values [1] of cellulose and polystyrene are 0.528 and 0.081 in the 0.5% LiCl/DMAc (wt/vol) system, respectively.

3. Results and discussion

3.1 Effect of cellulose concentration in NMMO·H_2O solution on the rheology-based results

Fig.1a, 1b, and 1c show *G'* and *G''* master curves of the pulp 1 at various cellulose concentrations in NMMO·H_2O solutions according to the time-temperature superposition theory [13]. In terms of the Tuminello diluted assumption theory [6-8, 14], these curves are converted to relative differential *MWD* curves of the pulp 1 at different cellulose concentrations in NMMO·H_2O solutions shown in Fig.2a. Using the same procedure, relative differential *MWD* curves of the pulp 2 and 3 are obtained at various cellulose concentrations in NMMO·H_2O solutions and presented in Fig.2b and 2c, respectively. Meanwhile, the calculated log $(1/\omega_p)$, σ and *PDI* values of the three pulps are given in Table 1.

Fig.2a, 2b, and 2c show that the relative differential *MWD* curves almost overlap with each other when the cellulose concentrations are 12% and 11%, respectively. However, when the cellulose concentration is 9%, the relative differential *MWD* curves shift slightly to the left.

The data in Table 1 clearly indicate that log $(1/\omega_p)$ are higher and *PDI* decrease with increasing concentration. However, the relative deviation of the cellulose concentration between 9% and 11% is up to 16%, and the relative deviation of the cellulose concentration between 11% and 12% is less than 0.9%. Obviously, the calculated log $(1/\omega_p)$ and *PDI* are approximately equal when the cellulose concentrations are 12% and 11%.

This phenomenon originates from the relaxation time (τ) of molecular chains and the steady-state recoverable compliance (J_e^0) in a polymer solution. With increasing concentration, the polymer chains are more strongly entangled with each other and get more and more

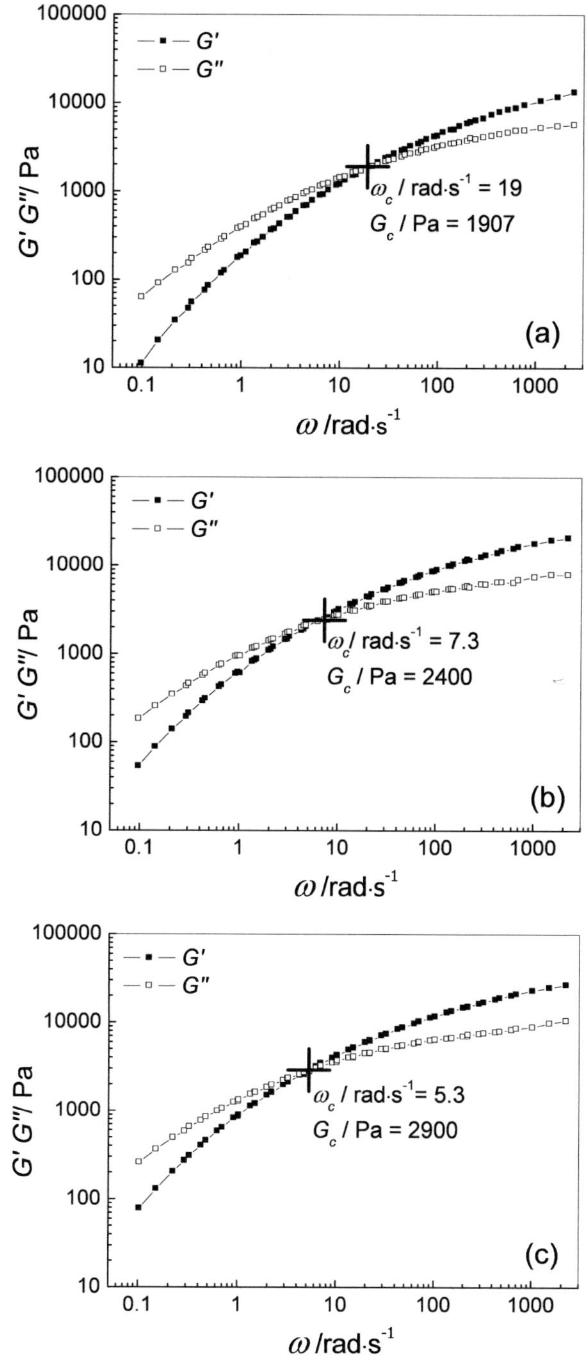

Fig. 1. Master curves of the pulp1 at different cellulose concentrations in NMMO·H_2O solutions: (a) 9%, (b) 11%, and (c) 12%.

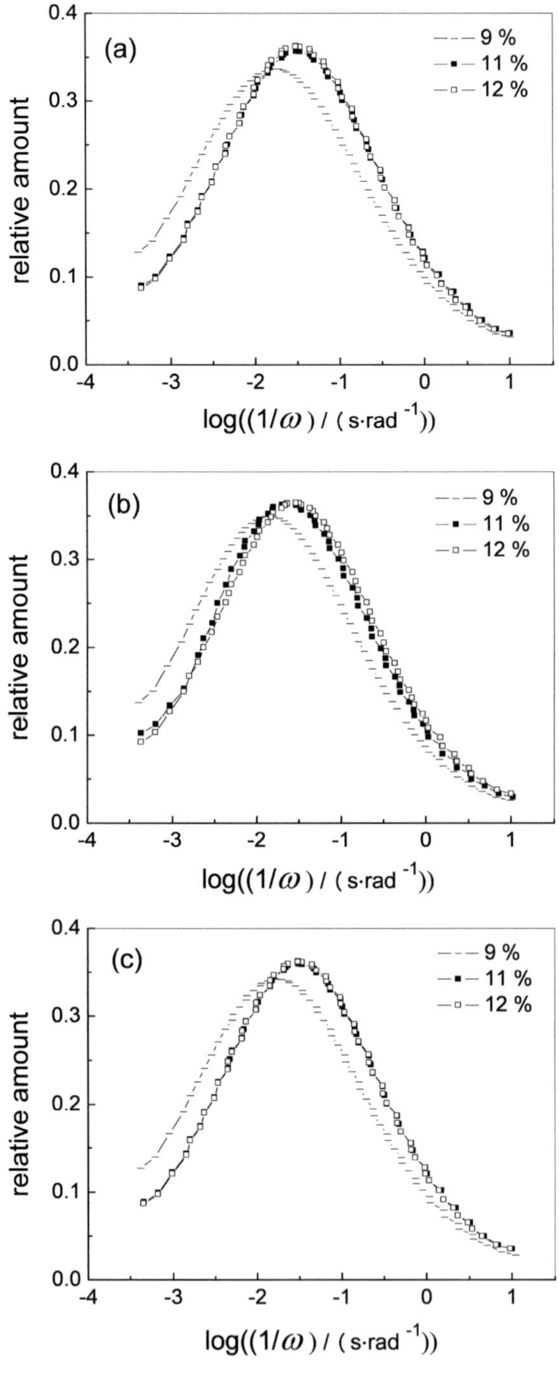

Fig. 2. Relative differential *MWD* curves of the three pulps at different cellulose concentrations in NMMO·H_2O solutions: (a) pulp 1, (b) pulp 2, and (c) pulp3.

Samples		Pulp1	Pulp 2	Pulp 3
	$\log(1/\omega_p)$	-1.5024	-1.5296	-1.4888
Concentration (12 %)	σ	1.1091	1.0956	1.1036
	PDI	3.4217	3.3211	3.3850
	$\log(1/\omega_p)$	-1.5099	-1.5437	-1.4982
Concentration (11 %)	σ	1.1118	1.0986	1.1069
	PDI	3.4424	3.3429	3.4046
	$\log(1/\omega_p)$	-1.7982	-1.8398	-1.7635
Concentration (9 %)	σ	1.1853	1.1410	1.1650
	PDI	4.0753	3.6759	3.8854

Table 1. Logarithm of relative molecular weight $\log((1/\omega_p)/(s \cdot rad^{-1}))$, standard deviation σ, and polydispersity index *PDI* of the three pulps calculated by the rheology-based method with different concentrations.

difficult to relax the entanglement network, leading to a longer τ. Meanwhile, with increasing concentration, the polymer solution shows a decreasing J_e^0 because the entangled polymers do not fully and freely stretch with rearranging configurationally.

However, when the polymer concentration reaches a certain value, τ and J_e^0 of the polymer solution will tend towards equilibrium because the gyration radius, the end-to-end distance, and the "degree" of mutual entanglement of the chains will reach critical points, respectively [17].

It is well known that the τ is a function of *MW* of the polymer [17, 18], and J_e^0 is correlated with *PDI* of the polymer [17, 18]. Therefore, *MW* and *PDI* of the polymer move towards stabilization with the polymer concentration reaches a critical point. Accordingly, in the rheology-based method, a high enough concentration of cellulose in NMMO·H_2O solution has to be used in order to obtain reliable and stable data.

3.2 Prediction of MW scale and MWD of cellulose using the rheology-based method

Methodology. Definition of the Rouse terminal relaxation time is well known as follows [14].

$$\tau_1 = \frac{6\eta_0 M}{\pi^2 \rho RT} = 1/\omega_{char} \tag{10}$$

where ρ is the density, η_0 the zero-shear viscosity, R the universal gas constant, M the molecular weight, T the temperature. ω_{char} is the 79th percentile point of the zero shear normalized flow curve and normally symbolized as $\omega_{.79}$. The $\omega_{.79}$ is the gradually-changed frequency from Newtonian to non-Newtonian behavior for a liquid of polymer solution, with which the apparent viscosity of the polymer solution decreases and the untangling effect is stronger than the entangling effect among molecular chains.

According to the Rouse terminal relaxation time theory [14], a polymer solution always has a corresponding characteristic relaxation frequency (ω_{char}). Here, the ω_{char} is the $\omega_{.79}$ corresponding to the point of maximum curvature of the flow curve. Therefore the calculated M is believed to be the peak *MW* (*Mp*). *Mp* indicates the maximum probability of molecular weight on the curve of *MWD*. In addition, it can be believed that the maximum probability of molecular weight begins to untangle, which would lead to a decreasing apparent viscosity of the polymer solution.

Accordingly, the peak *MW* scale is obtained by $Mp = M = \pi^2 \rho RT/(6\eta_0 \omega_{.79})$ and the *MWD* scale curve is obtained with shifting the abscissa (log *Mp* - log $(1/\omega_p)$) units. Here, the *Mp* and $(1/\omega_p)$ indicate the peak *MW* on the *MWD* scale curve and the relative *MWD* curve, respectively.

For the 79th percentile point of the normalized flow curve, the Vinogradov extrapolation leads to [14]:

$$\omega_{.79} = 0.3365 m_0^2/m_1 \tag{11}$$

where m_0 and m_1 are respectively the intercept and the slope on the curve of stress versus viscosity at the low frequency of the dynamic data. In the current paper, the m_0 and m_1 are obtained in the frequency range from 0.2 to 2 (*rad*·s^{-1}).

Results. Fig.2a, 2b, and 2c respectively represent relative *MWD* curves of the three pulps at various cellulose concentrations in NMMO·H_2O solutions. Choosing a concentration of 12%, these relative *MWD* curves are converted to the *MWD* scale curves of the three pulps shown in Fig.3, using the above-mentioned method. Furthermore, in terms of the *MWD* scale curve, log *Mp*, σ, and *PDI* of the three pulps are calculated and given in Table 2. From Fig.3 and Table 2, it can be found that the relation of peak *MW* is pulp 3 > pulp 1 > pulp 2. For the *MWD* of the three pulps, pulp 1 appears the broadest, next is pulp 3, and then pulp 2.

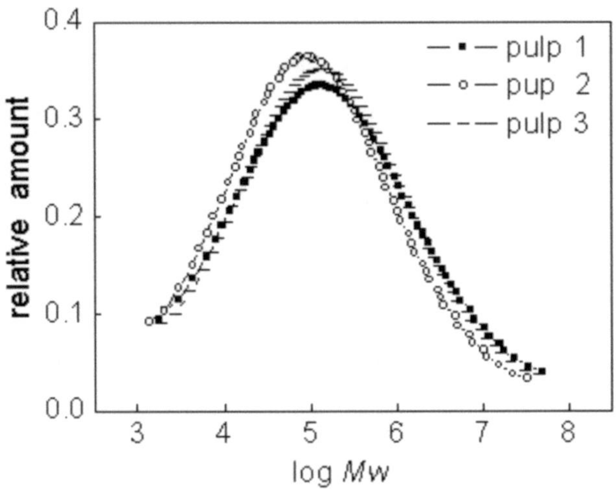

Fig. 3. *MW* scale curves of the three pulps by the rheology-based method with 12% cellulose concentration in NMMO·H_2O solution.

Samples	Pulp 1	Pulp 2	Pulp 3
log *Mp*	5.12	4.96	5.14
σ	1.109	1.096	1.104
PDI	3.42	3.32	3.38

Table 2. Logarithm of molecular weight scale log*Mp*, standard deviation σ, and polydispersity index *PDI* values of the three pulps calculated by the rheology-based method with 12% cellulose concentration in NMMO $\cdot H_2O$ solution.

3.3 Comparison of the results from the rheology-based method and the GPC method

Because of the lack of commercial cellulose standards with a narrow distribution, the narrow distribution polystyrenes standards are employed to measure *MW* and *MWD* of cellulose. The *MWD* curves of the three pulps measured by the *GPC* method are illustrated in Fig.4. Meanwhile, the calibrated *GPC* data are listed in Table 3.

Comparing the data of Table 2 with those of Table 3, one can observe that log *Mp* calculated by the rheology-based method is nearly equal to log *Mp*' determined by the *GPC* method. Therefore, it is feasible and reasonable that the calculated *M* with Eq. (10) is regarded as the peak *MW* (*Mp*) on the *MWD* curve. Consequently, the reciprocal of the frequency is converted to the *MW* scale in the rheology-based method.

The results of Table 3 show that *PDI*' of the three pulps is pulp 1 > pulp 3 > pulp 2, which are consistent with the results from the rheology-based method. Moreover, more information can also be obtained from Fig.4 by the GPC method than that from Fig.3 by the rheology-based method. For example, pulp 1 shows a symmetrical distribution, and moderate *MW* components are dominating. Pulp 3 shows a slightly asymmetrical distribution, moderate *MW* components are the major composition, and it has a little lower *MW*. For pulp 2, it is asymmetrical and slightly protuberant in the lower MW region, which indicates the presence of a higher low *MW* content. However, such useful information can not be reflected from the *MWD* scale curves obtained by the rheology-based method directly. It shows that further modification is still needed for the application of the rheology-based method.

Samples	Pulp 1	Pulp 2	Pulp 3
log *Mp*'	5.133	5.071	5.157
$\overline{Mw}$	16.07 ×10⁴	12.83 ×10⁴	15.59 ×10⁴
$\overline{Mn}$	3.989×10⁴	3.878 ×10⁴	4.355 ×10⁴
PDI'	4.030	3.309	3.581

Table 3. Logarithm of molecular weight log*Mp*', mass-average molecular weight $\overline{Mw}$, number-average molecular weight $\overline{Mn}$, and polydispersity index *PDI*' of the three pulps measured by the *GPC* method.

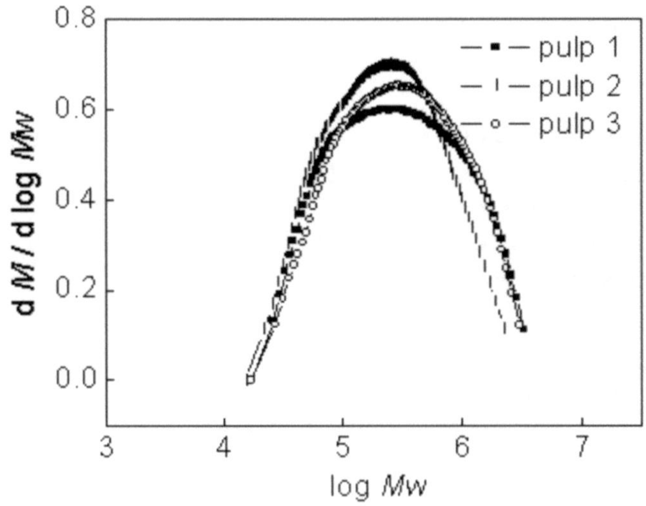

Fig. 4. *MWD* curves of the three pulps by the *GPC* method.

In the present work, the data obtained by the *GPC* method are relative values because of the use of polystyrenes standards, nevertheless the *GPC* method is an effective way for observing the differences of *MW* and *MWD* of cellulose. The relative data can not reflect the real *MW* characteristics of cellulose, so the data from *GPC* can not be used to calibrate the results from the rheology-based method. Even so, the comparison of the results from the two methods shows that it may be feasible to compare the *MW* and *MWD* of cellulose by the rheology-based method.

4. Conclusions

Prediction of *MW* scale and *MWD* of cellulose by means of a rheology-based method was developed. With this method, insignificant effect of cellulose concentration on predicting *MW* and *MWD* of cellulose was found using a rheology-based method when the cellulose concentration in the NMMO·H_2O solution is high enough. Furthermore, a method of calculating *PDI* of cellulose was established according to the Wesslan function which is the logarithm of the normal distribution function. For the cellulose/NMMO·H_2O solution, the cellulose *MW* values calculated by the Rouse terminal relaxation time can be considered as the peak *MW* on the *MWD* curves of cellulose. Consequently, the reciprocal of the frequency is converted to the *MW* scale, obtaining *MWD* scale curves of cellulose.

Meanwhile, the results obtained by the rheology-based method were compared with those measured by the *GPC* method. All obtained results from the two methods are only relative values. The comparison shows that the calculated peak *MW* are approximately equal, the calculated *PDI* have the same trends, but the shapes of the *MWD* curves do not match. *GPC* method is advantageous to depict finer characteristics of the *MWD* of cellulose. In spite of that, the rheology-based method is simple and fast. Therefore it is a useful and easy way to analyze the *MW* scale and *MWD* of cellulose in the fiber industry.

5. References

T. Bikova and A. Treimanis, *Carbohydr. Polym.*, 48, 23 (2002).
M. Strlic and J. Kolar, *Biochem. Biophys. Methods,* 56, 265 (2003).
Z. D. Xu and M.S. Song, *Macromolecules,* 14, 1597 (1981).
J. M. Evans, *Polym. Eng. Sci.*, 13(6), 401 (1973).
S. Wu, *Polym. Eng. Sci.*, 25, 122 (1985).
W. H. Tuminello, *Polym. Eng. Sci.*, 26(19), 1339 (1986).
W. H. Tuminello and N. C. Mauroux, *Polym. Eng. Sci.*, 31(10), 1496 (1991).
W. H. Tuminello, T. A. Treat, and A. D. English, *Macromolecules,* 21, 2606 (1988).
G. X. Gu, X. C. Hu, and H. L. Shao, *Sen'i Gakkaishi,* 57, 22 (2001).
H. H. Zhang, H. L. Shao, and X. C. Hu, *J. Appl. Polym. Sci.*, 94(2), 598 (2004).
H. R. Zhang, X. Y. Liu, D. S. Li, and R. X. Li, *Polym. Eng. Sci.*, 49(3), 554(2009).
H. R. Zhang, M. W. Tong, and R. X. Li, *Polym. Eng. Sci.*, 48(1), 102(2008).
G. Schramm and X. H. Li, *A Practical Approach to Rheology and Rheometry,* Petroleum Industry Press, Beijing (1998).
C. Lavallee and A. Berker, *J. Rheol.*, 41(4), 851 (1997).
D. Z. Ma, Z. D. Xu, and P. Z. He, *Structure and Property of Polymer,* Science Press, Beijing (2000).

C. R. Zheng, *Molecular Weight and Distribution of Polymer*, Chemistry Industry Press, Beijing (1986).
M. Kroger and H. Voigt, *Macromol. Theory Simul.*, 3, 639(1994).
B. J. Collier, M.Dever and S. Petrovan, *J. Polym. Environment*, 8(3), 151(2000).